NEUROSCIENCE RESEARCH PROGRESS

NEUROECOLOGY AND NEUROETHOLOGY IN MOLLUSCS

THE INTERFACE BETWEEN BEHAVIOUR AND ENVIRONMENT

NEUROSCIENCE RESEARCH PROGRESS

Additional books in this series can be found on Nova's website
under the Series tab.

Additional e-books in this series can be found on Nova's website
under the e-book tab.

NEUROSCIENCE RESEARCH PROGRESS

NEUROECOLOGY AND NEUROETHOLOGY IN MOLLUSCS

THE INTERFACE BETWEEN BEHAVIOUR AND ENVIRONMENT

ANNA DI COSMO
AND
WILLIAM WINLOW
EDITORS

New York

Library of Congress Cataloging-in-Publication Data

Neuroecology and neuroethology in molluscs : the interface between behaviour and environment / editors, Anna Di Cosmo and William Winlow (Dept. of Structural and Functional Biology, University of Naples "Federico II", Complesso Universitario Monte Sant'Angelo, Napoli, Italy and others).
pages cm
Includes index.
ISBN: 978-1-62948-970-4 (hardcover)
1. Mollusks--Nervous system. 2. Mollusks--Ecology. 3. Mollusks--Behavior. I. Di Cosmo, Anna. II. Winlow, W. (William)
QL431.2.N48 2014
564--dc23

2013048321

Published by Nova Science Publishers, Inc. † New York

Contents

Preface vii

List of Contributors ix

Chapter 1 Prologue: Neuroecology and the Molluscan Connection 1
Charles D. Derby and Richard K. Zimmer

Part 1. Molluscan Chemical Diversity 11

Chapter 2 Chemical Diversity in Molluscan Communities: From Natural Products to Chemical Ecology 13
Kirsten Benkendorff

Part 2. Predator-Prey Relationships and Reproductive Behaviour 43

Chapter 3 Olfaction in Gastropods 45
Scott F. Cummins and Russell C. Wyeth

Chapter 4 Communications between Predator and Prey 73
Matthew Nusnbaum

Chapter 5 Ecological Consequences of Chemically-Mediated Prey Perception in Molluscan Communities 89
Miranda Wilson and Marc Weissburg

Chapter 6 Cephalopods Meet Neuroecology: The Role of Chemoreception in Octopus vulgaris Reproductive Behaviour 117
Anna Di Cosmo and Gianluca Polese

Part 3. Evolutionary Conservation and Plasticity 133

Chapter 7 The Neuroendocrine-Immune Orchestra in Molluscs 135
Enzo Ottaviani

Chapter 8 A Neuroplastic Network Underlying Behaviour and Seasonal Change in Lymnaea stagnalis: A Neuroecological Standpoint 145
William Winlow and Gianluca Polese

Chapter 9 Evolutionary Sophistication of Aerial Respiratory Behaviour in the Freshwater Mollusc *Lymnaea stagnalis* **177**
Tara A. Janes and Naweed I. Syed

Chapter 10 Epilogue: Molluscs in Neuroecology **211**
Anna Di Cosmo and William Winlow

Index **215**

Preface

The term Neuroecology was first coined in the 1980s and describes the ways in which species adapt to their environment both in the short term and in evolutionary time. Here we focus on molluscan Neuroecology to show how it interfaces with Neuroethology and how animal behaviour adapts to environmental cues. Neuroecology is thus meeting point of ethology, neuroscience, evolution, ecology, physics and chemistry.

It has been suggested that molluscs are not major contributors to the structuring ecological communities and that molluscan research has functioned poorly in elucidating fundamental principles at higher levels of biological organisation beyond the level of whole behaving organisms. The contributions presented here correct this viewpoint and examine the concept of Neuroecology as applied to molluscs for the first time. The areas of chemical ecology and predator prey relationships are well covered as are neuroethology and evolution.

Significant neuroecological progress has been made with a number of molluscan groups in terms of neurotoxic peptides, complex defensive ink alarm pheromones, bioaccumulation of shellfish toxins and the evolution of chemical defence mechanisms in species with reduced physical protection. Many molluscs inhabit a world of olfactory cues and their perceptions of the outside world are largely chemically driven, leading to odorant driven behavioural responses. This is important to both molluscan predators and prey and this area is explored in some detail with particular reference to gastropod molluscs. It is also true for cephalopod molluscs which although they have excellent vision, rely on distance chemoreception particularly prey perception, thus demonstrating their significance in affecting marine populations and communities. Furthermore the olfactory lobes and olfactory organs of cephalopods play a crucial role in the chemical perception and integration of odour signals that allow the animal to balance their energy allocation between growth and reproduction. A better understanding the interactions of chemical cues between predator and prey molluscs will be important in future studies in the area of chemical ecology.

Within the immune and neuroendocrine systems of molluscs the primary signalling molecules are exceptionally well-conserved and the pool of molecules used is similar to those in advanced vertebrates. This conservation of molecular systems is important in neuroecology and underlies the conservation of neural mechanisms underlying behaviours as organ systems alter under evolutionary pressure. Indeed studies on pulmonates such as *Lymnaea stagnalis* have shown that behavioural networks remain virtually intact during these processes, although reflecting the evolutionary changes that adapt the animal to its environment. Such networks exhibit plastic responses to allow for behavioural selection and seasonal change.

However studies on respiratory behaviour lead to the conclusion that many molluscs are facing extinction because they cannot adapt their respiratory behaviour to the rapidly changing environmental conditions currently occurring particularly where hypoxia threatens the health of ecosysystems due to human activity. Such animals may prove to be sensitive models for studying the neuroecological effects of climate change with particular respect to invasive freshwater species such as predators, freshwater plans or microbial species.

The challenge for the future is to utilise molluscan species research to illuminate the effects of environmental change on the neuroecology of these species, using them wherever possible to evaluate the degree of ecological change taking place.

Anna Di Cosmo and Bill Winlow

List of Contributors

Kirsten Benkendorff. Marine Ecology Research Centre, School of Environmental Science and Engineering, Southern Cross University, PO Box 157, Lismore, NSW, Australia. Email: kirsten.benkendorff@scu.edu.au

Scott F. Cummins. University of the Sunshine Coast, Maroochydore, Queensland, Australia Email: scummins@usc.edu.au

Charles D. Derby. Neuroscience Institute and Department of Biology, Georgia State University, P.O. Box 5030, Atlanta, GA 30302-5030 USA. Email: cderby@gsu.edu

Anna Di Cosmo. Dipartimento di Biologia, Università degli Studi di Napoli, Federico II, 80126 Napoli, Italia.
Email: dicosmo@unina.it

Tara A. Janes. Dept. of Cell Biology and Anatomy and the Hotchkiss Brain Institute, Faculty of Medicine, University of Calgary, 3280 Hospital Dr. N.W., Calgary, Alberta, Canada T2N 4Z6.

Matthew Nusnbaum. Department of Biology Georgia State University, Atlanta, GA, USA. Email: m.nusnbaum@gmail.com

Enzo Ottaviani. Department of Biology, University of Modena and Reggio Emilia, Via Campi 231/D, 41125 Modena, Italy. E-mail: enzo.ottaviani@unimore.it

Gianluca Polese. Dipartimento di Biologia, Università degli Studi di Napoli, Federico II, 80126 Napoli, Italia.

Naweed I. Syed. Dept. of Cell Biology and Anatomy and the Hotchkiss Brain Institute, Faculty of Medicine, University of Calgary, 3280 Hospital Dr. N.W., Calgary, Alberta, Canada T2N 4Z6. Email: nisyed@ucalgary.ca

Marc Weissburg School of Biology, Georgia Institute of Technology, Atlanta, GA, USA. Email: marc.weissburg@biology.gatech.edu

Miranda Wilson. School of Biology, Georgia Institute of Technology, Atlanta, GA, USA

William Winlow. Visiting Professor, Departimento di Biologia, Università degli Studi di Napoli, Federico II, 80126 Napoli, Italia; Honorary Research Fellow, University of Liverpool, UK; Professor Emeritus of Neuroscience, University of Central Lancashire, Preston, Lancashire, UK. E-mail: bill@winlow.myzen.co.uk

Russell C. Wyeth. St. Francis Xavier University, Antigonish, Nova Scotia, Canada.

Richard K. Zimmer. Department of Ecology and Evolutionary Biology, Neurosciences Program, and Brain Research Institute, University of California, Los Angeles, CA 90095-1606 USA. Email: z@biology.ucla.edu

In: Neuroecology and Neuroethology in Molluscs
Editors: Anna Di Cosmo and William Winlow
ISBN: 978-1-62948-970-4

Chapter 1

Prologue: Neuroecology and the Molluscan Connection

***Charles D. Derby*[1*] *and Richard K. Zimmer*[2†]**
[1] Neuroscience Institute and Department of Biology,
Georgia State University, Atlanta, GA, US
[2] Department of Ecology and Evolutionary Biology, Neurosciences Program,
and Brain Research Institute, University of California, Los Angeles, CA, US

What Is Neuroecology?

Nervous systems are sculpted by evolution through natural selection. Many ecological interactions among individual organisms are mediated by behavioral processes. The outcomes of these organismal-level encounters have profound consequences for populations and communities (Zimmer and Butman, 2000; Paul et al., 2007; Hay, 2009; Derby and Aggio, 2011). Thus, neuroecology unifies principles from diverse disciplines, scaling from biophysical and molecular properties of nerve and muscle cells to community-wide impacts of behavior. It integrates neuroethological and ecological principles, embodying both the neural basis for behavior and the roles of behavior in establishing patterns of organismal abundances and species distributions (Zimmer and Derby, 2007, 2011). Figure 1 provides a graphic depiction of the span and aims of neuroecology.

Neuroecology does not take the place of sensory ecology or neuroethology. Rather, it creates bridges that unite such disciplines. Sensory ecology largely concerns how organisms detect and respond to external stimuli within environmental contexts (Dusenbery, 1992). Research in sensory ecology characterizes physical and chemical environmental stimuli, toward understanding how the information carried by them is encoded by sensory systems. In contrast, neuroecology considers sensory mechanisms, but also includes elucidations of central neural (such as learning and memory), neuroendocrine, and other higher-order processes that modulate information in ways that may, or may not, be tied directly to initial

[*] Email: cderby@gsu.edu.
[†] Email: z@biology.ucla.edu.

sensory inputs. Behavioral responses of animals ultimately are amalgamations of complex sensorimotor activities, and often cannot be described or predicted accurately from purely additive effects of applied environmental stimuli. A neuroecological approach casts a larger and more effective net than does sensory ecology, thereby linking the cellular basis for behavior, integrative organismal processes, and ecological functions.

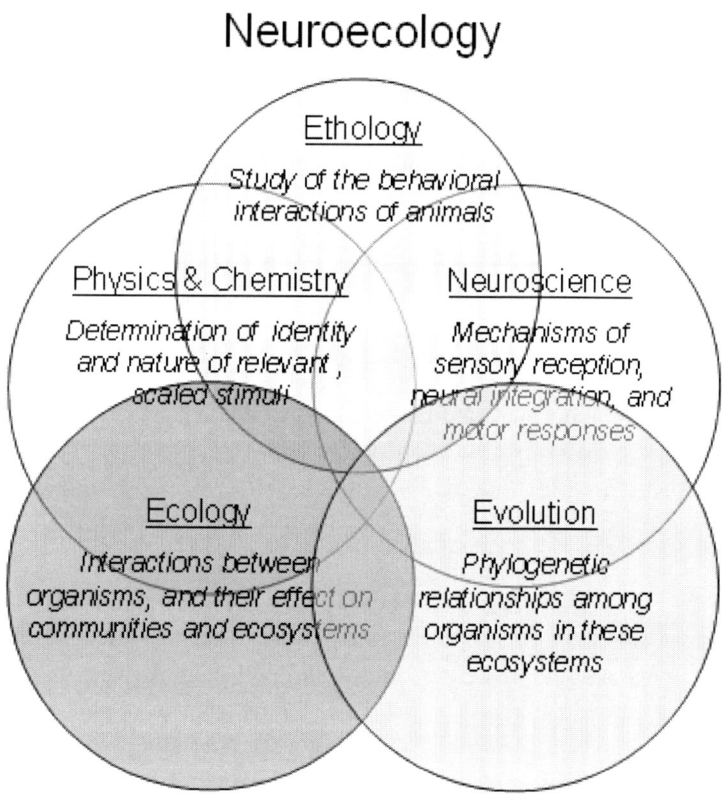

Figure 1. Venn diagram showing the neuroecological approach. Disciplinary approaches are shown in the circles.

By comparison, neuroethology scales from molecular mechanisms to processes that operate at the level of an individual organism (Zupanc, 2004). Much has been discovered about basic principles of information processing, sensorimotor coordination, memory, learning, and cognition. Neuroethological approaches have also promoted comparative research across phylogenetic lines to yield insights into proximate mechanisms, ultimate causes, and evolution of behavior. Neuroecology, however, treads where neuroethology has yet to travel by establishing causal linkages between sensory stimulus space, neural processes, behavior, and the demographic- and community-level consequences of decisions made by individuals.

The objective of neuroecology is to synthesize ideas and approaches with the goal of promoting a continued fusion of disciplines. Neuroecology, as it appears here, is a concept renovated and refocused from other applications (Zimmer and Derby, 2007, 2011; Derby,

2009; Derby and Zimmer, 2012). The term neuroecology first appeared in the neuroscience literature to describe interactions among neurons as they compete for limited resources within a developing nervous system (Purves and Lichtman, 1985). That usage has fallen out of favor and is seldom seen anymore. A second application of the term is quite different from all others. *Human neuro-ecology* describes the process by which couples are brought to "understand the neuro-ecological [brain] systems impacting attachment, survival and transactional reflexes" (E. Grey and R.D. Morrow, Institute of Human Neuro-Ecology®: www.rtpgh.com/Human-Neuro-Ecology-.html, 21 Oct 2012). A third contemporary definition of neuroecology encompasses the study of adaptive variation in the brain towards understanding evolution of animal cognition (Sherry, 2005; Smulders et al., 2010). Finally, Robertson (2004) employs a fourth application of the word towards elucidating the relationship between the functional organization of a nervous system and the ecological niche of an animal. Beginning in 2007, we co-opted the term neuroecology for a different use. As *neuroethology* recognizes the integration of two major disciplines, neurobiology and ethology, neuroecology is best employed towards promoting the fusion of neuroethology with ecology, thus scaling from molecules to community assemblages of species and even to entire ecosystems. A neuroecological approach seeks to identify the neural and behavioral basis for material transport throughout complex trophic webs, as well as the outcomes of biotic interactions that determine community structure and function. The time is ripe for such a broad synthesis.

The road connecting neuroethology and ecology is a two-way street. Much can be learned by integrating findings across higher and lower levels of biological organization. Population and community ecologists, behaviorists, and neuroscientists are encouraged to work together towards linking genes and ecosystems within an integrated framework of neural processing. Already, numerous species have been identified upon which the pillars of ecological communities are established and organized (Paine, 1969; Power et al., 1996; Kurzava and Morin, 1998; Davic and Welsh, 2004; Smith, 2006). In the past, many of these foundation species (such as corals, giant kelps, mussels, barnacles, prairie grasses, and pine trees) and keystone species (such as beavers, seastars, sea otters, grizzly bears, and red-banded newts) were not amenable to genomic sequencing or accessible to immediate physiological analyses. This situation, however, is changing rapidly. New tools have been developed and used routinely for studies of the major biomedical models such as mice (*Mus musculus*), fruit flies (*Drosophila melanogaster*), and nematode worms (*Caenorhabditis elegans*). For example, optogenetics allows using light to activate or inactivate specific types of cells in the nervous systems of behaving animals, allowing experimental exploration of neural circuits and their components (Deisseroth, 2011). RNA interference allows fine control of the levels of expression of genes of interest, allowing identification of the function of assorted molecules and pathways (Mello and Conte, 2004). These and other molecular techniques are now being applied to other species (Fenno et al., 2011; Kato et al., 2011), and they should be adaptable to species that are of vital ecological significance. With recent advancements in genetics and neurobiology, species once classified as non-models for molecular and cellular research are becoming amenable for study. Through improved integration across the tree of life, molecular biologists and physiologists can gain valuable new insights into the evolution (such as conservation or divergence between lineages) and diversity of basic biochemical and biophysical processes that drive electrically excitable cells. In turn, population and community ecologists can gain critical knowledge towards predicting behaviorally mediated interactions among individuals of the same, or different, species that determine demographic and community dynamics.

Examples of the Neuroecological Approach

Neuroecology is still an infant. No synthesis, as yet, has bridged the gap from molecules and nerve cells to individual behavior and organization of species assemblages. To date, some of the best examples come from the field of chemical ecology (reviewed in Derby, 2007; Paul et al., 2007; Zimmer and Ferrer, 2007; Nevitt, 2011; Riffell, 2011). Here, new conceptual frameworks are rallying around a neuroecological approach towards understanding the cross-cutting effects of select molecules. The keystone species concept, for example, is seminal in ecological theory. It defines a species whose impacts on communities are far greater than would be predicted from its relative abundance and biomass (Paine, 1969; Power et al., 1996). Similarly, certain neurotoxins function in keystone roles. They are rare within natural habitats but exert strong impacts on species interactions at multiple trophic levels (Ferrer and Zimmer, 2012).

Effects of two guanidine alkaloids, tetrodotoxin (TTX) and saxitoxin (STX), coalesce neurobiological and ecological perspectives (Zimmer and Ferrer, 2007). These molecules compose some of the most potent natural poisons ever described, and they are introduced into respective communities by one or only a few host species. Functioning as voltage-gated sodium channel blockers for nerve and muscle cells, TTX and STX serve in chemical defense (Catterall et al., 2007). Toxin-mediated processes that inhibit the generation of action potentials in nerve and muscle cells are well understood at molecular and cellular levels (Noda et al., 1989; Satin et al., 1992; Cestèle and Catterall, 2000). When co-opted by resistant consumer species, however, TTX and STX are used either in chemical defense against higher-order predators or as chemosensory cues that elicit courtship and mating, alarm, and predatory search (Kvitek and Bretz, 2004, 2005; Zimmer et al., 2006; Williams, 2010; Hara, 2011). Requisite to these multifunctional properties, biosynthetic capacity appears to evolve along with mechanisms for resistance to and/or storage of the toxins by primary producers. Subsequently, consumers acquire resistances or tolerances, and the toxins are transferred through food webs via trophic interactions (Ferrer and Zimmer, 2012). In consumers, mechanisms eventually evolve for neural recognition of toxins as feeding cues and, ultimately, as signals or pheromones in chemical communication within or between species. The molecular basis for neural recognition of TTX and STX, as well as the transduction events leading to chemosensory excitation, are still unknown. Well described, however, are cascading effects of TTX and STX that profoundly impact community-wide attributes, including species compositions and rates of energy and material exchange (Zimmer and Ferrer, 2007; Ferrer and Zimmer, 2012). A diverse array of genetic, molecular, and physiological properties, expressed differentially across many species, renders these substances fully functional as keystone molecules, with vast ecological consequences at multiple trophic levels.

Neuroecology and the Molluscs

This book begins a conversation on the neuroecology of molluscs. Within its chapters lie some of the best models and best examples of the power of the neuroethological and neuroecological approaches as led by some of the outstanding researchers in their respective fields. This volume describes work pertaining to important aspects of sensory biology, animal communication, comparative physiology, evolutionary biology, chemical ecology, and environmental science.

And yet, what is missing in the book indicates the vast potential for the interdisciplinary approach of neuroecology. For example, molecular and cellular processes that drive behavior have been elucidated largely with the gastropods and cephalopods, while important contributions to community-wide attributes also include the bivalves. A more integrative approach in the future will be needed towards interfacing neural mechanisms to community consequences across a common set of species. Currently, synthesis can be achieved only by assembling knowledge for many different species as investigated in many different ways. Distinct benefits would be gained through stronger coalescence of research from molecules to ecosystems for at least certain, key species.

Select molluscs have provided valuable models towards understanding the cellular basis of behavior. Topics of intensive study have included membrane biophysics and ion channels, intercellular communication using chemical and electrical synapses, and changes in intracellular signaling within neurons and of connections among neurons related to learning and memory. Two molluscan species have contributed more than others to the field of neuroscience: the squid *Doryteuthis* (formerly *Loligo*) *pealeii*, and the sea hare, *Aplysia californica*. Squid have been used since the 1930s to study the biophysical properties of membranes in excitable cells, because of their rapid, jetting escape behavior (Young, 1938). As with most rapid escape behaviors, jetting by squid is mediated by neurons with large axons, because rapid responses require fast conduction of axon potentials, and speed of conduction is related to size of axons. Thus, the giant axon of squid has allowed generations of neuroscientists to elucidated mechanisms of excitability. In fact, the 1963 Nobel Prize in Physiology or Medicine was awarded to John Eccles, Alan Hodgkin, and Andrew Huxley "for their discoveries concerning the ionic mechanisms involved in excitation and inhibition in the peripheral and central portions of the nerve cell membrane" ("The Nobel Prize in Physiology or Medicine 1963" Nobelprize.org. 28 Nov 2012, http://www.nobelprize.org/nobel_prizes/medicine/laureates/1963/), with the contribution of the team of Hodgkin and Huxley coming largely from their work on squid giant axons. Sea hares, especially *Aplysia californica*, have been used to examine the cellular and molecular bases of learning and memory. In 2000, the Nobel Prize in Physiology or Medicine was given to Arvid Carlsson, Paul Greengard, and Eric Kandel "for their discoveries concerning signal transduction in the nervous system" ("The Nobel Prize in Physiology or Medicine 2000" Nobelprize.org. 28 Nov 2012 http://www.nobelprize.org/nobel_prizes/medicine/laureates/2000/), with the contributions of Kandel coming from work on learning and memory in sea hares.

Neuroethological approaches have helped elucidate properties of neurons and circuits involved in behaviors such as feeding, swimming, crawling, and mating (e.g., Katz 2009; Molluscan Neuroscience Gateway: squishybrain.org, 28 Nov 2012). Their sensory abilities, especially involving the visual and chemical senses, and aspects of their intra- and interspecific communication are also relatively well studied and are represented in this volume. These include: the use of visual displays, color changes, and camouflage in interactions; the use of pheromones in recognizing potential mates, the perception of chemicals from substrates or conspecifics in determining when and where larval molluscs settle and metamorphose; and the use of chemicals as anti-predatory defenses.

Among the metazoans, the molluscs are second only to the arthropods in the number of extant species. Such incredible diversity in biological form and function is evident from species that inhabit every major biome on earth, exploit natural products chemistry to produce a hard exoskeleton or toxins in defense, and have modified a head and foot into one solid mass or into distinct tentacles or arms and cephalic region. Given tremendous opportunities for comparative research, it is surprising that so many molecular and cellular studies have focused on only a few of the available species. This approach has yielded significant findings regarding fundamental principles of electrical conduction, learning, memory, and other subjects. Moreover, the few chosen species often have been endowed by unique and experimentally exploitable morphological traits (for example, neurons with giant axons or large and accessible somata), or simple, yet distinctive, behaviors (for example, gill withdrawal reflex, food ingestion reflex, escape responses) that lend themselves to effective analyses at molecular, cellular, and organismal levels.

Many of these model animals, however, are not major contributors structuring ecological communities. To our knowledge, there is no evidence that predation or herbivory by any of these major molluscan models, such as squids (e.g., *Doryteuthis*), *Octopus*, sea hares (*Aplysia* spp.), sea slugs (e.g., *Pleurobranchaea californica, Hermissenda crassicornis, Tritonia diomedea*), or the great pond snail (*Lymnaea stagnalis*), regulates species abundances or distributions in any natural community. They also lack competitive dominance for space, food, or other critical resources. Similarly, their actions fail to facilitate significantly positive biotic interactions towards establishing and maintaining biodiversity. Although effective for understanding basic molecular, cell, and organismal processes, these species function rather poorly towards elucidating fundamental principles at higher levels of biological organization.

The ultimate goal of neuroecology is to span from molecules to communities. Exciting opportunities thus await investigators who work collaboratively on common species to cross broad intellectual expanses. Given recent advances, it now is possible to target ecologically vital organisms that may have escaped investigations on molecular and cellular neural processes in the past. Bivalvia is the least understood of all major molluscan classes with respect to peripheral and central neural processing. New findings, however, are challenging established theory on such basic functions as the role of chemosensation in food detection (Pales Espinosa et al., 2009, 2010). Mussels (*Mytilus* spp., *Modiolus/Geukensia* spp., and *Bathymodiolus thermophilus*), oysters (*Crassostrea* spp. and *Ostrea* spp.), and clams (*Calyptogena magnifica* and *Mercenaria mercenaria*) are amongst the most important foundation species on earth (Witman, 1985; Stachowicz, 2001; Bruno et al., 2003; Kimbro and Grosholz, 2006; Norkko et al., 2006). They dominate space while serving as critical ecosystem engineers within muddy embayments, along rocky, wave-swept shores, and amid deep-sea hydrothermal vents. For Pacific oysters (*Crassostrea gigas*), the entire genome is sequenced and annotated (Zhang et al., 2012). This resource would facilitate greatly studies at the molecular and cellular levels. Many bivalves exhibit complex behavioral and ecological interactions, even though natural selection has promoted small brain size and a reduction in the fusion of central neural elements over evolutionary time. Surprising discoveries undoubtedly will emerge at the interface between sensory activation, behavioral response, and ecological consequence.

Many gastropod species similarly play seminal roles in structuring communities. They often are numerically abundant, voracious carnivores. Predation by whelks (*Nucella* spp., *Busycon* spp., *Cominella adspersa*, *Morula marginalba*, *Urosalpinx cinerea*, and *Stramonita haemastoma*) modifies competitive hierarchies and drives prey population dynamics (Connell, 1972; Lively, 1986; Hughes and Burrows, 1993; Navarrete, 1997; Ferner et al., 2009). When occurring on foundation species prey, such predation creates gaps and enables habitat colonization by sub-dominant competitors to increase biodiversity (Connell, 1970). Moreover, ingestive conditioning and prey switching behavior promote prey population stability and species coexistence (Murdoch, 1969). These distinctive behavioral traits and strong community-wide impacts make excellent candidates for studies that integrate sensory biology with trophic ecology, and scale from molecules to entire species assemblages.

It is our hope that this volume not only provides an overview of what is known about the neuroecology of molluscs, but that it also serves as an inspiration for a more comprehensive exploration of the breadth of neuroecological issues.

References

Bruno, J.F., J.J. Stachowicz, and M.D. Bertness. 2003. Inclusion of facilitation into ecological theory. *Trends Ecol. Evol.* 18: 119-125.

Catterall, W.A., S. Cestèle, V. Yarov-Yarovoy, F.H. Yu, K. Konoki, and T. Scheuer. 2007. Voltage-gated ion channels and gating modifier toxins. *Toxicon* 49: 124-141.

Cestèle, S., and W.A. Catterall. 2000. Molecular mechanisms of neurotoxin action on voltage-gated sodium channels. *Biochimie* 82: 883-892.

Connell, J.H. 1970. A predator-prey system in the marine intertidal region. I. *Balanus glandula* and several predatory species of *Thais*. *Ecol. Monogr.* 40: 49-78.

Connell, J.H. 1972. Community interactions on marine rocky intertidal shores. *Annu. Rev. Ecol. Syst.* 3: 169-191.

Davic, R.D., and H.H. Welsh, Jr. 2004. On the ecological roles of salamanders. *Annu. Rev. Ecol. Evol. Syst.* 35: 405-434.

Deisseroth, K. 2011. Optogenetics. *Nature Methods* 8: 26-29.

Derby, C.D. 2007. Escape by inking and secreting: marine molluscs avoid predators through a rich array of chemicals and mechanisms. *Biol. Bull.* 213: 274-289.

Derby, C.D. 2009. Symposium overview: neuroecology of chemical senses. Proceedings of the International Symposium on Olfaction and Taste, ed., Thomas Finger. *Ann. N.Y. Acad. Sci.* 1170: 447-449.

Derby, C.D., and J.F. Aggio. 2011. The neuroecology of chemical defenses. *Integr. Comp. Biol.* 51: 771-780.

Derby, C.D., and R.K. Zimmer. 2012. Neuroecology of predator-prey interactions. Pp. 158-171 in C. Brönmark and L.-A. Hansson, eds. Chemical Ecology in Aquatic Systems. Oxford University Press, London.

Dusenbery, D.B. 1992. Sensory Ecology. Freeman Press, New York.

Fenno, L., O. Yizhar, and K. Deisseroth. 2001. The development and application of optogenetics. *Annu. Rev. Neurosci.* 34: 389-412.

Ferner, M.C., D.L. Smee, and M.J. Weissburg. 2009. Habitat complexity alters lethal and non-lethal olfactory interactions between predators and prey. *Mar. Ecol. Prog. Ser.* 374: 13-22.

Ferrer, R.P., and R.K. Zimmer. 2012. Community ecology and the evolution of molecules of keystone significance. *Biol. Bull.* 223: 167-177.

Hara, T.J. 2011. Gustatory detection of tetrodotoxin and saxitoxin, and its competitive inhibition by quinine and strychnine in freshwater fishes. *Mar. Drugs* 9: 2283-2290.

Hay, M.E. 2009. Marine chemical ecology: chemical signals and cues structure marine populations, communities, and ecosystems. *Annu. Rev. Mar. Sci.* 1: 193-212.

Hughes, R.N., and M.T. Burrows. 1993. Predatory behaviour of the intertidal snail, *Nucella lapillus*, and its effect on community structure. Pp. 63-83 in H. Kawanabe, J.E. Cohen, and K. Iwasaki, eds. Mutualism and Community Organization: Behavioural, Theoretical, and Food-Web Approaches. Oxford University Press, Oxford.

Kato, Y., Y. Shiga, K. Kobayashi, S. Tokishita, H. Yamagata, T. Iguchi, and H. Watanabe. 2011. Development of an RNA interference method in the cladoceran crustacean *Daphnia magna*. *Devel. Genes Evol.* 220: 337-345.

Katz, P.S, ed. 2009. Molluscan Neuroscience: Recent Advances and New Vistas. *Brain Behav. Evol.* 74: 155-246.

Kimbro, D.L., and E.D. Grosholz. 2006. Disturbance influences oyster community richness and evenness, but not diversity. *Ecology* 87: 2378-2388.

Kurzava, L.M., and P.J. Morin. 1998. Tests of functional equivalence: complementary roles of salamanders and fish in community organization. *Ecology* 79: 477-489.

Kvitek, R.G., and C. Bretz. 2004. Harmful algal bloom toxins protect bivalve populations from sea otter predation. *Mar. Ecol. Prog. Ser.* 271: 233-243.

Kvitek, R.G., and C. Bretz. 2005. Shorebird foraging behavior, diet, and abundance vary with harmful algal bloom toxin concentrations in invertebrate prey. *Mar. Ecol. Prog. Ser.* 293: 303-309.

Lively, C.M. 1986. Predator-induced shell dimorphism in the acorn barnacle *Chthamalus anisopoma*. *Evolution* 40: 232-242.

Mello, C.C., and D. Conte, Jr. 2004. Revealing the world of RNA interference. *Nature* 431: 338-342.

Murdoch, W.W. 1969. Switching in general predators: experiments on predator specificity and stability of prey populations. *Ecol. Monogr.* 39: 335-359.

Navarrete, S.A. 1997. Variable predation: effects of whelks on a mid-intertidal successional community. *Ecol. Monogr.* 66: 301-321.

Nevitt, G.A. 2011. The neuroecology of dimethyl sulfide: a global-climate regulator turned marine infochemical. *Integr. Comp. Biol.* 51: 819-825.

Noda, M., H. Suzuki, S. Numa, and W. Stühmer. 1989. A single point mutation confers tetrodotoxin and saxitoxin insensitivity on the sodium channel II. *FEBS Lett.* 259: 213-216.

Norkko, A., J.E. Hewitt, S.E. Thrush, and G.A. Funnell. 2006. Conditional outcomes of facilitation by a habitat-modifying subtidal bivalve. *Ecology* 87: 226-234.

Paine, R.T. 1969. A note on trophic complexity and community stability. *Am. Nat.* 103: 91-93.

Pales Espinosa, E., M. Perrigault, J.E. Ward, S.E. Shumway, and B. Allam. 2009. Lectins associated with the feeding organs of the oyster *Crassostrea virginica* can mediate particle selection. *Biol. Bull.* 217: 130-142.

Pales Espinosa, E., D. Hassan, J.E. Ward, S.E. Shumway, and B. Allam. 2010. Role of epicellular molecules in the selection of particles by the blue mussel, *Mytilus edulis*. *Biol. Bull.* 219: 50-60.

Paul, V.J., K.E. Arthur, R. Ritson-Williams, C. Ross, and K. Sharp. 2007. Chemical defenses: from compounds to communities. *Biol. Bull.* 213: 226-251.

Power, M.E., D. Tilman, J.A. Estes, B.A. Menge, W.J. Bond, L.S. Mills, G. Daily, J.C. Castilla, J. Lubchenko, and R.T. Paine. 1996. Challenges in the quest for keystones. *BioScience* 46: 609-620.

Purves, D., and J.W. Lichtman. 1985. Principles of Neural Development. Sinauer Associates, Sunderland, MA.

Riffell, J.A. 2011. The neuroecology of a pollinator's buffet: olfactory preferences and learning in insect pollinators. *Integr. Comp. Biol.* 51: 781-793.

Robertson, R.M. 2004. Neuroecology. In G. Adelman and B.H. Smith, eds. Encyclopedia of Neuroscience, 3rd edition. Elsevier, Amsterdam.

Satin, J., J.W. Kyle, M. Chen, P. Bell, L.L. Cribbs, H.A. Fozzard, and R.B. Rogart. 1992. A mutant of TTX-resistant cardiac sodium channels with TTX-sensitive properties. *Science* 256: 1202-1205.

Sherry, D.F. 2005. Neuroecology. *Annu. Rev. Psychol.* 57: 167-197.

Smith, K.G. 2006. Keystone predators (eastern newts, *Notophthalmus viridescens*) reduce the impacts of an aquatic invasive species. *Oecologia* 148: 342-349.

Smulders, T.V., K.L. Gould, and L.A. Leaver. 2010. Using ecology to guide the study of cognitive and neural mechanisms of different aspects of spatial memory in food-hoarding animals. *Phil. Trans. R. Soc. B* 365: 883-900.

Stachowicz, J.J. 2001. Mutualism, facilitation, and the structure of ecological communities. *BioScience* 51: 235-246.

Williams, B.L. 2010. Behavioral and chemical ecology of marine organisms with respect to tetrodotoxin. *Mar. Drugs* 8: 381-398.

Witman, J.D. 1985. Refuges, biological disturbance, and rocky subtidal community structure in New England. *Ecology* 55: 421-445.

Young, J.Z. 1938. The functioning of the giant nerve fibres of the squid. *J. Exp. Biol.* 15: 170-185.

Zhang, G., X. Fang, X. Guo, L. Li, R. Luo, F. Xu, P. Yang, L. Zhang, and 78 additional co-authors. 2012. The oyster genome reveals stress adaptation and complexity of shell formation. *Nature* 490: 49-54.

Zimmer, R.K., and C.A. Butman. 2000. Chemical signaling processes in the marine environment. *Biol. Bull.* 198: 168-187.

Zimmer, R.K., and C.D. Derby. 2007. Biological Bulletin Virtual Symposium: Neuroecology of chemical defenses. *Biol. Bull.* 213: 205-207.

Zimmer, R.K., and C.D. Derby. 2011. Neuroecology and the need for broader synthesis. *Integr. Comp. Biol.* 51: 751-755.

Zimmer, R.K., and R.P. Ferrer. 2007. Neuroecology, chemical defense, and the keystone species concept. *Biol. Bull.* 213: 208–225.

Zimmer, R.K., D.W. Schar, R.P. Ferrer, P.J. Krug, L.B. Kats, and W.C. Michel. 2006. The scent of danger: tetrodotoxin (TTX) as an olfactory cue of predation risk. *Ecol. Monogr.* 76: 585-600.

Zupanc, G.H.K. 2004. Behavioral Neurobiology: An Integrative Approach. Oxford University Press, Oxford.

Part 1. Molluscan Chemical Diversity

In: Neuroecology and Neuroethology in Molluscs
Editors: Anna Di Cosmo and William Winlow
ISBN: 978-1-62948-970-4

Chapter 2

Chemical Diversity in Molluscan Communities: From Natural Products to Chemical Ecology

Kirsten Benkendorff
Marine Ecology Research Centre, School of Environmental Science and Engineering, Southern Cross University, Lismore, NSW, Australia

Abstract

Marine natural products chemistry commenced in the early 1900s when the first marine pigment was isolated and identified from a predatory neogastropod. Growing interest in marine secondary metabolites throughout the 1960-80s led to the initiation of Faulkner's annual reviews of marine natural products, which have since been continued by Blunt and co workers. These reviews provide an excellent starting point for investigating the current state of knowledge on molluscan natural products. From these reviews, data from 828 publications encompassing 277 molluscan species and reporting 1145 distinct secondary metabolites was collated. This was used to examine the global distribution of molluscan natural products research, with respect to climate, geographical location, habitats and depth. The diversity, bioactivity and biosynthetic origin of molluscan secondary metabolites are also examined with respect to trophic niche and morphological defense. It was found that the majority of species subject to natural products research have been collected from subtidal rocky reefs in tropical and warm temperate locations and represent grazing gastropods with reduced physical protection (mostly opisthobranchs). Some interesting chemical defense strategies in these species include: the alarm pheromones produced by mucus trail following Cephalaspideans; the multifunctional defensive ink of Anapsidean sea hares; defensive autonomy and compounds that stimulate the regeneration of autonomous cerata in some Saccoglossans and nudibranchs; kleptocnides and the associated loss of chemical defense in aeolid nudibranchs; and aposeomatic colouration in many nudibranch species that contain toxic compounds. Some interesting neurotoxic compounds have also been identified in shelled predatory caenogastropods, and a few cephalopods with venom glands, to subdue prey. Further studies are required to investigate the ecological role of most molluscan natural products and to expand to field of search to encompass representative molluscs with unique niches in poorly studied marine communities.

Introduction

As the second largest animal phylum, molluscs currently represent 23% of all named marine species (52,000 and many more unnamed, (Bouchet, 2006). They are the dominant fauna in many marine communities, particularly rocky reefs, where they are often used as indicators for marine species diversity (Benkendorff and Davis, 2002). Molluscs are typically slow moving or sessile and occupy a wide range of ecological niches. With a long history of marine evolution, natural selection has led to an arsenal of chemical compounds that enable molluscs to gain a competitive edge in this pathogen and predator rich environment. Investigation into the diversity and function of molluscan secondary metabolites is a vital step towards developing a comprehensive understanding of how these chemicals might help structure marine populations, communities and ecosystems.

Marine natural products research provides an insight into the diversity of compounds produced or sequestered by molluscs. Since the 1960s, a significant amount of research has focused on molluscan chemistry (Faulkner, 2000a), with a total of 828 papers summarised in the annual reviews of marine natural products (Blunt et al., 2008, Blunt et al., 2007b, Blunt et al., 2012, Blunt et al., 2011, Blunt et al., 2010, Blunt et al., 2006, Blunt et al., 2005, Blunt et al., 2004, Blunt et al., 2003, Faulkner, 2002, Faulkner, 2000b, Faulkner, 1996, Faulkner, 2001, Faulkner, 1986, Faulkner, 1984a, Faulkner, 1984b, Blunt et al., 2007a, Blunt et al., 2009). On average 30 papers have been produced per year (since 1987) from studies undertaken on between 13-22 molluscs (Figure 1a). Despite this consistent interest in secondary metabolites from marine molluscs, still less than 1% of potential species in this diverse Phylum have been examined by natural products researchers (Benkendorff, 2010). Nevertheless, an accumulated total of 1145 unique secondary metabolites have been reported from molluscs in the annual reviews of marine natural products (Figure 1c). Each year a proportion of publications also report on previously known compounds from different species (Figure 1b). Drawing on this significant body of chemical research, much progress has been made towards understanding the relationship between molluscan secondary metabolite production and phylogenetic diversity (Cimino and Gavagnin, 2006, Benkendorff, 2010).

In recent years, molluscan natural products research has expanded to include in depth studies detailing the bioactivity of the reported compounds (Figure 2a). However, only 48% of all molluscan metabolites reported in the annual reviews have yet been tested for any form of bioactivity (Figure 2). The majority of these studies are driven by human health interests, for example shellfish toxins, or cytotoxins as targets for anticancer agents, mostly from soft-bodied Opisthorbanchs (Figure 2b &c). The “other” bioactivities reported for molluscan metabolites (Figure 2b) primarily involve receptor binding activity, investigated for anticancer potential. As shellfish toxins and anticancer compounds from molluscs are regularly reviewed (Cimino and Gavagnin, 2006, Watkins et al., 2008), these will not be considered further here. By comparison, very little is known about the natural function of most secondary metabolites isolated from marine molluscs.

The multi-disciplinary field of neuroecology has developed to provide an holistic understanding of the role of chemicals in facilitating interactions between species and structuring communities through higher level trophic interactions (Derby and Aggio, 2011).

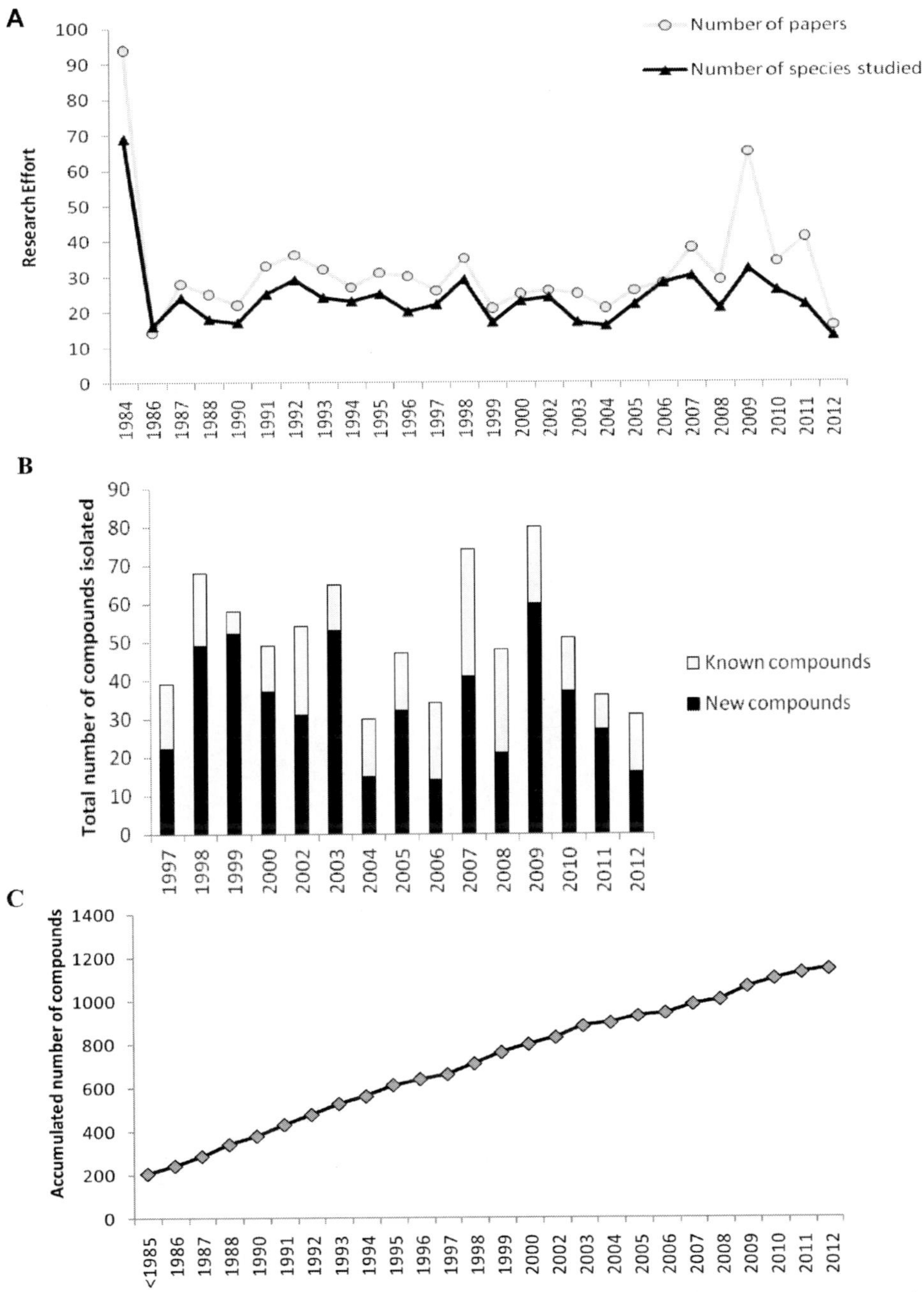

Figure 1. The number of: A) molluscan species studies and publications on molluscan natural products; B) novel and previously known compounds isolated from marine molluscs; and C) the accumulated number of distinct secondary metabolites reported from molluscs in the annual reviews of marine natural products. Data complied from Faulkner 1984-2002; Blunt et al. 2003-2012.

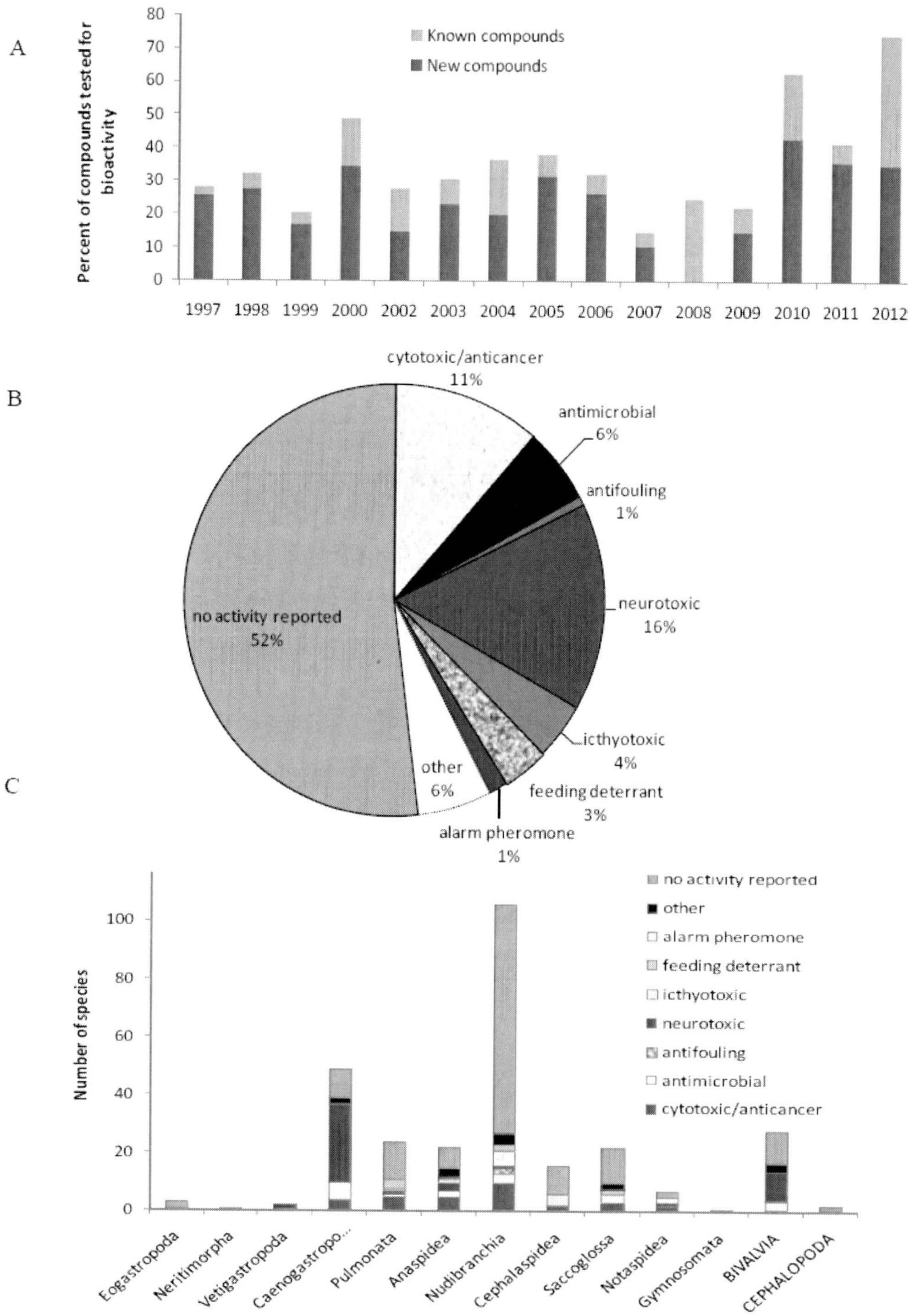

Figure 2. Bioactivity assessment of molluscs; A) the percent of novel and known molluscan secondary metabolites tested for bioactivity; B) the proportion of all species with compounds that are active in various bioassays; and C) in the number of species in the different molluscan taxa (Classes or Orders of Gastropoda) with reported bioactivity.

This extends traditional research in the fields of marine natural products and chemical ecology, by detailed investigation into the mechanistic explanations, at the cell and molecular level (Derby and Aggio, 2011). Thus neuroecology effectively integrates research across all scales of biology, from the molecule through to cell, individual organism, species, population, community and ecosystem level interactions. Recently, a number of secondary metabolites found in molluscs have been identified as "molecules of keystone significance" (Derby and Aggio, 2011, Zimmer and Ferrer, 2007, Williams, 2010) and these provide good models for neuroecological studies into the multi-functional interactions involved in some marine chemical defenses systems. However, as neuroecology is only a relatively recent development, this review focuses on synthesizing the current information on molluscan chemical diversity and chemical ecology, examining biases in the literature and identifying future targets for neuroecological research. Information on habitat, geographical distribution, morphological defense and diet, for all species represented in the annual review of marine natural products, was obtained from various sources (Cimino and Gavagnin, 2006, Cimino and Ghiselin, 2009, Rudman, 2010, Appeltans et al., 2012).

Ecologically Relevant Bioactivity

Amongst the marine invertebrates, molluscs are considered relatively well studied for multiple aspects of their chemical ecology (Paul et al., 2011), despite the limited number of publications to date. Understanding of marine chemical ecology was initially derived largely from parallel investigations in terrestrial models (Paul, 1992), which has led to a focus on defense against consumers. Many of the earlier studies investigating the defensive role of molluscan metabolites used fresh water fish to test for ichthyotoxicity (e.g., Thompson et al., 1982, Hochlowski et al., 1983, Ichiba and Higa, 1986, Cimino et al., 1988, Bobzin and Faulkner, 1989, Miyamoto et al., 1995, Hellou et al., 1982). These types of studies are not ecologically relevant and toxic metabolites do not always deter predators (Pawlik, 1993). Because different predators will respond differently to chemical defences, a combination of field based bioassays and aquarium bioassays with ecologically relevant predators are typically required for comprehensive assessment (e.g., Pawlik et al., 1986, Rogers et al., 1995, Pennings, 1994, Pennings et al., 1999, Neves et al., 2009). Other limitations in previous feeding trials include the use of extracts or secondary metabolites at much higher concentrations than they exist naturally and adding these to an attractive food source that may be higher in nutritional value, thus masking the unpalatability. Only ~3% of molluscan metabolites have been tested for feeding deterrent activity in ecologically relevant experiments (Figure 2b). A number of additional studies have examined the feeding deterrent activity of mucus secretions from molluscs against relevant predators (Bancala, 2009, San-Martin et al., 2009), but without isolation and identification of the active factors.

In the last decade, significant advances have been made in specific areas of marine chemical communication including predator–prey and seaweed–herbivore interactions, defenses against microbial and fouling organisms, competitive interactions and settlement cues (Paul et al. 2011). When investigated in detail, chemically mediated interactions between species are usually found to be a lot more complex that the often assumed toxic or feeding deterrent role of secondary metabolites. For example, investigations in seahares (*Aplysia* spp.)

have revealed that dietary accumulation of some algal secondary metabolites comes at a cost, with significant impacts on fitness, including egg production and growth (Rogers et al., 2002). Furthermore some algae compounds, such as palisadin A are selectively transferred to mucous and opaline secretions, but were ineffective against natural assemblages of predatory fishes or sea spiders (Rogers et al., 2000a, Rogers et al., 2002, Rogers et al., 2000b, Rogers et al., 1995). Thus overall, the benefit from acquiring these toxic dietary metabolites is unclear, but it is assumed they form part of a broader defensive strategy including chemical camouflage on host algae, making the sea hares less likely to be detectable by scent. Increasingly sophisticated experiments are being developed to investigate the behavior of predators and prey in response to chemical cues. A good example of this is the detailed investigation into the ecological function of Aplysia ink and opaline secretions, which examine sensory detection, in addition to aversion behaviors in predators (Sheybani et al., 2009, Aggio and Derby, 2008, Shabani et al., 2007). Similar studies have been undertaken on Cephalopod ink (Wood et al. 2010) and there is evidence that melanin containing ink can invoke an alarm response in conspecific squid (Wood et al., 2008).

Few studies on molluscan natural products report pheromone activity (Figure 2). In Cephalaspideans, alkyl phenols and alkaloids (haminols) are reported as alarm pheromones (Figure 2c) (Della Sala et al., 2007, Cimino et al., 1991). These gastropods follow conspecific mucus trails and when threatened they release haminols that elicit an escape response in conspecifics (Cimino and Ghiselin, 2009, Cimino et al., 1991). Water borne, peptide based pheromones stimulating mate attraction have been isolated from *Aplysia* spp. (Cummins et al., 2008, Cummins and Degnan, 2010, Painter et al., 2003, Painter et al., 1998). Reproductive behavior in both gastropods and cephalopods is also thought to be regulated by soluble peptide and protein pheromones (Susswein and Nagle, 2004, Kuanpradit et al., 2010, Kuanpradit et al., 2011). Other pheromones regulating behaviors in molluscs include prostaglandins and eicosanoids, which play roles in egg production, spawning, neurophysiology and defense (Avila, 2006, Moulis, 2006).

In a detailed review, Hay (2009) outlines the important role of chemical signaling in the functioning of marine ecosystems. In addition to significant advances in understanding marine chemical communication in predator–prey and seaweed–herbivore interactions, the field of marine chemical ecology has progressed with investigation into defense against microbial and fouling organisms, competitive interactions, and settlement cues (Paul et al., 2011). However, in the marine natural products literature, little progress has been made in assessing chemical interactions between molluscs and microorganisms. All of the natural products reported to have antimicrobial activity in the annual reviews of marine natural products (Figure 2b) were tested against human pathogenic bacteria, fungi and viruses. Nevertheless, antimicrobial activity in the extracts of egg masses from seven mollusc species has been demonstrated against some marine pathogens (Benkendorff et al., 2001b). The active factors were found to be brominated indoles in the leathery egg capsules from the Muricidae family of Neogastropods (Benkendorff et al., 2001a, Benkendorff et al., 2000), whereas combinations of polyunsaturated fatty acids modeled on the lipid composition extracted from gelatinous egg masses were found to contribute significantly towards the antimicrobial activity against marine pathogens (Benkendorff et al., 2005).

Throughout their life cycle, molluscs rely on an innate immune system (Hooper et al., 2007a), which includes the constitutive biosynthesis of antimicrobial defence factors (Li et al., 2009, Mitta et al., 2000a, Mitta et al., 2000b, Mitta et al., 2000c, Cellura et al., 2007).

Infection by microbial pathogens represents a type of biotic stress, which can lead to inducible defense reactions. For example, some antimicrobial peptides and proteins produced by commercial bivalves are induced by bacterial infection (Mitta et al., 2000b). Assays for antibacterial activity against marine pathogens and antiviral activity against a model herpes virus have recently been developed to screen the hemolymph plasma of abalone and oysters in response to a variety of environmental stressors (Dang et al., 2011a, Dang et al., 2011b). Further studies are required to identify the active factors in abalone and other gastropods. Chemical interactions with marine micro-organisms are of particular interest considering the recent global increases in the frequency and intensity of disease in marine mollusc populations (Marcogliese, 2008). Notable examples of mass mortalities in commercial molluscs are the range expansion of oyster diseases (Cook et al., 1998, Ford, 1996, Ford and Smolowitz, 2007), viral outbreaks (Hooper et al., 2007b, Chang et al., 2005) and escalating summer mortality in bivalves and abalone (Sauvage et al., 2009). Increased incidence and prevalence of marine disease is likely to have substantial ecological consequences for species persistence and ecosystem function (Hoegh-Guldberg and Bruno, 2010) and is sign of significant deterioration of ocean ecosystem health.

Assessment of anti-fouling activity in marine natural products typically focuses on soft-bodied sessile marine invertebrates and has only been reported for compounds from two nudibranch species (Figure 2c). However, any undefended substrate in the marine environment is a potential site for colonization and the formation of biofilms which could then facilitate pathogenic infiltration, proliferation and invasion. Ramasamy and Murugan, (2005) reported antimicrobial activity in a broad range of molluscs against a panel of 40 biofilm bacteria. Przeslawski and Benkendorff, (2005) have demonstrated that significantly higher algal biofilm formation and colonization by protists occurred on gelatinous molluscan egg capsules, compared to the leathery egg masses of Neogastropods. This suggests some antifouling defense in the leathery egg capsules, which may include a combination of physical and chemical mechanisms (Lim et al., 2007). Anti-fouling activity against mussel byssus formation was demonstrated in extracts from the egg capsules of three Murcidae species (Ramasamy and Murugan, 2007, Ramasamy and Murugan, 2005). Peters et al., (2012) have also demonstrated that the biofilms on the surface of gelatinous *Siphonaria diemenensis* egg masses differed significantly from intertidal rocky substrate biofilms. One epibiotic bacteria was isolated which had antibacterial activity (Peters et al. 2012), thus suggesting complex chemical interactions can occur in molluscan associated biofilm communities. Further investigation of molluscan compounds, secretions and epibionts that interfere with bacterial signaling and biofilm formation, would also be valuable because such compounds could significantly influence the structure and development of fouling communities.

The Global Distribution of Molluscan Natural Products Research

Molluscan natural products research has generally focused on species from tropical and temperate locations (Figure 3a). The uneven distribution in collecting efforts across different oceanic regions (Figure 3b) in part reflects convenience to some of the major research centres in marine natural products.

A

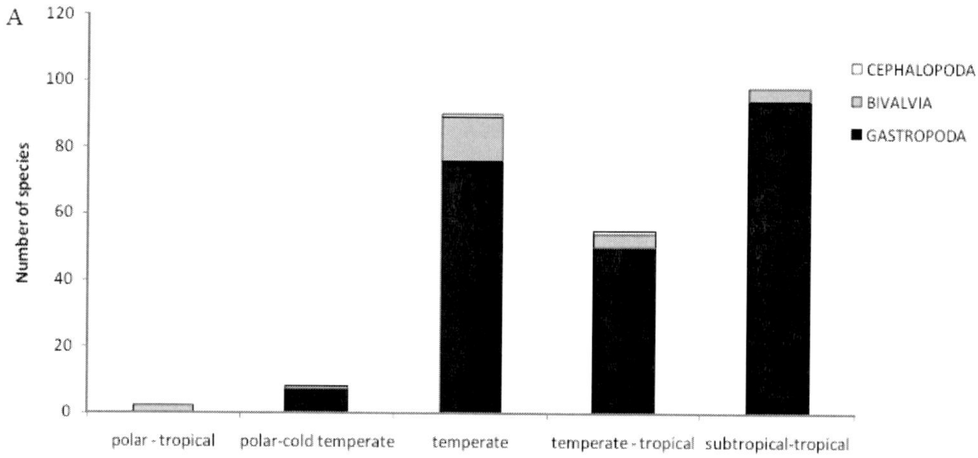

B

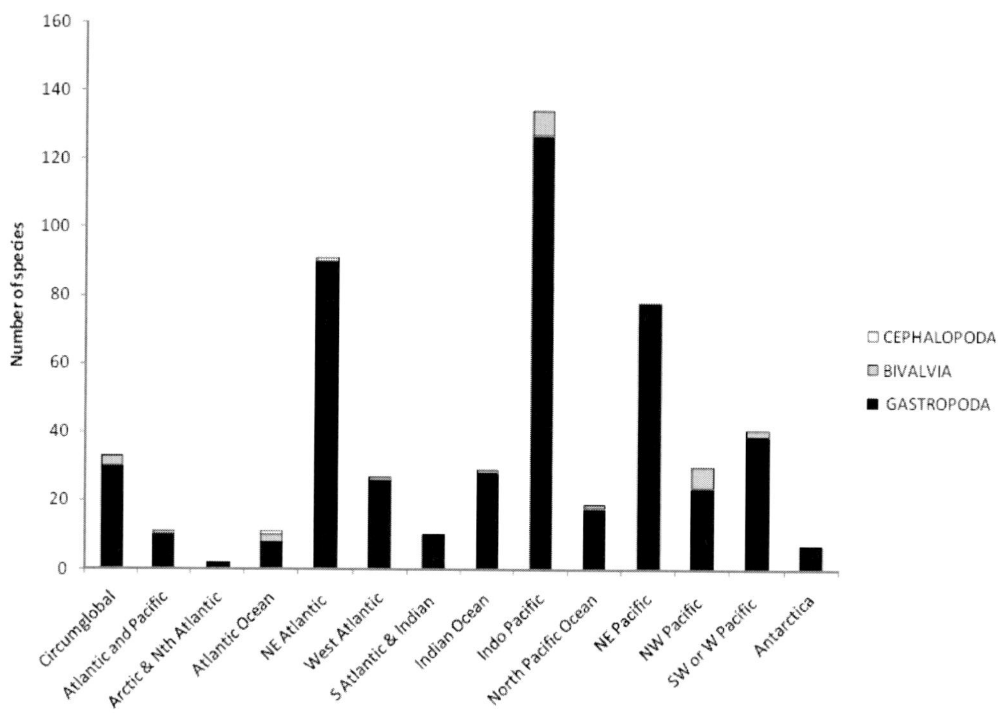

Figure 3. The geographic distribution of marine molluscs subject to natural products research, as reported in the annual reviews of marine natural products (Faulkner 1997-2002; Blunt et al. 2003-2012): A) distribution according to major climate zones; B) oceanic distribution with major oceans combined and divided into smaller regions to account for broadly distributed and regionally restricted species. Species from the Mediterranean Sea and Black Sea are included under NE Atlantic, species from the Red Sea and Persian Gulf are included under Indian Ocean, species from the Gulf of Mexico are included under West Atlantic.

For example, significant research has been undertaken on Mediterranean & North East Atlantic molluscs by Cimino and co-workers (Cimino and Gavagnin, 2006, Cimino and Ghiselin, 2009, Cimino, 2010). Andersen et al., (2006) provide a detailed review of nudibranch skin chemistry from the west coast of North America (NE Pacific, Figure 3b), reporting over 100 compounds from 27 species. This is based primarily on studies from Faulkner and coworkers at the Scripps Institute of Oceanography, as well as Andersen's group at the University of British Columbia. Significant research has also been undertaken on Japanese opisthobranchs (NW Pacific, Figure 3b), by several leading marine natural products researchers, as reviewed by Miyamoto, (2006). The intense research effort on Indo-Pacific molluscs, in part, reflects the huge diversity of marine species from this area, which have lead to a number of international collaborative investigations. For example, Wahidullah et al., (2006) report on some of these collaborative investigations into less-studied areas of the Indian, Chinese and Egyptian coast, reporting 99 compounds from 20 species. Some Indo-Pacific species have also been studied in Australia, although the majority of Australian studies have focused on temperate species from the south west Pacific. Garson, (2006) provides a review of chemical and ecological studies from Australia and New Zealand, reporting 89 compounds from a range of opisthobranchs, pulmonates and one caenogastropod.

Very few species have been collected from Antarctica or cold southern oceans (Figure 3b), with secondary metabolites from only eight species found in cold temperate or polar regions reported in the annual reviews of marine natural products. Davies-Coleman, (2006) provide a review of secondary metabolites from Antarctica, South Africa and South America, listing 100 metabolites from 21 species, most of which are shell-less opisthobranchs. A review of chemical ecological studies on Antarctic opisthobranchs is provided by McClintock and Baker, (1997). As discussed by Avila et al., (2000) benthic communities in Antarctica are considered to be controlled mainly by biological factors, such as predation and competition, which along with the slow growth rates and long life spans, provides significant selective pressures for the evolution of defensive metabolites in Antarctic marine invertebrates. Sea stars are the main predators in these communities and indeed Antarctic molluscs do show significant feeding deterrence against predatory sea stars (Mc Clintock and Baker, 1997; Avila et al., 2000; Davis and Coleman, 2006). Further examination of the role of molluscan secondary metabolites in structuring Antarctic marine communities would expand our knowledge of ecosystem structure in polar regions.

A small proportion of the molluscan species examined to date have broad oceanic distributions, with 15 listed as circumglobal and two of these are bivalves that range from cold temperate to tropical zones (Figure 3). A further 55 species have distributions that occur from temperate to tropical regions (Figure 3a). These species provide an opportunity for comparative investigation into the importance of chemical defense in different ecosystems. For example, there is limited and conflicting evidence for the importance of temperature generating latitudinal gradients in marine secondary metabolites. In most cases, the studies undertaken are confounded by inter-specific comparison between tropical and temperate regions. However, a strong decreasing trend in the sponge metabolite salicylihalamide A was found with increasing temperature (Abdo et al., 2007). Conversely, Becerro et al., (2003) found no overall latitudinal difference in sponge defenses when comparing multiple phylogenetically paired species of temperate and tropical sponges. The interest in latitudinal gradients in secondary metabolism has centred on the hypothesis that higher species diversity in the tropics leads to great competition and predation pressure than in temperate regions. This could imply a greater need for defense at higher latitudes. However, recent studies on Antarctic molluscs clearly demonstrate a role for chemical defense at low latitudes (Mc

Clintock and Baker, 1997). Geographic differences in chemical defense against microbial pathogens also require consideration. Fuhrman et al., (2008) demonstrated that a latitudinal gradient occurs in marine planktonic bacteria species richness and the virulence of marine pathogenic Vibrios has been correlated with elevated temperatures (Harvell et al., 2002, Harvell et al., 2004, Marcogliese, 2008, Vezzulli et al., 2011). Thieltges et al., (2009) demonstrated a negative correlation between latitude and trematode parasite richness and infection rates in crustaceans. These examples suggest that latitudinal gradients in molluscan secondary metabolites could occur in response to pathogen infection and virulence at higher temperatures, but well designed ecologically relevant experiments are required to test this hypothesis.

Habitat and Molluscan Natural Products Research

As highlighted by Benkendorff (2010) several molluscan classes have been completely overlooked by natural products chemists and this is in part influenced by accessibility according to habitat. Of the neglected molluscan classes, the Polyplacophora are large and common enough to deserve further attention in neuroecological studies. These chitons are often abundant on intertidal rock platforms (Beesley et al., 1998) and many have a similar ecological niche to limpets (Eogastropoda) and false limpets (Siphonaridae, Pulmonata). Also like poyplacophora, these gastropods only have a shell on their dorsal surface, which may have given rise to some interested associated chemistry. Over 30 polyproprionates (Davies-Colman and Garson, 1998) and eight novel fatty acid derivatives have been isolated from 14 *Siphonaria* spp. (Carballeira et al., 2001), some of which have been tested and shown to contain icthyotoxic properties. In the eogastropods 23 fatty acid derivatives have been isolated from just two species (*Cellana grata* and *Collisella dorsuosa,* (Kawashima, 2005, Kawashima and Ohnishi, 2006), whilst in a third species *Lottia (Collisella) limatula*, a terpene with anti-feedant properties has been isolated (Pawlik et al., 1986). Further investigation into this ancestral lineage of gastropods, along with the class Polyplacophora could reveal key chemical interactions shaping rocky intertidal communities. In total, natural products from only 30 gastropods and a further 4 bivalves that live exclusively in the intertidal zone have been investigated to date (Figure 4a). Additional studies on the egg masses of 46 intertidal gastropods have revealed a range of mycosporine-like amino acids (MAAs) that are likely to provide protection to the developing embryos against UV radiation (Przeslawski et al., 2005).

Most of the species subjected to natural products research are associated with subtidal environments (Figure 4), reflecting the primary collection effort by scuba diving. Only a small number of deep sea molluscs have been investigated (6 gastropods and a single bivalve, Figure 4a) most likely reflecting difficulty in accessing these communities. To increase the diversity of deep sea molluscs in natural products research, dredges could be used; for example species in the neglected ancestral classes Aplacophora and Monoplacophora have been reported from deep sea dredges (Beesley et al., 1998). Furthermore, targeted collections using remotely operated vehicles around deep sea vents could be worthwhile. Deep sea vent communities have evolved with chemoautotrophy (Stewart et al., 2005) and bivalves with chemosymbionts for example, could provide a unique niche for exploring these interesting chemical interactions. Only three chemical studies have focused on pelagic or open water molluscs (Figure 4). More studies on pelagic cephalopods would be of particular interest given their predatory behavior, size and apparent dominance in deep sea pelagic communities.

By far the majority of natural products research has focused on molluscs from rocky or coral reef environments (Figure 4b). A subset of gastropods and bivalves, as well as the cephalopods, have habitats that range across reef and soft sediment, including seagrass. Only 36 molluscs that live exclusively in soft-sediment have been studied, including 17 bivalves (clams) and 19 gastropods, mostly from the order Cephalaspidea (Figure 4b).

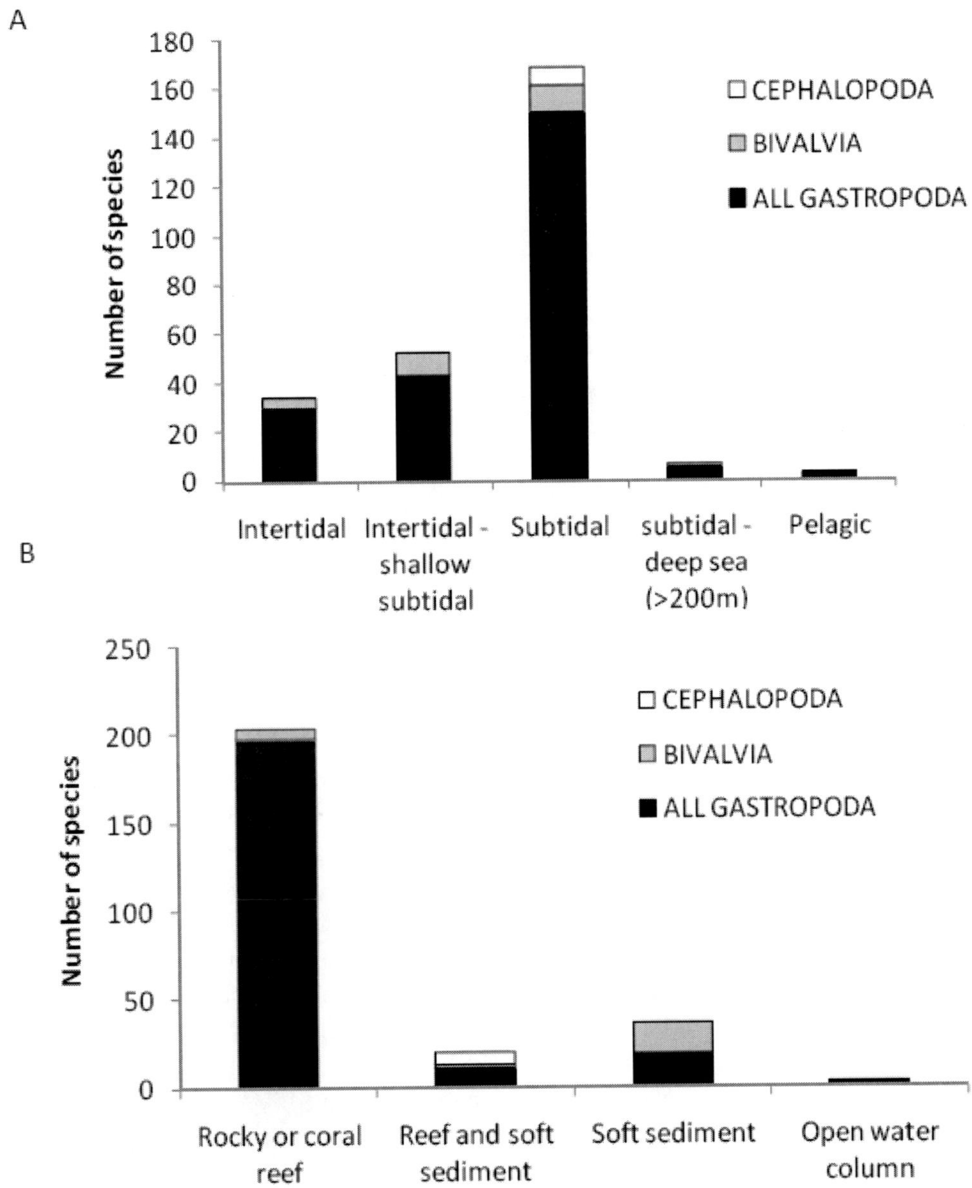

Figure 4. The habitat of marine molluscs subject to natural products research, as reported in the annual reviews of marine natural products (Faulkner 1997-2002; Blunt et al. 2003-2012): A) distribution according to tidal zone; and B) distribution according to primary substrate.

The class Scaphopoda has been neglected to date, but these species live buried in sediment and although small and patchy in distribution, they can often be quite abundant when found. They possess a tubular shell that is open at both ends and many are microcarnivores that probe the sand for meiofauna (Beesley et al., 1998) The taxonomic distinctness of scaphopods, along with their unique lifestyle, may be associated with some novel chemistry worthy of investigation for understanding of the role of chemicals in structuring meiofaunal communities.

Trophic Niche and the Dietary Origin of Molluscan Metabolites

The trophic niche of species subject to chemical investigations is an important consideration because this may influence the types of biological activity that could be expected and their potential dietary origin. For example, a survey of MAAs in intertidal gastropod egg masses revealed that both phylogeny and adult diet influence the chemical composition, with significantly higher levels of MAAs in herbivores than predators (Przeslawski et al., 2005), as might be expected based on closer links to the *de novo* algal source. In general, much less research on molluscan natural products has focused on active predators, relative to algal grazers or grazing carnivores (Figure 5). However, active predators are more likely to produce neurotoxic venoms to sedate or kill their prey than grazers. In recognition of this, the amount of research into neurotoxic peptides produced in the venom glands of predatory species in the family Conidae has greatly increased (Halai and Craik, 2009, Lewis et al., 2012, Myers et al., 1993). There are ~700 species in the Conidae family and each species is thought to possess over 1000 conopeptides, with less than 0.1% characterized to date. Those that have been pharmacologically characterized target a range of membrane proteins with high potency and specificity (Lewis et al., 2012). This chemical diversity and specificity is likely to be the result of strong evolutionary pressure for improved prey-capture and defense, making conotoxins an excellent model for future neuroecological studies.

Another interesting family of predatory marine Neogastropods that have been subjected to natural products research is the Muricidae. Muricids are the most diverse family of predatory marine gastropods, comprising ten recognized subfamilies with about 1600 extant species (Barco et al., 2010). The secondary metabolites have only been examined in ~ 10 species to date, but they have attracted a long history of interest. Marine natural products research commenced in the early 1900s, with the identification of 6,6'dibromoindigo from the Muricidae, as the traditional source of the highly valued pigment Tyrian Purple (Freidlander, 1909) (Baker, 1974). Over half a century later, Baker and colleagues identified the precursor compound, which is stored in these molluscs as a brominated indole-choline ester salt (Baker and Duke, 1973, Baker and Duke, 1976, Baker and Sutherland, 1968). Subsequently, at least five choline esters, 16 indole derivatives, several indolequinones and two brominated imidazoles have been isolated from the Muricidae family (Fukuyama et al., 1998, Benkendorff et al., 2001a, Benkendorff et al., 2000, Benkendorff et al., 2004, Roseghini et al., 1970, Roseghini et al., 1996, Cooksey, 2001).

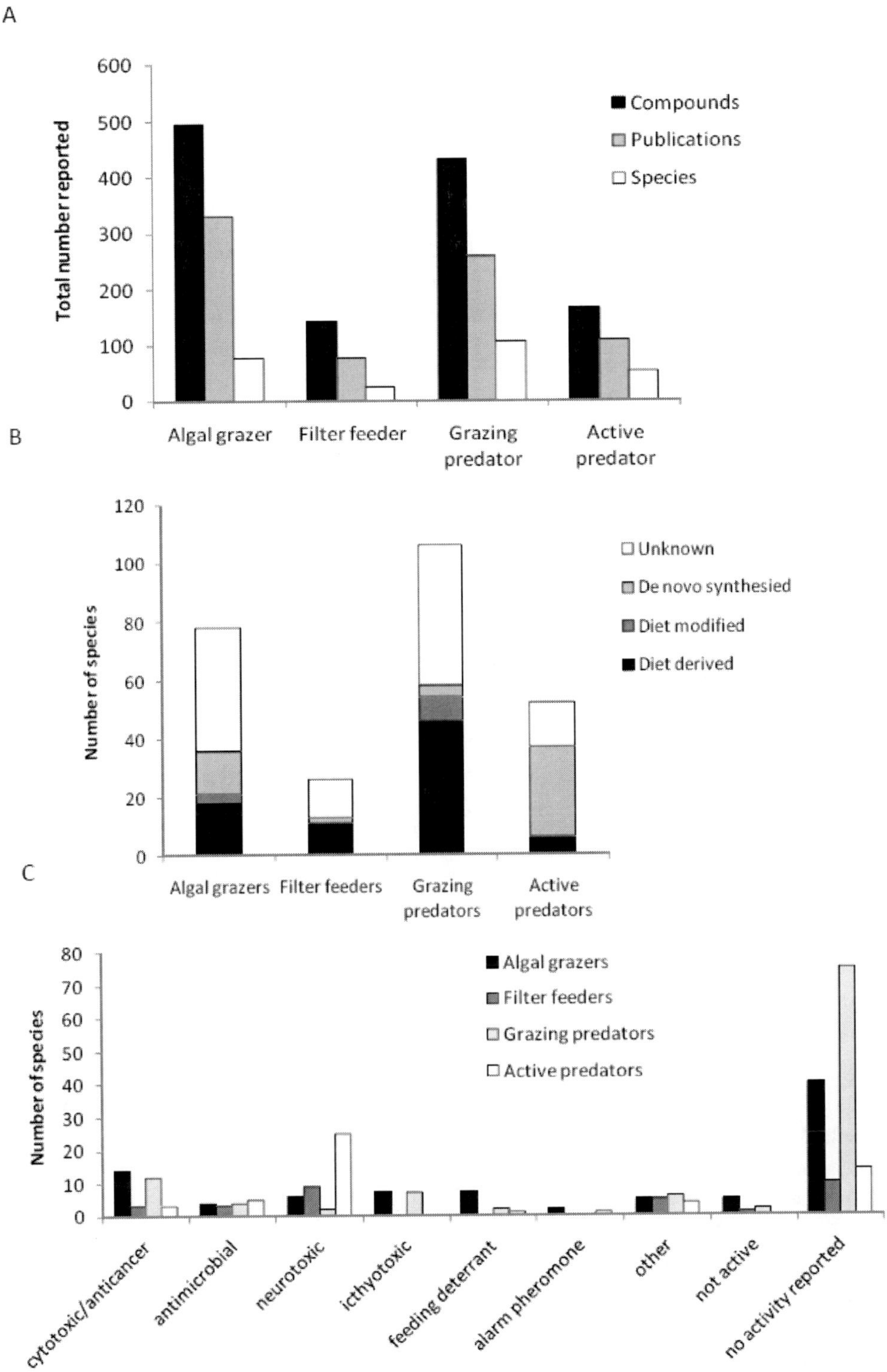

Figure 5. The research effort on marine molluscs grouped according to trophic niche; A) the total number of publications, compounds isolated and species studied; B) the proportion of compounds of with known dietary origin; and c) the proportion of compounds tested for biological activity.

Antibacterial and anticancer activity is associated with the brominated compounds from the Muricidae (Westley et al., 2010b, Magiatis and Skaltsounis, 2006, Meijer et al., 2003, Benkendorff et al., 2001a, Benkendorff et al., 2004, 2011, 2013, Edwards et al., 2012), whereas the choline esters have muscle relaxing activity (Roseghini et al., 1996, Erspamer and Glasser, 1957). Nevertheless, the functional role of these dye precursor and pigments in the molluscs remains uncertain. Benkendorff (2013) and Westley et al. (2006) review the possible role of these compounds in the reproduction, defense and feeding behavior of Muricidae. Further studies in well designed experiments are required to elucidate the full range of ecological roles of the purple secretion and associated compounds in Muricidae.

Only three studies on predatory Cephalopods have featured in the annual reviews of marine natural products to date. However, more papers appear in the toxicology and physiology literature. Studies on the venom and salivary glands of octopus and cuttlefish have revealed several proteins (Cariello and Zanetti, 1977). These proteins have diverged through gene duplication events in different Cephalopod species (Fry et al., 2009b) and some show convergent structure with a range of other animal venoms (Fry et al., 2009a). Tetradodoxin has also been isolated as a neurotoxic compound from the venom gland and egg masses of the blue ringed octopus (Sheumack et al., 1984, Sheumack et al., 1978) and is mostly likely produced by symbiotic bacteria. Tetradodoxin has also been isolated from a range of predatory neogastropods and depending on species, it appears to mediate a range of different roles including predatory venom, antipredator defense, and intra and interspecific communication (Williams, 2010). The broad species distribution and functional roles of tetradotoxin suggests it is a molecule of keystone significance that can affect the ecology of entire communities (Williams, 2010).

Comparatively few studies have been undertaken on filter feeding bivalves (Figure 5a). Yet accounting for the different research effort in the various trophic groups (Figure 5a), the average number of compounds isolated per species per publication is higher in filter feeders (2.3 ± 1.8) and algal grazers (2.4 ± 3.6), than in grazing and active predators (2 ± 1.6 & 1.8 ± 1.2 respectively). This may be due in part to the number of bioaccumulated compounds in algal grazing gastropods and filter feeding bivalves. However, the relative proportion of known dietary derived compounds is highest in the grazing predators (Figure 5b). Dietary derived compounds in the filter feeding bivalves are predominately shellfish toxins and carotenoids, whereas a huge range of dietary derived metabolites have been reported from grazing gastropods. One example is the sequestration of dietary alkaloids from sponges in the Notaspidean *Tylodina perversa* (Thoms et al., 2006). Some of the dietary derived compounds acquired by opisthobranchs are selectively accumulated and a small proportion are biotransformed (Figure 5b) into more or less toxic compounds (e.g., Fontana, 2006b).

Across all grazing gastropods and filter feeding bivalves, the biosynthetic origin is not known for over 45% of compounds (Figure 5b). By comparison, the biosynthetic origin is known for over 70% of natural products from active predators. This is because this group is dominated by conopeptides which can be identified from gene transcriptomes (Lewis et al., 2012, Halai and Craik, 2009). Also amongst the de novo synthesized secondary metabolites from active predators are the brominated compounds from Muricidae. Bromoperoxidase and aryl sulfatase activity in the biosynthetic hypobranchial gland of these snails supports their biosynthetic origin (Westley and Benkendorff, 2008, Westley and Benkendorff, 2009, Jannun and Coe, 1987, Jannun et al., 1981, Westley et al., 2010a). A number of compounds that are known to be de novo synthesized are also found across the other tropic groups, including polyproprionates from herbivorous pulmonates (Garson, 1993, Garson et al., 1994, Manker et al., 1988), polyketide derived pyrones in herbivorous saccoglossans and some terpenes in carnivourous grazing nudibranchs (reviewed by (Fontana, 2006a). Some peptides and proteins are de novo synthesized across all groups, including filter feeders.

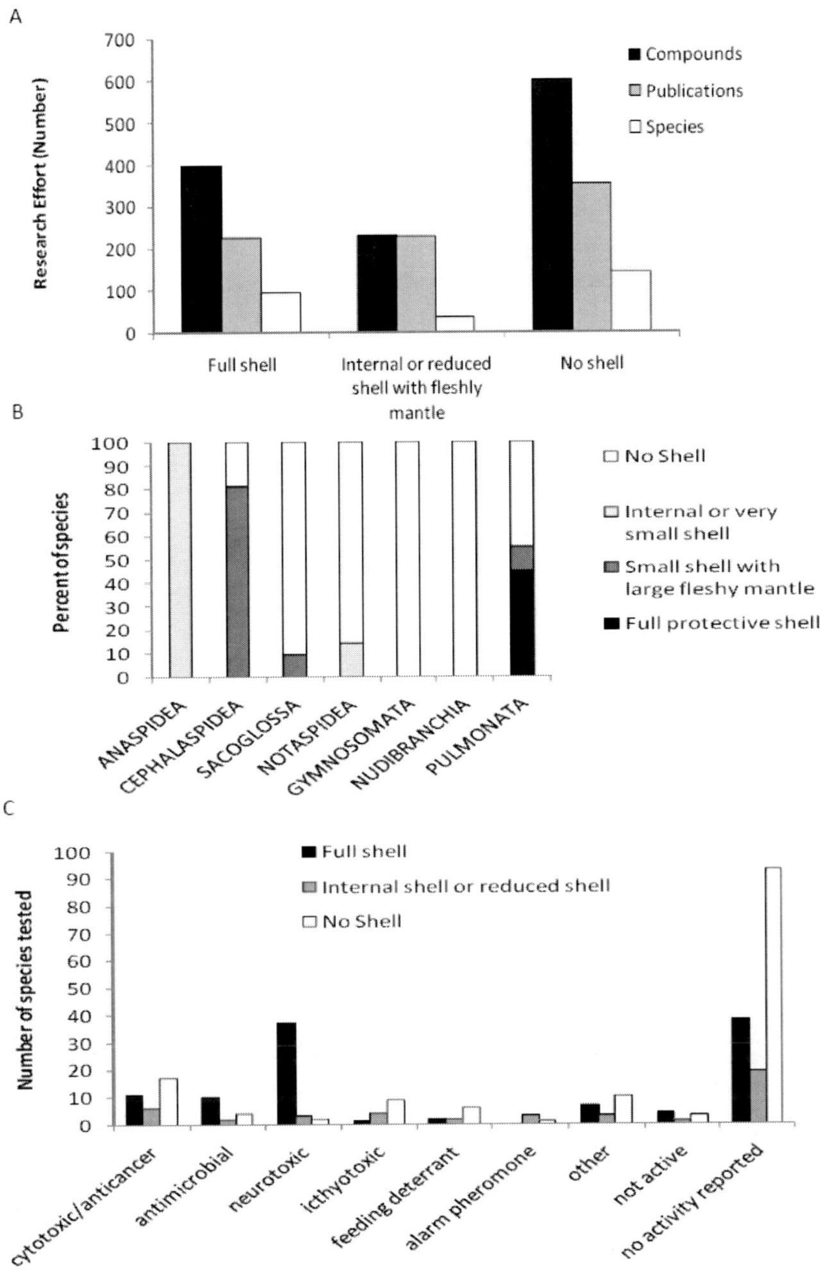

Figure 6. The distribution of research effort on marine molluscs according to the presence of physical defense: A) the total number of compounds isolated, publications and species studied with differing levels of physical protection; B) proportion of species with different levels of physical protection that have been studied within the Heterobranchia; and C) the reported biological activity of compounds isolated from molluscs with different levels of physical protection.

With regards to the influence of trophic niche on bioactivity (Figure 5c), as expected the highest proportion of neurotoxic compounds have been isolated from active predators (venoms) and filter feeding bivalves (shellfish toxins). Icthyotoxicity and feeding deterrence has been most frequently assessed in grazing gastropods (Figure 5c). The infrequency of testing for defense functions in filter feeders and active predatory gastropods is most likely due to their physical protection via the shell (Figure 6c).

Morphological and Chemical Defense

The study of marine natural products has developed alongside chemical ecology with an initial focus on defense against predators (Avila, 2006, Hay, 1996, Paul et al., 2007). Consequently, there has been a strong bias towards investigating shell-less molluscs, which have no apparent physical protection against predators (Avila, 1995, Avila, 2006, Benkendorff, 2010). Although shelled molluscs are more diverse in terms of species richness and abundance, as seen in Figure 6a, a greater number of publications have focused on species without a shell. This is despite the fact that nearly 400 distinct natural products have been isolated from <100 shelled species (Figure 6a). These compounds from shelled molluscs also show an interesting range of bioactivity (Figure 6c). In addition to the neurotoxins mentioned previously, shelled species are associated with anticancer, antimicrobial and even some feeding deterrent and ichthyotoxic properties, thus demonstrating the limitations in predicting interesting chemistry from a lack of physical defense. Nevertheless, research into the natural products from shell-less gastropods has certainly been rewarding with over 600 compounds reported from 145 species (Figure 6a).

Amongst the Heterobranchia, there are a number of distinct extant lineages that represent intermediate stages between shelled "snails" and fully shell-less "slugs" (Cimino and Ghiselin, 2009). The repeated loss of shell in the evolution of these gastropods most likely represents a trade-off between the physical protection of a shell and the cost of producing one. Optimal defense theory predicts that organisms will allocate energy towards defense mechanisms that maximize fitness (Orians and Janzen, 1974). Chemical defense may be less energetically expensive than a shell, but it is not without costs. For de novo synthesis, secondary metabolites are produced from the same chemical precursors as primary metabolites, thus generating a physiological trade-off at the biochemical level (Ivanisevic et al., 2011, Rogers et al., 2002). Even dietary accumulated secondary metabolites would require some resource allocation for transport and storage, and can have potential impacts on fitness (Rogers et al., 2002). Considering all the trade-offs involved in the evolution of chemical defense and the loss of a shell, one interesting question is which came first? By extensive examination of lineages of gastropods that include shelled, shell-less and intermediate forms, Cimino and Ghiselin, (2009) conclude that chemical defense must have originated when the ancestors of slugs still had a shell. This is supported by a proportionally high number of compounds isolated from species with a partial or internal shell (>230 from just 36 species, Figure 6b), including de novo synthesized feeding deterrents and alarm pheromone (Figure 6c).

The first chemical investigation of "shell-less" marine gastropods was actually undertaken on the sea hare *Aplysia kurodai*, which has a small internal shell. From this algal grazer, Yamamura and Hirata, (1963) isolated several brominated terpenes, presumed to be dietary derived. Subsequently, over 200 compounds have been isolated from Anaspidean seahares (Benkendorff, 2010), including many more dietary terpenes, as well as peptides, alkaloids, macrolides, polyproprionate, acetogenin, pyrones, lactones, sterols and fatty acid derivatives. The bioactive molecules from Anapsidea seahares has been reviewed by Kamiya et al., (2006). More recently, the ink produced by seahares has attracted significant interest as a chemical defense strategy. Derby, (2007) highlights the fact that a diversity of chemicals in the ink help mediate a diversity of effects that together help protect the sea hares from a range of different predators. Compounds in the ink, such as Aplysioviolin can act directly to deter predators, whereas indirect effects are conferred by nucleic acids and nucleosides (Kicklighter et al., 2007), as well as an oxidase enzyme (escapin), which act as intra-specific alarm cues (Derby and Aggio, 2011). Free amino acids in the ink also appear to cause sensory disruption (Derby et al., 2007), and along with the acidity of the ink, could potentiate the defensive effects through phagomimicry (Derby and Aggio, 2011). This highlights the complexity of chemical interactions that can contribute to antipredator defense mechanisms in sea hares. Nevertheless, other researchers provide evidence that ink and opaline do not always deter ecologically relevant predators, especially fishes (Pennings et al., 2001, Pennings, 1994, Rogers et al., 2002).

Certainly the absence of a protective shell does appear to necessitate alternative defense mechanisms (Putz et al., 2010). In addition to the sequestration and de novo synthesis of secondary metabolites, marine molluscs with a reduced or no shell use a range of alternative morphological defense mechanisms. These include parapodial flaps that may facilitate a swimming escape response in the Anaspidea and Gymnosomata. Dorsal processes called cerrata are found in some shell-less Saccoglossa and these are often associated with defensive autonomy. Interestingly, de novo synthesised polyproprionates and pyrones in some Saccoglossa appear to stimulate regeneration of the autonomised cerrata (Cimino and Ghiselin, 2009, Di Marzo et al., 1993). However, in at least one species that does not autonomise, *Placida dendritica*, the polyproprionate pyrones do not stimulate regeneration and appear to play a purely chemical defensive role (Cimino and Ghiselin, 2009). Several species in the Order Nudibranchia also possess cerrata that can be autonomised and a few species undergo autonomy of the mantle tissue. In nudibranchs, such as *Tethys fimbra* defensive metabolites derived from prostaglandins are implicated in the regeneration of autonomised cerrata (Cimino and Ghiselin, 2009).

Some other nudibranchs contain cerrata with cnidosacs that have been acquired from their cnidarians prey without discharging (Greenwood, 2009). This defense strategy, known as kleptocnides, has evolved in the Aeolidoidea nudibranchs which do not have the defensive glands found in other Cladobranchia (Putz et al., 2010). Only a few pigments and some sterols have been isolated from Aeolids containing cnidocytes, suggesting that they may rely entirely on kleptocnides for defense. Nevertheless, there is limited evidence that the dietary acquired cnidocytes provide defense in the absence of secondary metabolites (Greenwood, 2009, Penney et al., 2010). A few species of Aeolids that do not contain functional cnidocysts sequester deterrent chemicals from their octocoral prey (Putz et al., 2010), suggesting diverse defense strategies have evolved as part of the predator-prey interactions within this clade. Other nudibranchs sequester spicules from their sponge diets, which could contribute to a combined chemical and physical defense strategy. Other morphological features found in some nudibranchs and notaspideans include tubercles or hairy papillae. These may facilitate textural camouflage that mimics the surface of their sessile invertebrate prey such as sponges and ascidians. However, in some nudibranchs the tubercles have become brightly pigmented and appear to contribute to aposematic colouration.

Cryptic appearance is an important defense strategy providing protection against predation in many species (Putz et al., 2010, Becerro et al., 2006, Fahey and Carroll, 2007). Cryptic colouration that matches the primary substrate and/or food source is found across all groups of molluscs that have reduced physical defense (Figure 6b). Disruptive colouration is also found in some nudibranch species, which may help them blend in the complex habitats created by sessile invertebrates of differing colours. However, in some Cephalaspidea, Saccoglossa and Nudibranchia bright colouration is often associated with the edge of the parapodia or mantle. The reason for this type of coloration is not entirely clear, but could indicate a concentration of chemicals around the vulnerable mantle edge. In the Chromodoridadae nudibranchs and at least one Saccoglossan, mantle dermal formations have evolved as specialized structures to store toxic compounds in the external part of the body (Avila, 2006). A more extreme form of warning coloration, involving bright contrasting colours, is seen in a significant proportion of opisthobranch species that contain secondary metabolites (e.g., Beccero et al., 2006, Fahey and Carroll, 2007). Aposematism increases the effectiveness of chemical defense and as noted by Cimino and Ghiselin, (2009), this enables some well defended nudibranch species to venture forth into open habitats in broad daylight. However, in most cases further studies are required to confirm the role of aposematism in molluscan chemical defense.

Conclusion

The annual reviews of marine natural products (Faulkner, 1984-2002; Blunt et al. 2003-2012) provide a comprehensive resource for exploring the diversity of molluscan secondary metabolites. These reviews cover over 800 publications on molluscs reporting over 1000 secondary metabolites from 277 species. This represents only 0.53% of all named marine molluscan species. Across the globe, there is an uneven distribution of molluscs collected for natural products research, with a paucity of studies on polar and cold temperate species, in particular. Most of the molluscs studied come from subtidal rocky reef habitats. There would be much value in expanding the studies on intertidal rocky shores to encompass species in the class Polyplacophora. In addition further studies on deep sea and pelagic molluscs would be worthwhile, including predatory Cephalopods and chemosymbiotic bivalves from hydrothermal vent communities. Chemosymbiotic bivalves could also be investigated from shallow subtidal sandy habitats, along with meiofaunal carnivores from the Class Scaphopoda.

To date, less that 50% of the species subject to natural products research have been tested for any form of biological activity and much of the testing that has been done is not ecologically relevant. Nevertheless, significant progress has been made towards understanding the neuroecology of some molluscan groups, including 1) predatory Neogastropoda (Conus sp.) with respect to their neurotoxic peptides; 2) Anaspidea (sea hares) and their complex defensive ink; 3) Cephalaspideans (bubble shells) and their alarm pheromones; and 4) filter feeding bivalves and their bioaccumulated shellfish toxins.

Nudibranchs and other Heterobranchs with reduced physical protection have provided valuable insights into the evolution of chemical defense. The large amount of literature on *de novo* synthesized and dietary derived secondary metabolites from nudibranchs provides a good starting point for future investigations into ecological chemical interactions and holds much potential for uncovering further molecules of keystone significance.

Acknowledgments

Funding by the Marine Ecology Research Centre for teaching relief to facilitate preparation this Chapter is greatly appreciated. Thanks to Dr Steve Whalan & Dr Cary Rogers for constructive feedback which improved this manuscript.

References

Abdo, D. A., Motti, C. A., Battershill, C. N. & Harvey, E. S. (2007) Temperature and spatiotemporal variability of salicylihalamide A in the sponge *Haliclona* sp. *Journal of Chemical Ecology,* 33, 1635-1645.

Aggio, J. F. & Derby, C. D. (2008) Hydrogen peroxide and other components in the ink of sea hares are chemical defenses against predatory spiny lobsters acting through non-antennular chemoreceptors. *Journal of Experimental Marine Biology and Ecology,* 363, 28-34.

Andersen, R. J., Desjardine, K. & Woods, K. (2006) Skin chemistry of nudibranchs from the West Coast of North America. In Cimino, G. & Gavagnin, M. (Eds.) *Molluscs: From Chemo-ecological Study to Biotechnolgical Application.* Heidelberg, Springer-Verlag.

Appeltans, W., Bouchet, P., Boxshall, G., De Broyer, C., De Voogd, N., Gordon, D., Hoeksema, B., Horton, T., Kennedy, M., Mees, J., Poore, G., Read, G., Stöhr, S., Walter, T. & Costello, M. (2012) World Register of Marine Species. Accessed 2012 from http://www.marinespecies.org/.

Avila, C. (1995) Natural Products of Opisthobranch Molluscs: A Biological Review. *Oceanography and Marine Biology: an Annual Review,* 33, 487-559.

Avila, C. (2006) Molluscan natural products as biological models: Chemical ecology, histology, and laboratory culture. In Cimino, G. & M., G. (Eds.) *Molluscs: From Chem-Ecological Study to Biotechnological Application.* Heidelburg, Spinger Verlag.

Avila, C., Iken, K., Fontana, A. & Cimino, G. (2000) Chemical ecology of the Antarctic nudibranch *Bathydoris hodgsoni* Eliot, 1907: defensive role and origin of its natural products. *Journal of Experimental Marine Biology and Ecology,* 252, 27-44.

Baker, J. & Duke, C. (1973) Chemistry of the indoleninones. II. Isolation from the hypobranchial glands of marine molluscs of 6-bromo-2,2-dimethylthioindolin-3-one and 6-bromo-2-methylthioindoleninone as alternative precursors to Tyrian purple. *Australian Journal of Chemistry,* 26, 2153-2157.

Baker, J. & Duke, C. (1976) Isolation of choline and choline ester salts of tyrindoxyl sulphate from the marine molluscs *Dicathais orbita* and *Mancinella keineri. Tetrahedron Letters,* 15, 1233-1234.

Baker, J. & Sutherland, M. (1968) Pigments of marine animals VIII. Precursors of 6,6'-dibromoindigotin (Tyrian purple) from the mollusc *Dicathais orbita* (Gmelin). *Tetrahedron Letters,* 1, 43-46.

Baker, J. T. (1974) Tyrian purple: an ancient dye, a modern problem. *Endeavour,* 33, 11-17.

Bancala, F. (2009) Function of mucus secretion by lamellose ormer, *Haliotis tuberculata lamellosa*, in response to starfish predation. *Animal Behaviour,* 78, 1189-1194.

Barco, A., Claremont, M., Reid, D. G., Houart, R., Bouchet, P., Williams, S. T., Cruaud, C., Couloux, A. & Oliverio, M. (2010) A molecular phylogenetic framework for the Muricidae, a diverse family of carnivorous gastropods. *Molecular Phylogenetics and Evolution,* 56, 1025-1039.

Becerro, M. A., Starmer, J. A. & Paul, V. J. (2006) Chemical defenses of cryptic and aposematic gastropterid molluscs feeding on their host sponge *Dysidea granulosa. Journal of Chemical Ecology,* 32, 1491-1500.

Becerro, M. A., Thacker, R. W., Turon, X., Uriz, M. J. & Paul, V. J. (2003) Biogeography of sponge chemical ecology: comparisons of tropical and temperate defenses. *Oecologia,* 135, 91-101.

Beesley, P., Ross, G. & Wells, A. (1998) *Mollusca: the Southern Synthesis. Fauna of Australia. Vol. 5,* Melbourne, CSIRO Publishing.

Benkendorff, K. (2013) Natural product research in the Australian marine invertebrate *Dicathais orbita. Marine Drugs* 11, 1370-1398.

Benkendorff, K. (2010) Molluscan biological and chemical diversity: secondary metabolites and medicinal resources produced by marine molluscs. *Biological Reviews,* 85, 757-775.

Benkendorff, K., Bremner, J. & Davis, A. (2001a) Indole derivatives from the egg masses of muricid molluscs. *Molecules,* 6, 70-78.

Benkendorff, K., Bremner, J. B. & Davis, A. R. (2000) Tyrian purple precursors in the egg masses of the Australian muricid, Dicathais orbita: A possible defensive role. *Journal of Chemical Ecology,* 26, 1037-1050.

Benkendorff, K. & Davis, A. R. (2002) Identifying hotspots of molluscan species richness on rocky intertidal reefs. *Biodiversity and Conservation,* 11, 1959-1973.

Benkendorff, K., Davis, A. R. & Bremner, J. B. (2001b) Chemical defense in the egg masses of benthic invertebrates: An assessment of antibacterial activity in 39 mollusks and 4 polychaetes. *Journal of Invertebrate Pathology,* 78, 109-118.

Benkendorff, K., Davis, A. R., Rogers, C. N. & Bremner, J. B. (2005) Free fatty acids and sterols in the benthic spawn of aquatic molluscs, and their associated antimicrobial properties. *Journal of Experimental Marine Biology and Ecology,* 316, 29-44.

Benkendorff, K., Mciver, C. M. & Abbott, C. A. (2011) Bioactivity of the Murex homeopathic remedy and of extracts from an Australian muricid mollusc against human cancer cells. *Evidence-Based Complementary and Alternative Medicine,* 2011, art. no. 879585.

Benkendorff, K., Pillai, R. & Bremner, J. B. (2004) 2,4,5-Tribromo-1H-Imidazole in the egg masses of three muricid molluscs. *Natural Product Research,* 18, 427-431.

Blunt, J. W., Copp, B. R., Hu, W.-P., Munro, M. H., Northcote, P. T. & Prinsep, M. R. (2007a) Marine natural products. *Natural Product Reports,* 24, 31 - 86.

Blunt, J. W., Copp, B. R., Hu, W.-P., Munro, M. H. G., Northcote, P. T. & Prinsep, M. R. (2007b) Marine natural products. *Natural Product Reports,* 24, 31-86.

Blunt, J. W., Copp, B. R., Hu, W.-P., Munro, M. H. G., Northcote, P. T. & Prinsep, M. R. (2008) Marine natural products. *Natural Product Reports,* 25, 35-94.

Blunt, J. W., Copp, B. R., Hu, W.-P., Munro, M. H. G., Northcote, P. T. & Prinsep, M. R. (2009) Marine natural products. *Natural Product Reports,* 26, 170-244.

Blunt, J. W., Copp, B. R., Keyzers, R. A., Munro, M. H. G. & Prinsep, M. R. (2012) Marine natural products. *Natural Product Reports,* 29, 144-222.

Blunt, J. W., Copp, B. R., Munro, M. H. G., Northcote, P. T. & Prinsep, M. R. (2003) Marine natural products. *Natural Product Reports,* 20, 1-48.

Blunt, J. W., Copp, B. R., Munro, M. H. G., Northcote, P. T. & Prinsep, M. R. (2004) Marine natural products. *Natural Product Reports,* 21, 1-49.

Blunt, J. W., Copp, B. R., Munro, M. H. G., Northcote, P. T. & Prinsep, M. R. (2005) Marine natural products. *Natural Product Reports,* 22, 15-61.

Blunt, J. W., Copp, B. R., Munro, M. H. G., Northcote, P. T. & Prinsep, M. R. (2006) Marine natural products. *Natural Product Reports,* 23, 26-78.

Blunt, J. W., Copp, B. R., Munro, M. H. G., Northcote, P. T. & Prinsep, M. R. (2010) Marine natural products. *Natural Product Reports,* 27, 165-237.

Blunt, J. W., Copp, B. R., Munro, M. H. G., Northcote, P. T. & Prinsep, M. R. (2011) Marine natural products. *Natural Product Reports,* 28, 196-268.

Bobzin, S. & Faulkner, D. (1989) Diterpenes from the marine sponge *Aplysilla polrhaphis* and the dorid nudibranch *Chromordis norrisi. Journal of Organic Chemistry,* 54, 3902-3907.

Bouchet, P. (2006) The magnitude of marine biodiversity. In C.M., D. (Ed.) *The Exploration of Marine Biodiveristy, Scientific and Technological Challenges.* France, Fundacion BBVA www.fbbva.es.

Carballeira, N. M., Cruz, H., Hill, C. A., De Voss, J. J. & Garson, M. (2001) Identification and total synthesis of novel fatty acids from the siphonarid limpet *Siphonaria denticulata. Journal of Natural Products,* 64, 1426-1429.

Cariello, L. & Zanetti, L. (1977) Alpha-cephalotoxin and beta-cephalotoxin - 2 paralyzing proteins from posterior salivary-glands of *Octopus vulgaris. Comparative Biochemistry and Physiology C-Pharmacology Toxicology & Endocrinology,* 57, 169-173.

Cellura, C., Toubiana, M., Parrinello, N. & Roch, P. (2007) Specific expression of antimicrobial peptide and HSP70 genes in response to heat-shock and several bacterial challenges in mussels. *Fish & Shellfish Immunology,* 22, 340-350.

Chang, P. H., Kuo, S. T., Lai, S. H., Yang, H. S., Ting, Y. Y., Hsu, C. L. & Chen, H. C. (2005) Herpes-like virus infection causing mortality of cultured abalone *Haliotis diversicolor supertexta* in Taiwan. *Diseases of Aquatic Organisms,* 65, 23-27.

Cimino, G. (2010) From sepiamelanin to chemical ecology of opisthobranchs and diatoms: an exciting 40 years game in the field of marine natural products. *Phytochemistry Reviews,* 9, 547-556.

Cimino, G., Crispino, A., Spinella, A. & Sodano, G. (1988) 2 Ichthyotoxic diacylglycerols from the opisthobranch mollusk *Umbraculum mediterraneum. Tetrahedron Letters,* 29, 3613-3616.

Cimino, G. & Gavagnin, M. (Eds.) (2006) *Molluscs: From Chemo-ecological Study to Biotechnological Application,* Heidleburg, Springer-Verlag

Cimino, G. & Ghiselin, M. (2009) Chemical defense and evolution of opisthobranch gastropods. *Proceedings of the California Academy of Sciences,* 60, 175-422.

Cimino, G., Passeggio, A., Sodano, G., Spinella, A. & Villani, G. (1991) Alarm pheromones from the mediterranean opisthobranch *Haminoea navicula*. *Experientia*, 47, 61-63.

Cook, T., Folli, M., Klinck, J., Ford, S. & Miller, J. (1998) The relationship between increasing sea-surface temperature and the northward spread of *Perkinsus marinus* (Dermo) disease epizootics in oysters. *Estuarine Coastal and Shelf Science*, 46, 587-597.

Cooksey, C. J. (2001) Tyrian purple: 6,6 '-dibromoindigo and related compounds. *Molecules*, 6, 736-769.

Cummins, S., Degnan, B. & Nagle, G. (2008) Towards an understanding of pheromone communication in molluscs. *Journal of Shellfish Research*, 27, 1000-1000.

Cummins, S. F. & Degnan, B. M. (2010) Sensory sea slugs: Towards decoding the molecular toolkit required for a mollusc to smell. *Communicative & integrative biology*, 3, 423-6.

Dang, V. T., Benkendorff, K. & Speck, P. (2011a) In vitro antiviral activity against herpes simplex virus in the abalone Haliotis laevigata. *Journal of General Virology*, 92, 627-637.

Dang, V. T., Li, Y., Speck, P. & Benkendorff, K. (2011b) Effects of micro and macroalgal diet supplementations on growth and immunity of greenlip abalone, Haliotis laevigata. *Aquaculture*, 320, 91-98.

Davies-Coleman, M. T. (2006) Secondary metabolites from the marine gastropod molluscs of Antarctica, southern Africa and South America. In Cimino, G. & Gavagnin, M. (Eds.) *Molluscs: From Chem-Ecological Study to Biotechnological Application*. Heidelburg, Spinger-Verlag.

Davies-Colman, M. T. & Garson, M. J. (1998) Marine polypropionates. *Natural Product Reports*, 15, 477-493.

Della Sala, G., Cutignano, A., Fontana, A., Spinella, A., Calabrese, G., Coll, A. D., D'ippolito, G., Della Monica, C. & Cimino, G. (2007) Towards the biosynthesis of the aromatic products of the Mediterranean mollusc *Scaphander lignarius*: isolation and synthesis of analogues of lignarenones. *Tetrahedron*, 63, 7256-7263.

Derby, C. D. (2007) Escape by inking and secreting: Marine molluscs avoid predators through a rich array of chemicals and mechanisms. *Biological Bulletin*, 213, 274-289.

Derby, C. D. & Aggio, J. F. (2011) The neuroecology of chemical defenses. *Integrative and Comparative Biology*, 51, 771-780.

Derby, C. D., Kicklighter, C. E., Johnson, P. M. & Zhang, X. (2007) Chemical composition of inks of diverse marine molluscs suggests convergent chemical defenses. *Journal of Chemical Ecology*, 33, 1105-1113.

Di Marzo, V., Marin, A., Vardaro, R. R., Depetrocellis, L., Villani, G. & Cimino, G. (1993) Histological and biochemical bases of defense mechanisms in 4 species of Polybranchioidea-Ascoglossan mollusks. *Marine Biology*, 117, 367-380.

Edwards, V., Benkendorff, K. & Young, F. (2012) Marine compounds selectively induce apoptosis in female reproductive cancer cells but not in primary-derived human reproductive granulosa cells. *Marine Drugs*, 10, 64-83.

Erspamer, V. & Glasser, A. (1957) The pharmacological actions of murexine (urocanylcholine). *British journal of pharmacology and chemotherapy*, 12, 176-84.

Fahey, S. J. & Carroll, A. R. (2007) Natural products isolated from species of Halgerda Bergh, 1880 (Mollusca : Nudibranchia) and their ecological and evolutionary implications. *Journal of Chemical Ecology*, 33, 1226-1234.

Faulkner, D. J. (1984a) Marine natural products - metabolites of marine-algae and herbivorous marine mollusks. *Natural Product Reports,* 1, 251-280.

Faulkner, D. J. (1984b) Marine natural products - metabolites of marine-invertebrates. *Natural Product Reports,* 1, 551-598.

Faulkner, D. J. (1986) Marine natural products. *Natural Product Reports,* 3, 1-33.

Faulkner, D. J. (1996) Marine natural products. *Natural Product Reports,* 13, 75-125.

Faulkner, D. J. (2000a) Highlights of marine natural products chemistry (1972-1999). *Natural Product Reports,* 17, 1-6.

Faulkner, D. J. (2000b) Marine pharmacology. *Antonie Van Leeuwenhoek International Journal of General and Molecular Microbiology,* 77, 135-145.

Faulkner, D. J. (2001) Marine natural products. *Natural Product Reports,* 18, 1-49.

Faulkner, D. J. (2002) Marine natural products. *Natural Product Reports,* 19, 1-48.

Fontana, A. (2006a) Biogenetic proposals and biosynthetic studies on secondary metabolites of opisthobranch molluscs. In Cimino, G. & Gavagnin, M. (Eds.) *Molluscs: From Chem-Ecological Study to Biotechnological Application.* Heidelburg, Springer Verlag.

Fontana, A. (2006b) Biogenetical proposals and biosynthetic studies on secondary metabolites of opisthobranch molluscs. In Cimino, G. & Gavagnin, M. (Eds.) *Molluscs: From Chemo-ecological Study to Biotechnolgical Application.* . Heidelberg, Springer-Verlag.

Ford, S. E. (1996) Range extension by the oyster parasite *Perkinsus marinus* into the northeastern United States: Response to climate change? *Journal of Shellfish Research,* 15, 45-56.

Ford, S. E. & Smolowitz, R. (2007) Infection dynamics of an oyster parasite in its newly expanded range. *Marine Biology,* 151, 119-133.

Freidlander, P. (1909) Ueber den farbstoff des antiken purpura aus *Murex brandaris. Chemische Berichte* 42, 765-770.

Fry, B. G., Roelants, K., Champagne, D. E., Scheib, H., Tyndall, J. D. A., King, G. F., Nevalainen, T. J., Norman, J. A., Lewis, R. J., Norton, R. S., Renjifo, C. & De La Vega, R. C. R. (2009a) The toxicogenomic multiverse: Convergent recruitment of proteins into animal venoms. *Annual Review of Genomics and Human Genetics.*

Fry, B. G., Roelants, K. & Norman, J. A. (2009b) Tentacles of venom: Toxic protein convergence in the Kingdom Animalia. *Journal of Molecular Evolution,* 68, 311-321.

Fuhrman, J. A., Steele, J. A., Hewson, I., Schwalbach, M. S., Brown, M. V., Green, J. L. & Brown, J. H. (2008) A latitudinal diversity gradient in planktonic marine bacteria. *Proceedings of the National Academy of Sciences of the United States of America,* 105, 7774-7778.

Fukuyama, Y., Iwatsuki, C., Kodama, M., Ochi, M. & Kataoka, K. (1998) Antimicrobial indolequinones from the mid-intestinal gland of the muricid gastropod *Drupella fragum. Tetrahedron,* 54, 10007-10016.

Garson, M. (2006) Marine molluscs from Australia and New Zealand: Chemical and ecological studies. In Cimino, G. & Gavagnin, M. (Eds.) *Molluscs: From Chemo-ecological Study to Biotechnolgical Application.* Heidelberg, Springer-Verlag.

Garson, M. J. (1993) The biosynthesis of marine natural products. *Chemical Reviews,* 93, 1699-1733.

Garson, M. J., Goodman, J. M. & Paterson, I. (1994) A configurational model for siphonariid polypropionates derived from structural and biosynthetic considerations. *Tetrahedron Letters,* 35, 6929-6932.

Greenwood, P. G. (2009) Acquisition and use of nematocysts by cnidarian predators. *Toxicon,* 54, 1065-1070.

Halai, R. & Craik, D. J. (2009) Conotoxins: natural product drug leads. *Natural Product Reports,* 26, 526-536.

Harvell, C. D., Mitchell, C. E., Ward, J. R., Altizer, S., Dobson, A. P., Ostfeld, R. S. & Samuel, M. D. (2002) Ecology - Climate warming and disease risks for terrestrial and marine biota. *Science,* 296, 2158-2162.

Harvell, D., Aronson, R., Baron, N., Connell, J., Dobson, A., Ellner, S., Gerber, L., Kim, K., Kuris, A., Mccallum, H., Lafferty, K., Mckay, B., Porter, J., Pascual, M., Smith, G., Sutherland, K. & Ward, J. (2004) The rising tide of ocean diseases: unsolved problems and research priorities. *Frontiers in Ecology and the Environment,* 2, 375-382.

Hay, M. E. (1996) Marine chemical ecology: What's known and what's next? *Journal of Experimental Marine Biology and Ecology,* 200, 103-134.

Hay, M. E. (2009) Marine chemical ecology: Chemical signals and cues structure marine populations, communities, and ecosystems. *Annual Review of Marine Science.*

Hellou, J., Andersen, R. J. & Thompson, J. E. (1982) Terpenoids from the dorid nudibranch *Cadlina luteomarginata. Tetrahedron,* 38, 1875-1879.

Hochlowski, J. E., Faulkner, D. J., Matsumoto, G. K. & Clardy, J. (1983) The denticulatins, two polypropionate metabolites from the pulmonate *Siphonaria denticulata. Journal of the American Chemical Society,* 105, 7413-7415.

Hoegh-Guldberg, O. & Bruno, J. F. (2010) The impact of climate change on the world's marine ecosystems. *Science,* 328, 1523-1528.

Hooper, C., Day, R., Slocombe, R., Handlinger, J. & Benkendorff, K. (2007a) Stress and immune responses in abalone: Limitations in current knowledge and investigative methods based on other models. *Fish and Shellfish Immunology,* 22, 363-379.

Hooper, C., Hardy-Smith, P. & Handlinger, J. (2007b) Ganglioneuritis causing high mortalities in farmed Australian abalone (*Haliotis laevigata* and *Haliotis rubra*). *Australian Veterinary Journal,* 85, 188-193.

Ichiba, T. & Higa, T. (1986) New cuparene-derived sesquiterpenes with unprecedented oxygenation patterns from the sea hare *Aplysia dactylomela. Journal of Organic Chemistry,* 51, 3364-3366.

Ivanisevic, J., Thomas, O. P., Pedel, L., Penez, N., Ereskovsky, A. V., Culioli, G. & Perez, T. (2011) Biochemical trade-offs: Evidence for ecologically linked secondary metabolism of the sponge *Oscarella balibaloi. Plos One,* 6.

Jannun, R. & Coe, E. L. (1987) Bromoperoxidase from the marine snail, *Murex trunculus. Comparative Biochemistry and Physiology B-Biochemistry & Molecular Biology,* 88, 917-922.

Jannun, R., Nuwayhid, N. & Coe, E. (1981) Biological bromination - bromoperoxidase activity in the murex sea-snail. *Federation Proceedings,* 40, 1774-1774.

Kamiya, K., Sakai, R. & Jimbo, M. (2006) Bioactive Molecules From Sea Hares. In Cimino, G. & Gavagnin, M. (Eds.) *Molluscs: From Chemo-ecological Study to Biotechnolgical Application.* . Heidelberg, Springer-Verlag.

Kawashima, H. (2005) Unusual minor nonmethylene-interrupted di-, tri-, and tetraenoic fatty acids in limpet gonads. *Lipids,* 40, 627-630.

Kawashima, H. & Ohnishi, M. (2006) Occurrence of novel nonmethylene-interrupted C-24 polyenoic fatty acids in female gonad lipids of the limpet *Cellana grata. Bioscience Biotechnology and Biochemistry,* 70, 2575-2578.

Kicklighter, C. E., Germann, M., Kamio, M. & Derby, C. D. (2007) Molecular identification of alarm cues in the defensive secretions of the sea hare *Aplysia californica. Animal Behaviour,* 74, 1481-1492.

Kuanpradit, C., Cummins, S. F., Degnan, B. M., Sretarugsa, P., Hanna, P. J., Sobhon, P. & Chavadej, J. (2010) Identification of an attractin-like pheromone in the mucus-secreting hypobranchial gland of the abalone *Haliotis asinina* Linnaeus. *Journal of Shellfish Research,* 29, 699-704.

Kuanpradit, C., Stewart, M. J., York, P. S., Degnan, B. M., Sobhon, P., Hanna, P. J., Chavadej, J. & Cummins, S. F. (2011) Characterization of mucus-associated proteins from abalone (Haliotis) - candidates for chemical signaling. *FEBS Journal,* 279, 437-450.

Lewis, R. J., Dutertre, S., Vetter, I. & Christie, M. J. (2012) Conus venom peptide pharmacology. *Pharmacological Reviews,* 64, 259-298.

Li, C.-H., Zhao, J.-M. & Song, L.-S. (2009) A review of advances in research on marine molluscan antimicrobial peptides and their potential application in aquaculture. *Molluscan Research,* 29, 17-26.

Lim, N. S. H., Everuss, K. J., Goodman, A. E. & Benkendorff, K. (2007) Comparison of surface microfouling and bacterial attachment on the egg capsules of two molluscan species representing Cephalopoda and Neogastropoda. *Aquatic Microbial Ecology,* 47, 275-287.

Magiatis, P. & Skaltsounis, A.-L. (2006) *From Hexaplex trunculus to new kinase inhibitory indirubins.*

Manker, D. C., Garson, M. J. & Faulkner, D. J. (1988) De novo biosynthesis of polypropionate metabolites in the marine pulmonate *Siphonaria denticulata. Journal of the Chemical Society-Chemical Communications,* 1061-1062.

Marcogliese, D. J. (2008) The impact of climate change on the parasites and infectious diseases of aquatic animals. *Revue Scientifique Et Technique-Office International Des Epizooties,* 27, 467-484.

McClintock, J. B. & Baker, B. J. (1997) A review of the chemical ecology of Antarctic marine invertebrates. *American Zoologist,* 37, 329-342.

Meijer, L., Skaltsounis, A. L., Magiatis, P., Polychronopoulos, P., Knockaert, M., Leost, M., Ryan, X. Z. P., Vonica, C. A., Brivanlou, A., Dajani, R., Crovace, C., Tarricone, C., Musacchio, A., Roe, S. M., Pearl, L. & Greengard, P. (2003) GSK-3-selective inhibitors derived from Tyrian purple indirubins. *Chemistry & Biology,* 10, 1255-1266.

Mitta, G., Hubert, F., Dyrynda, E. A., Boudry, P. & Roch, P. (2000a) Mytilin B and MGD2, two antimicrobial peptides of marine mussels: gene structure and expression analysis. *Developmental and Comparative Immunology,* 24, 381-393.

Mitta, G., Vandenbulcke, F., Noel, T., Romestand, B., Beauvillain, J. C., Salzet, M. & Roch, P. (2000b) Differential distribution and defence involvement of antimicrobial peptides in mussel. *Journal of Cell Science,* 113, 2759-2769.

Mitta, G., Vandenbulcke, F. & Roch, P. (2000c) Original involvement of antimicrobial peptides in mussel innate immunity. *Febs Letters,* 486, 185-190.

Miyamoto, T. (2006) Selected bioactive compounds from Japanese Anaspideans and nudibranchs. In Cimino, G. & Gavagnin, M. (Eds.) *Molluscs: From Chemo-ecological Study to Biotechnolgical Application.* Heidelberg, Springer-Verlag.

Miyamoto, T., Ebisawa, Y. & Higuchi, R. (1995) Aplyparvunin, a bioactive acetogenin from the sea hare *Aplysia parvula. Tetrahedron Letters,* 36, 6073-6074.

Moulis, A. (2006) The action of RFamide neuropeptides on molluscs, with special reference to the gastropods *Buccinum undatum* and *Busycon canaliculatum. Peptides,* 27, 1153-1165.

Myers, R. A., Cruz, L. J., Rivier, J. E. & Olivera, B. M. (1993) Conus peptides as chemical probes for receptors and ion channels. *Chemical Reviews,* 93, 1923-1936.

Neves, R., Gaspar, H. & Calado, G. (2009) Does a shell matter for defence? Chemical deterrence in two cephalaspidean gastropods with calcified shells. *Journal of Molluscan Studies,* 75, 127-131.

Orians, G. & Janzen, D. (1974) Why are embryos so tasty? *American Naturalist,* 108, 581-592.

Painter, S. D., Clough, B., Black, S. & Nagle, G. T. (2003) Behavioral characterization of attractin, a water-borne peptide pheromone in the genus Aplysia. *Biological Bulletin,* 205, 16-25.

Painter, S. D., Clough, B., Garden, R. W., Sweedler, J. V. & Nagle, G. T. (1998) Characterization of Aplysia attractin, the first water-borne peptide pheromone in invertebrates. *Biological Bulletin,* 194, 120-131.

Paul, V. J., Arthur, K. E., Ritson-Williams, R., Ross, C. & Sharp, K. (2007) Chemical defenses: From compounds to communities. *Biological Bulletin,* 213, 226-251.

Paul, V. J., Ritson-Williams, R. & Sharp, K. (2011) Marine chemical ecology in benthic environments. *Natural Product Reports,* 28, 345-387.

Pawlik, J. R. (1993) Marine invertebrate chemical defenses. *Chemical Reviews,* 93, 1911-1922.

Pawlik, J. R., Albizati, K. F. & Faulkner, D. J. (1986) Evidence of a defensive role for limatulone, a novel triterpene from the intertidal limpet *Collisella limatula. Marine Ecology-Progress Series,* 30, 251-260.

Penney, B. K., Laplante, L. H., Friedman, J. R. & Ocasio Torres, M. (2010) A noninvasive method to remove kleptocnidae for testing their role in defence. *Journal of Molluscan Studies,* 76, 296-298.

Pennings, S. C. (1994) Interspecific variation in chemical defenses in the sea hares (Opisthobranchia, Anaspidea). *Journal of Experimental Marine Biology and Ecology,* 180, 203-219.

Pennings, S. C., Nastisch, S. & Paul, V. J. (2001) Vulnerability of sea hares to fish predators: importance of diet and fish species. *Coral Reefs,* 20, 320-324.

Pennings, S. C., Paul, V. J., Dunbar, D. C., Hamann, M. T., Lumbang, W. A., Novack, B. & Jacobs, R. S. (1999) Unpalatable compounds in the marine gastropod *Dolabella auricularia*: Distribution and effect of diet. *Journal of Chemical Ecology,* 25, 735-755.

Peters, C., Collins, G. M. & Benkendorff, K. (2012) Characterisation of the physical and chemical properties influencing bacterial epibiont communities on benthic gelatinous egg masses of the pulmonate Siphonaria diemenensis. *Journal of Experimental Marine Biology and Ecology,* In press.

Przeslawski, R. & Benkendorff, K. (2005) The role of surface fouling in the development of encapsulated gastropod embryos. *Journal of Molluscan Studies,* 71, 75-83.

Przeslawski, R., Benkendorff, K. & Davis, A. (2005) A Quantitative Survey of Mycosporine-Like Amino Acids (MAAS) in Intertidal Egg Masses from Temperate Rocky Shores. *Journal of Chemical Ecology,* 31, 2417-2438.

Putz, A., Koenig, G. M. & Waegele, H. (2010) Defensive strategies of Cladobranchia (Gastropoda, Opisthobranchia). *Natural Product Reports,* 27, 1386-1402.

Ramasamy, M. & Murugan, A. (2007) Fouling deterrent chemical defence in three muricid gastropod egg masses from the Southeast coast of India. *Biofouling,* 23, 259-265.

Ramasamy, M. S. & Murugan, A. (2005) Potential antimicrobial activity of marine molluscs from tuticorin, southeast coast of India against 40 biofilm bacteria. *Journal of Shellfish Research,* 24, 243(9).

Rogers, C. N., De Nys, R., Charlton, T. S. & Steinberg, P. D. (2000a) Dynamics of algal secondary metabolites in two species of sea hare. *Journal of Chemical Ecology,* 26, 721-744.

Rogers, C. N., De Nys, R. & Steinberg, P. D. (2000b) Predation on juvenile Aplysia parvula and other small anaspidean, ascoglossan, and nudibranch gastropods by pycnogonids. *Veliger,* 43, 330-337.

Rogers, C. N., De Nys, R. & Steinberg, P. D. (2002) Effects of algal diet on the performance and susceptibility to predation of the sea hare Aplysia parvula. *Marine Ecology-Progress Series,* 236, 241-254.

Rogers, C. N., Steinberg, P. D. & Denys, R. (1995) Factors associated with oligophagy in 2 species of sea hares (Mollusca, Anaspidea). *Journal of Experimental Marine Biology and Ecology,* 192, 47-73.

Roseghini, M., Erspamer, V., Ramorino, L. & Gutierrez, J. E. (1970) Choline esters, their precursors and metabolites in the hypobranchial gland of prosobranchiate molluscs, *Concholepas concholepas* and *Thais chocolata. European journal of biochemistry FEBS,* 12, 468-73.

Roseghini, M., Severini, C., Erspamer, G. F. & Erspamer, V. (1996) Choline esters and biogenic amines in the hypobranchial gland of 55 molluscan species of the neogastropod Muricoidea superfamily. *Toxicon,* 34, 33-55.

Rudman, B. (2010) Sea Slug Forum. Australian Museum accessed 2012 from http://www.seaslugforum.net/.

San-Martin, A., Rovirosa, J., Gaete, K., Olea, A. & Ampuero, J. (2009) Mantle defensive response of marine pulmonate *Trimusculus peruvianus. Journal of Experimental Marine Biology and Ecology,* 376, 43-47.

Sauvage, C., Pepin, J. F., Lapegue, S., Boudry, P. & Renault, T. (2009) Ostreid herpes virus 1 infection in families of the Pacific oyster, *Crassostrea gigas*, during a summer mortality outbreak: Differences in viral DNA detection and quantification using real-time PCR. *Virus Research,* 142, 181-187.

Shabani, S., Yaldiz, S., Vu, L. & Derby, C. D. (2007) Acidity enhances the effectiveness of active chemical defensive secretions of sea hares, *Aplysia californica*, against spiny lobsters, *Panulirus interruptus. Journal of Comparative Physiology A - Neuroethology Sensory Neural and Behavioral Physiology,* 193, 1195-1204.

Sheumack, D. D., Howden, M. E. H. & Spence, I. (1984) Occurrence of a tetrodotoxin-like compound in the eggs of the venomous blue-ringed octopus (*Hapalochlaena maculosa*). *Toxicon,* 22, 811-812.

Sheumack, D. D., Howden, M. E. H., Spence, I. & Quinn, R. J. (1978) Maculotoxin - neurotoxin from venom glands of octopus *Hapalochlaena maculosa* identified as tetrodotoxin. *Science,* 199, 188-189.

Sheybani, A., Nusnbaum, M., Caprio, J. & Derby, C. D. (2009) Responses of the sea catfish *Ariopsis felis* to chemical defenses from the sea hare *Aplysia californica. Journal of Experimental Marine Biology and Ecology,* 368, 153-160.

Stewart, F. J., Newton, I. L. G. & Cavanaugh, C. M. (2005) Chemosynthetic endosymbioses: adaptations to oxic-anoxic interfaces. *Trends in Microbiology,* 13, 439-448.

Susswein, A. J. & Nagle, G. T. (2004) Peptide and protein pheromones in molluscs. *Peptides,* 25, 1523-1530.

Thieltges, D. W., Fredensborg, B. L., Studer, A. & Poulin, R. (2009) Large-scale patterns in trematode richness and infection levels in marine crustacean hosts. *Marine Ecology-Progress Series,* 389, 139-147.

Thompson, J. E., Walker, R. P., Wratten, S. J. & Faulkner, D. J. (1982) A chemical defense-mechanism for the nudibranch *Cadlina luteomarginata. Tetrahedron,* 38, 1865-1873.

Thoms, C., Ebel, R. & Proksch, P. (2006) Sequestration and possible role of dietary alkaloids in the sponge-feeding mollusk *Tylodina perversa.* In Cimino, G. & Gavagnin, M. (Eds.) *Molluscs: From Chem-Ecological Study to Biotechnological Application.* Heidleberg, Springer-Verlag.

Vezzulli, L., Brettar, I., Pezzati, E., Reid, P. C., Colwell, R. R., Hoefle, M. G. & Pruzzo, C. (2011) Long-term effects of ocean warming on the prokaryotic community: evidence from the vibrios. *Isme Journal,* 6, 21-30.

Wahidullah, S., Guo, Y. W., Fakhr, I. M. I. & Mollo, E. (2006) Chemical diversity in opisthobranch molluscs from scarcely investigated Indo-Pacific areas. *Molluscs: From Chem-Ecological Study to Biotechnological Application,* 43, 175-198.

Watkins, S. M., Reich, A., Fleming, L. E. & Hammond, R. (2008) Neurotoxic shellfish poisoning. *Marine Drugs,* 6, 431-455.

Westley, C. & Benkendorff, K. (2008) Sex-specific Tyrian purple genesis: Precursor and pigment distribution in the reproductive system of the marine mollusc, *Dicathais orbita. Journal of Chemical Ecology,* 34, 44-56.

Westley, C. & Benkendorff, K. (2009) The distribution of precursors and biosynthetic enzymes required for Tyrian purple genesis in the hypobranchial gland, gonoduct, an egg masses of Dicathais orbita (Gmelin, 1791) (Neogastropoda: Muricidae). *Nautilus,* 123, 148-153.

Westley, C. B., Lewis, M. C. & Benkendorff, K. (2010a) Histomorphology of the hypobranchial gland in Dicathais orbita (Gmelin, 1791) (Neogastropoda: Muricidae). *Journal of Molluscan Studies,* 76, 186-195.

Westley, C. B., Mciver, C. M., Abbott, C. A., Le Leu, R. K. & Benkendorff, K. (2010b) Enhanced acute apoptotic response to azoxymethane-induced DNA damage in the rodent colonic epithelium by Tyrian purple precursors: A potential colorectal cancer chemopreventative. *Cancer Biology & Therapy,* 9, 371-379.

Williams, B. L. (2010) Behavioral and chemical ecology of marine organisms with respect to tetrodotoxin. *Marine Drugs,* 8, 381-398.

Zimmer, R. K. & Ferrer, R. P. (2007) Neuroecology, chemical defense, and the keystone species concept. *Biological Bulletin,* 213, 208-225.

Part 2. Predator-Prey Relationships and Reproductive Behaviour

In: Neuroecology and Neuroethology in Molluscs
Editors: Anna Di Cosmo and William Winlow
ISBN: 978-1-62948-970-4

Chapter 3

Olfaction in Gastropods

***Scott F. Cummins*[1] *and Russell C. Wyeth*[2]**
[1]University of the Sunshine Coast, Maroochydore, Queensland, Australia
[2]St. Francis Xavier University, Antigonish, Nova Scotia, Canada

Abstract

Within the Phylum Mollusca, gastropods encompass a large and diverse group of aquatic and terrestrial animals that inhabit areas of the deep or shallow sea, estuaries, lakes, and arid environments. They have no acoustic sense and, for many, little to no visual recognition of objects. Instead, their world is chemically driven so that water-soluble and volatile odorants are key mediators of many physiological and behavioral events. Olfaction, conventionally referred to as the sense of smell, serves to detect odorants at a distance from their source and provides the sensory afference leading to odorant-driven behavioral responses. These exogenous chemicals convey a remarkable amount of information critical to aggregation, habitat selection, defense, prey localization, courtship, and mating. To understand the role of olfaction in behavior, studies have focused on how olfactory systems recognize and interpret information from chemical signals. Researchers have made excellent progress toward understanding gastropod olfaction at behavioral and morphological levels and, more recently, at neural and biochemical levels. In this chapter, we review recent studies and provide insight into the olfactory neurobiology of aquatic (*Aplysia*, *Tritonia,* and *Lymnaea*) and terrestrial (*Achatina, Cantareus,* and *Limax*) gastropods. The three aquatic model systems are distantly related, with previous studies of each tending to have different foci, thereby providing distinct but complimentary contributions to the synthesis of gastropod olfaction research. By contrast, the three terrestrial model systems are more closely related, and their research histories are far more integrated. We continue this integration, taking care to consider the differences within and between aquatic and terrestrial systems in our synthesis effort.

Aquatic Gastropods

Water (including marine and freshwater) covers over 70% of the earth's surface (Ellis 2003), yet our knowledge of how aquatic animals detect water-borne odors in this vast environment is far from complete. This gap in our knowledge is concerning given that many aquatic species dwell on the bottom of bodies of water, where vision and other sensory modalities are at best limited by turbidity or the physical properties of water. Instead, odor cues impact all aspects of aquatic animal life, directly affecting chances of survival and reproduction. Aquatic gastropods exist in marine, freshwater, and brackish water and provide excellent models in which to investigate basic principles of the physiological, biochemical, and molecular aspects of olfaction in aquatic environments.

Aplysia

Aplysia, which are marine opisthobranch gastropods within the family Aplysiidae, are unique among molluscs in that they are highly characterized at multiple levels and therefore fulfill many of the requirements necessary for investigating the biological basis of olfaction. Their primary cephalic sensory organs are two pairs of tentacles located in the head region. All Aplysiidae possess a pair of short retractable rhinophores and oral tentacles (Klussmann-Kolb 2004; Cummins et al. 2009a) (Figure 1), allowing them to locate and discriminate food (Audesirk 1975), conspecific pheromones (Levy et al. 1997), and predators. Upon stimulation of the cephalic sensory organs, animals may orient and move towards the source (Leonard and Lukowiak 1986; Painter et al. 2004). Rhinophores were first described as olfactory organs by early nudibranch researchers (Agersborg 1925) and more accurately defined through the discovery that their ablation reduces, although does not abolish, food-finding (Audesirk 1975). Rhinophores are required for long-distance chemoreception, with oral tentacles generally thought to be close-range chemoreceptors (Boudko et al. 1999), although rhinophores appear to be as effective as oral tentacles in detecting food odors (Preston and Lee 1973). Other proposed functions of rhinophores and oral tentacles in other gastropods are photoreception and mechanoreception (Bicker et al. 1982; Keunzi and Carew 1991), the latter which allows for orientation within water currents.

Prior to the year 2000, there were few ultrastructural studies of the cephalic sensory organs of any opisthobranch, although it was established that sensory cells are the principal cell type, which may be of primary or secondary origin (Emery 1992). Light microscope histological analysis revealed two receptor cell types in the epithelium of the rhinophore groove, while a small rhinophore ganglion is present at the base of each darkly pigmented rhinophore groove (Figure 2) (Emery and Audesirk 1978). More recent studies describe that *Aplysia punctata* possesses three different cell types at the epidermis of the rhinophore groove: supporting cells, ciliated cells, and mucus cells (Gobbeler and Klussmann-Kolb 2007). The surface of the rhinophore groove differs markedly from the relatively smooth outer surface (Gobbeler and Klussmann-Kolb 2007; Cummins et al. 2009b). Scanning electron micrographs of both the rhinophore groove and oral tentacles demonstrate the presence of long cilia (Figure 1), with ciliated cells of the rhinophore occurring once every 5-

15 cells (Gobbeler and Klussmann-Kolb 2007). Ciliated cells may be required for rapid turnover and continual sampling of seawater (Emery and Audesirk 1978).

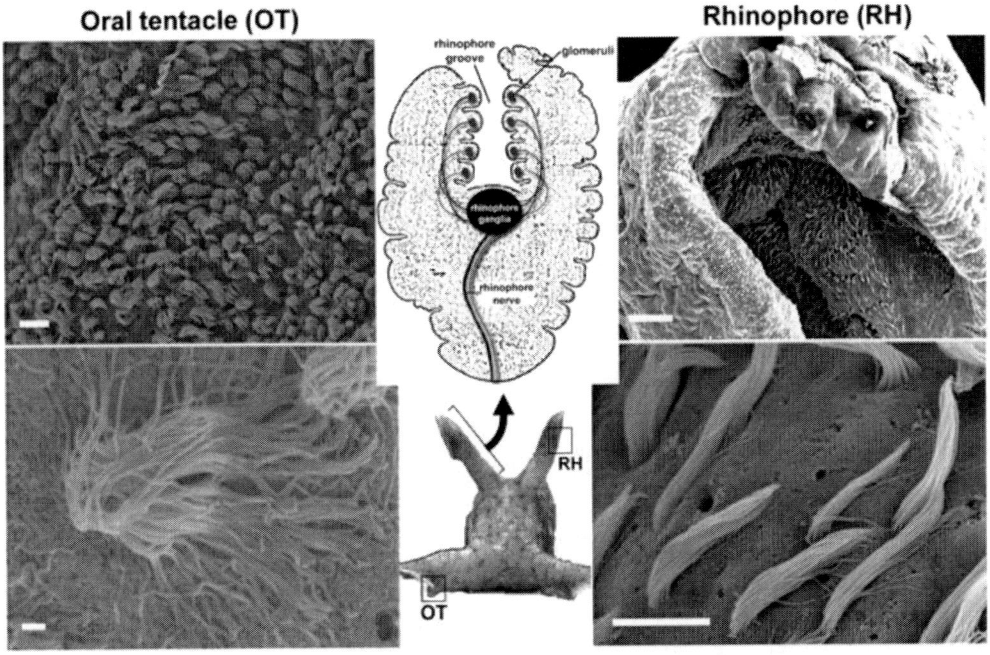

Figure 1. Scanning electron micrographs of *Aplysia* cephalic sensory organs. RH, rhinophore; OT, oral tentacle.

Neuromodulators such as tyrosine hydroxylase, serotonin, nitric oxide, FMRFamides (Phe-Met-Arg-Phe-NH_2), and FMRF-related peptides were identified within rhinophore preparations using immunolocalization techniques. Tyrosine hydroxylase-like immunoreactivity appears in both small and large somata beneath the epithelium of the tentacles and rhinophores, respectively (Croll 2001). Serotonin immunoreactivity, on the other hand, is distributed only within cellular processes and not in somata, suggesting that serotonergic neurons are extrinsic to rhinophores (Wertz et al. 2006). The neuronal messenger nitric oxide is also present in *Aplysia* peripheral sensory regions (foot, mouth, tentacles, and rhinophores), as determined by intense nicotinamide adenine dinucleotide phosphate-diaphorase (NADPH-d) labeling (Moroz 2006). Similar NADPH-d staining patterns are observed in other gastropod grazers, such as *Lymnaea stagnalis* (Moroz et al. 1993; Moroz 1994a, b), as well as the nudibranch mollusc *Melibe leonine* (Newcomb and Watson 2001). In contrast, no peripheral NADPH-d staining is observed in predators such as the pteropod molluscs *Clione limacina* and *Pleurobranchaea californica* (Moroz and Gillette 1996; Moroz et al. 2000).

In *Aplysia*, the functional neuroanatomy of the rhinophore has also been described by employing immunohistochemical and calcium imaging methods (Wertz et al. 2006). Using these techniques, the olfactory pathway was found to include sensory neurons, glomerulus-like structures, a rhinophore ganglion located at the base of the rhinophore groove, and a rhinophore nerve (Figure 1). In vertebrates and insects, spherical olfactory glomeruli in the central nervous system (CNS) act as functional units for neuronal processing of external chemical information (Pinching and Powell 1971; Kosaka et al. 1998; Martin et al. 2011). In

Aplysia, glomerulus-like structures are located immediately underneath the epithelium, consisting of large tangles of nerve fibers surrounded by glia-like processes (Moroz 2006; Wertz et al. 2006; Gobbeler and Klussmann-Kolb 2007). Estimations from a single *Aplysia punctata* rhinophore predict the existence of approximately 36 glomeruli with diameters varying between 25 and 135 μm (Wertz et al. 2006) or 30 and 90 μm (Gobbeler and Klussmann-Kolb 2007). Precisely how rhinophore groove sensory cells project to these glomerulus-like structures and onward to rhinophore or cerebral ganglia is currently unknown. Alterations in calcium levels in rhinophore ganglion neurons following stimulation by individual amino acids, such as alanine, provides further evidence for the chemosensory nature of rhinophores (Wertz et al. 2006).

Aplysia oral tentacles are also known as labial or cephalic tentacles; they consist of one pair, situated at the front margins of the head, lateral to the mouth. Although less studied, their position suggests that they function as contact chemoreceptors (Jahan-Parwar 1972; Boudko et al. 1999; Croll 2003), and their epithelium resembles that of rhinophores (Cummins et al. 2009a) (Figure 1). Another chemosensory organ, the osphradium, is implicated in chemoreception in some gastropods. In *Aplysia,* the osphradium is at the posterior right side of the mantle, in close proximity to the gills (Klussmann-Kolb 2004). The closeness of the osphradium to the gills predicts that its function is to sample incoming water prior to movement into the respiratory organs. However, there is currently no evidence for a role of the osphradium of *Aplysia* in olfaction (Kupfermann and Weiss 1976; Weiss et al. 1989).

Although the majority of these anatomical investigations used adult specimens, there are also some excellent insights into the ontogeny of juvenile *Aplysia californica* cephalic sensory organs. In this species, the first sensory organs to develop are the labial tentacles and the lip (stage 9 of development), followed by the rhinophore outgrowths that become prominent during stage 10 of development (Figure 3) (Wollesen et al. 2007) (stages are based on Kriegstein 1977). A serotoninergic meshwork is clearly observed in developing rhinophores, possibly innervated by a common labial nerve. Similar to adults, no serotonergic cell somata are observed within peripheral regions (Wollesen et al. 2007). Immunocytochemical comparisons of cellular distribution patterns with neurotransmitters such as serotonin and FMRFamide indicate a functional role for sensory epithelia.

Odor detection requires that sensory neurons are present within the olfactory epithelium and express receptors devoted to binding environmental odorants and activating intracellular pathways. In mammals, a large family of seven-transmembrane domain rhodopsin and metabotropic G-protein coupled receptors (GPCRs) are expressed on the surface of sensory neurons. An investigation of the *Aplysia* draft genome (http://www.ncbi.nlm.nih.gov/) with algorithms that identify genes encoding multi-transmembrane proteins (Cummins et al. 2009a) enabled the first detailed genetic analysis of candidate odorant-sensing receptors in a mollusc. A large multi-gene receptor family encoding rhodopsin-like GPCRs was observed within sensory epithelia of rhinophores and oral tentacles (Figure 4). Expression of at least some of these receptors is specific to the rhinophore groove. Other molecular components necessary for intracellular transduction, including G proteins, an IP_3 receptor, and a phopholipase C, are also located in the olfactory sensory epithelium of rhinophores (Cummins et al. 2009b).

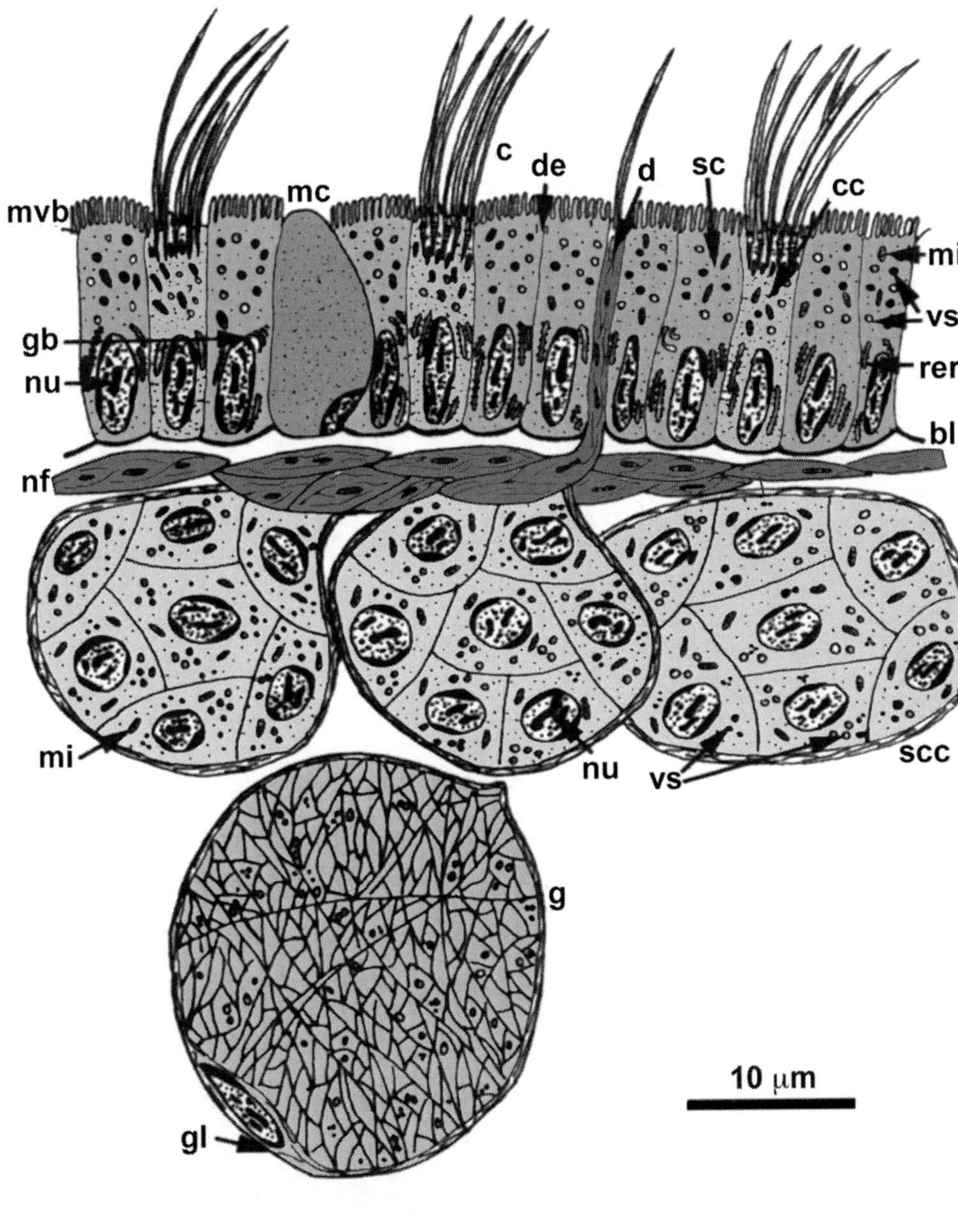

Figure 2. Semi-schematic illustration showing cellular composition of sensory epithelia in rhinophore of *Aplysia punctata* (modified from, Gobbeler and Klussmann-Kolb 2007. bl = basal lamina, c = cilia, cc = ciliated cell, d = dendrite, de = desmosome, g = glomerulus, gb = Golgi body, gl = glia cell, mc = mucus cell, mi = mitochondrium, mvb = microvillious border, nf = nerve fibre, nu = nucleus, rer = rough endoplasmatic reticulum, sc = supporting cell, vs = vesicle.

In insects, ionotropic receptors (IRs) on sensory neurons detect odorant molecules, an adaptation thought to enhance processing speed (Benton et al. 2009). Insect olfactory systems respond to a smaller range of odorants than terrestrial vertebrates, and their genomes, in turn, encode a smaller family of receptors (Robertson et al. 2003). *Aplysia* appear to have their own set of IRs, some of which, including the *IR25a* orthologue, are found predominantly in olfactory organs (Figure 4). The presence of both GPCRs and IRs within *Aplysia* olfactory organs suggests that molluscs may represent a "hybrid" organism in olfactory system evolution. Further research is essential for bridging the gap in our understanding of the evolution of animal olfaction.

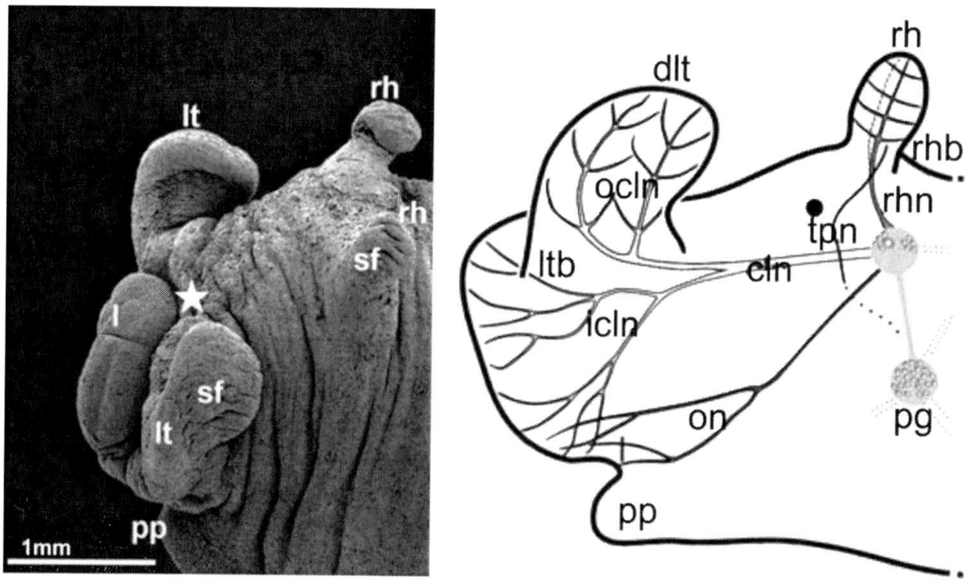

Figure 3. Ontogenesis of the *Aplysia californica* cephalic sensory organs at stage 12. Left, Lateral view scanning electron micrograph of juvenile *A. californica*. rh = rhinophore; l = lip, pp = propodium, sf = sensory fold, It, star = medial base of the labial tentacles. Right, serotonergic innervation of the cephalic region of A. californica. Cg = cerebral ganglion, l = lip, pg = pedal ganglion, on = oral nerve, ocln = outer branch of the common labial nerve, tpn = third pedal nerve, dlt = distal labial tentacle, ltb = labial tentacle base, rh = rhinophore, rhb = rhinophoral base. Modified from Wollesen et al. 2007.

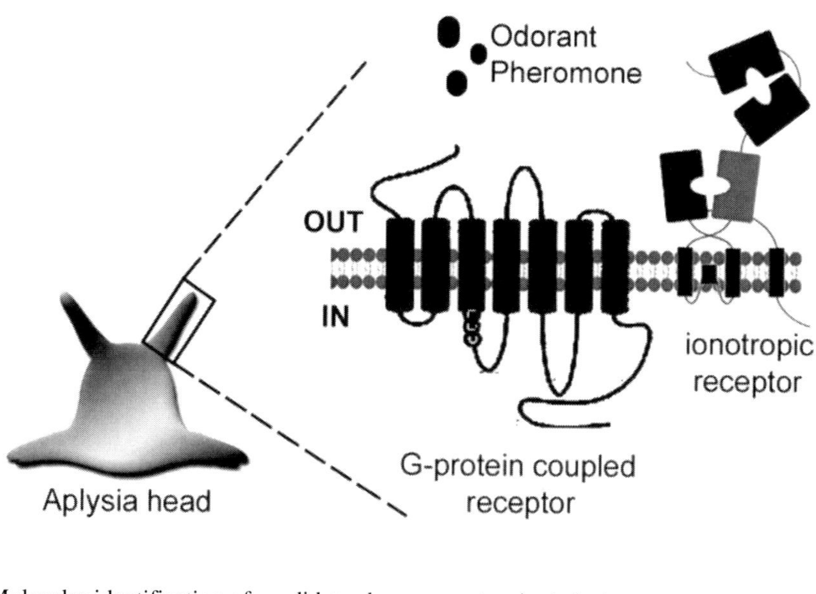

Figure 4. Molecular identification of candidate chemoreceptors in *Aplysia*.

Pheromones are chemical signals that have been well studied in the context of *Aplysia* mate attraction (Painter et al. 2004; Cummins et al. 2005). These studies were initiated following the observation that water surrounding *Aplysia* egg mass was able to increase conspecific respiratory pumping (Levy et al. 1997). Molecular investigation and experimental attraction assays revealed that a sex attraction pheromone consisting of at least four small proteins (attractin, enticin, temptin, and seductin) is released from egg-laying animals (Cummins et al. 2004). When released together, attraction pheromones are powerful cues that not only induce attraction but also stimulate mating. This discovery is the first example of a multi-component attraction pheromone used by a mollusc (or, indeed, any marine invertebrate). Rhinophores are responsible for the detection of *Aplysia* attraction pheromones; their ablation results in a decrease in time spent mating and egg laying (Levy et al. 1997). The identity of the pheromone receptors within sensory organs remains to be determined. Of similar interest is the investigation of pheromone-induced receptor neuron activation and subsequent processing in glomerulus-like regions and the cerebral ganglion.

Tritonia

Olfaction in the nudibranch *Tritonia diomedea* has been studied primarily in the context of navigational behavior. These nudibranchs live in shallow subtidal beds of two species of soft corals, *Ptlosarcus gurneyi* or *Virgularia* sp. The soft corals are the slugs' only prey. The slugs also share their habitat with a relatively small number of other community members including their only major predator, the sea star *Pycnopodia helianthoides* (Birkeland 1974). Field and laboratory studies demonstrate that movements toward prey and mates and away from predators depend on the detection of odors carried in flow (Wyeth and Willows 2006a,b; Wyeth et al. 2006). The available evidence suggests that slugs use positive odor-gated rheoaxis to move upstream inside odor plumes to locate attractive targets. This is complemented by downstream turns inside predator odor plumes, resulting in downstream movement away from predators. Thus, at least for large-scale movements, the *Tritonia* olfactory system is primarily dedicated to detecting and identifying odors rather than odor gradients, which would normally be eradicated in the slugs' natural habitat due to turbulent flow (see also Weissburg and Wilson, Chapter 5).

As in *Aplysia, Tritonia* rhinophores are olfactory organs. Rhinophores are necessary and sufficient for navigation inside odor plumes, and chemosensory afference occurs in response to odors isolated from prey, mates, and predators (Wyeth and Willows 2006; McCullagh 2013). The second cephalic sensory organ, the oral veil, is thought to be chemosensory, at least with respect to predator odors (Field and Macmillan 1973). However, it has not been properly established whether distance or contact chemoreceptors, or those for prey and mate odours, are present within this organ. Nonetheless, these chemosensory functions seem likely given the oral veil's role in prey approaches, predatory strikes, and mating behavior (Wyeth and Willows 2006). Preliminary behavioral observations combined with sensory organ lesions indicate the oral veil is not used for navigation based on odors trapped in the boundary layer but is used to detect prey and predator odors (McCullagh and Wyeth, unpublished). Moreover, the oral veil is required for normal predatory strikes although not strictly necessary for predation, as modified strikes were possible. Finally, unlike *Aplysia* and other aquatic gastropods, the osphradium has been lost in the Nudipleura (Wagele and Willan 2000). Thus, the available evidence indicates that rhinophores are the primary olfactory organ in *Tritonia*.

Very little is known about the neural substrates underlying olfaction in *Tritonia*. Initial studies of cephalic sensory organ neuroanatomy have just begun (Cain et al. 2010; Gallagher et al. 2011). Preliminary data indicate that rhinophores and the oral veil contain numerous catecholaminergic peripheral sensory cells (Kehoe et al. 2012) similar to those found in other gastropods (Baratte and Bonnaud 2009; Wyeth and Croll 2011), which may or may not be chemosensory. Meanwhile, no clear data yet demonstrate that *Tritonia* rhinophores contain glomeruli (Wyeth et al. 2012). Glomerulus-like structures in nudibranch *Phestilla sibogae* rhinophores were described (Croll et al. 2001), but no glomeruli were observed in another sea slug, *Archidoris pseudoargus* (Wertz et al. 2007). Further work in *Tritonia* is clearly needed to identify the peripheral and central sensory cell types present in both the rhinophores and oral veil, settle the question of whether glomeruli are present, and explore sensory projection patterns to the rhinophore and central ganglia.

Beyond extracellular recordings from isolated rhinophores, which reveal chemosensory responses in the rhinophore nerve, neurophysiological studies of olfactory sensory organs have been limited to two types of identified neurons. Audesirk and Audesirk (Audesirk and Audesirk 1980a; Audesirk and Audesirk 1980b) describe central primary mechanoreceptors that receive synaptic input from unidentified chemoreceptors. The balance of evidence suggests that these chemoreceptors are peripheral sensory cells with peripheral synapses on mechanoreceptors (or other neurons that form synapses with mechanoreceptors). A separate group of multimodal afferent neurons (S-cells) that respond to predatory chemosensory stimuli have also been identified (Getting 1976; Frost et al. 2001; Megalou et al. 2009). Although studies demonstrate a role of S-cells in contact chemoreception, they may also be involved in distance chemoreception, as we now know that escape swim can be stimulated at a distance (Wyeth and Willows 2006). These cells also respond to mechanical stimuli, and preliminary work on their receptive fields suggests they form a somatotopic map spread across the body of the animals. As their sensory endings are not concentrated in cephalic sensory organs, other neurons are most likely also involved in olfactory pathways. Finally, evidence suggesting S-cells are sensory cells with central somata does not fully exclude the possibility that they have peripheral synapses with other primary sensory neurons. Overall, our limited knowledge of the proximate mechanisms of olfaction in *Tritonia* is consistent with the idea that olfactory processing largely occurs with input from cephalic sensory organs, with possibly central or peripheral primary chemoreceptors in those organs as well as throughout the body wall.

In summary, despite the rather sparse neurophysiological study of olfaction in *Tritonia* to date, this model system is nonetheless suited for further study of the proximate mechanisms underlying olfactory-based navigation. The slugs have the benefit of relatively simple navigational behaviours based on just three odor source types: a narrow selection of prey, one predator and potential mates (with no sex differences). Moreover, *Tritonia* workers have established a suite of neurophysiological techniques, including the semi-intact preparation (Willows et al. 1973; Murray et al. 1992), fine wire recordings from crawling animals (Redondo and Murray 2005), and ensemble optical recordings (Hill et al. 2010) that can all be used to study the neural mechanisms that generate the different responses to odors (Murray et al. 2006). Future work can thus explore the central processing of olfactory inputs that generate orientation and movements relative to odor sources.

Lymnaea

Lymnaea stagnalis is an air-breathing aquatic pulmonate found in freshwater habitats across Europe, North America, and parts of Asia (Benjamin 2008 and see chapters 8 and 9). Little is known about *Lymnaea* distance chemoreception. Their cephalic sensory organs include the tentacles and lips, whereas the osphradium is found outside the mantle cavity close to the pneumostome (through which air is breathed). The cephalic sense organs are likely involved in both contact and distance chemoreception, whereas the osphradium also detects changes in respiratory gas concentrations (Wedemeyer and Schild 1995; Kamardin et al. 1999; Kamardin et al. 2001; Bell et al. 2007; Karnik et al. 2012). An early study concluded that the snails could find only animal flesh but not algae and plants in a Y-maze, although plants and algae are their natural diet in the wild (Bovbjerg 1968). We suggest this conclusion warrants verification given the number of closely related aquatic pulmonate snails that successfully use olfaction to find algal and plant prey (Townsend 1973; Lombardo et al. 1991; Madsen 1992; Fink et al. 2006). The bulk of subsequent work in *Lymnaea* has focused on sugar chemoreception and its involvement in the neural control of feeding (Jager 1971; Goldschmeding and Jager 1973; Kemenes et al. 1986). Because the primary focus has been on understanding characteristics of feeding central pattern generators, the details of chemoreceptor inputs have been left largely unexplored. Moreover, because these experiments stimulated the lips (or both lips and tentacles) with sugar solutions, their results likely provide information primarily on taste rather than true olfaction. Thus, we have little understanding of how olfaction is involved in the neural control of behavior in *Lymnaea*.

Several studies have explored the sensory neuroanatomy of the tentacles and lips in *Lymnaea*. Earlier work used both transmission and scanning electron microscopy to categorize sensory cells or their superficial endings (Zylstra 1972; Zaitseva and Bocharova 1981). More recently, both immunohistochemistry and histochemistry targeting specific neurotransmitters was used to describe putative sensory cells (Serfozo et al. 1998; Croll et al. 1999; Hegedus et al. 2004; Hatakeyama et al. 2007). Both peripheral and central sensory cells are evidently present, but integrating the various results to form a comprehensive picture is difficult due to the diverse array of cells described, the range of methods used, and the diversity of tissues studied. Moreover, ascribing modality to any of the morphologically categorized cell types has been largely speculative.

Recently, Wyeth and Croll (2011) completed a more comprehensive study of peripheral sensory cells in *Lymnaea* cephalic sensory organs. Using a combination of nonspecific vital dyes, backfills, histochemistry, and immunohistochemistry, four types of peripheral sensory cell were identified (Figure 5). These include catecholaminergic, histaminergic, and nitrergic cells with dendrites that penetrate sensory epithelia and axons that presumably project centrally. The fourth cell type are unipolar nitrergic cells with a single process that penetrates sensory epithelia. If the unipolar cells are indeed sensory, they must synapse with other peripheral neurons. Further complexity is added by the possibility that central sensory cells also innervate these structures (Nelson and Audesirk, 1986). Thus, with multiple cell types, most of which are distributed in structures with both chemosensory and mechanosensory functions, it is not yet possible to generate specific hypotheses regarding their different modalities. Physiological tests are now needed to verify both the putative neurotransmitters and the modality of all cell types.

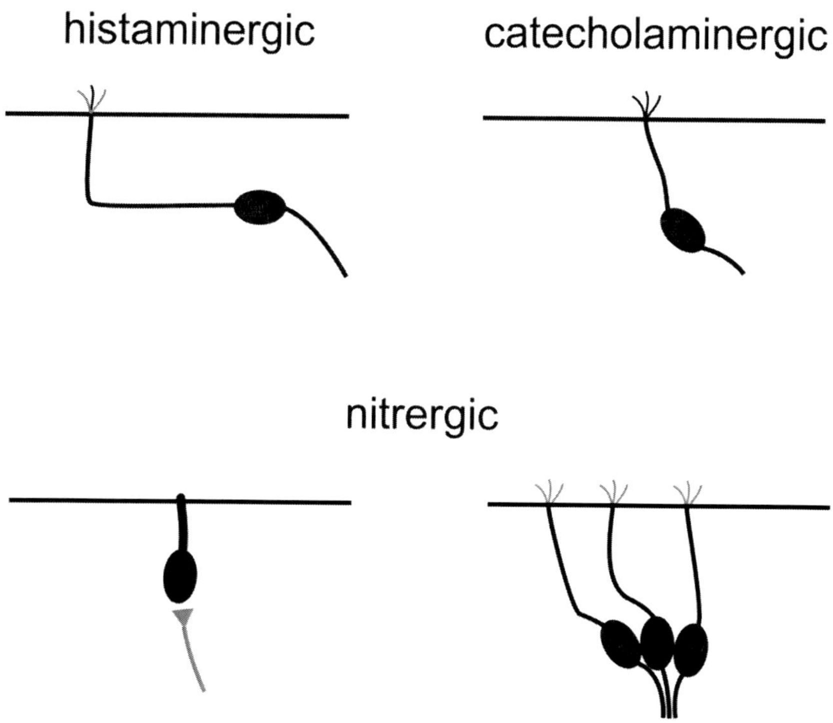

Figure 5. Peripheral sensory cell types identified in the lips and tentacles of *Lymnaea stagnalis* include bipolar catecholaminergic, histaminergic and nitrergic cells (the latter found in clusters), as well as unipolar nitrergic cells. Both sensory function and neurotransmitter content need to be verified with physiological experimentation. Some aspects of the neuroanatomy remain uncertain (grey structures), particularly the extent of ciliation in some cell types and somatic synapses with other peripheral neurons for the unipolar cells.

Physiological experiments in *Lymnaea* cephalic sensory organs have largely been limited to sugar chemosensation without addressing other odors or making a distinction between distance and contact chemoreception. Earlier work demonstrated the snails possess chemosensory cells that detect sugar and control feeding responses (Jager, 1971). Subsequently, both the median (Goldschmeding and Jager 1973; Kemenes et al. 1986) and superior (Nakamura et al. 1999a; Nakamura et al. 1999b) lip nerves were identified as sources of chemosensory afference. In addition, the tentacle was found to contain chemoreceptors that respond to amino acids (Zaitseva 1994). In all of these experiments, it is possible that the chemoreceptors are peripheral neurons either with central projections themselves or peripheral synapses with other peripheral neurons that project centrally. Finally, there is strong physiological evidence that a set of bipolar nitrergic chemosensory neurons in the lips projects centrally (Elphick et al., 1995). These cells detect sugar and use nitric oxide at central synapses that control the feeding motor program. The variability in both labeling quality and reliability that affects histochemical and immunohistochemical identification of putative nitrergic cells (Cooke et al. 1994; Moroz et al. 1994a) likely explains why Wyeth and Croll (2011) were unable to find the same population of cells. It is also possible that the cells identified by Elphick et al. 1995, were in fact unipolar, with peripheral synapses on other centrally-projecting nitrergic neurons. Regardless, at least one population of chemosensory cells in *Lymnaea* lips responds to sugars and nitric oxide is a neurotransmitter in the peripheral chemosensory pathway.

In addition to cephalic sensory organs, the osphradium is also implicated in several *Lymnaea* chemosensory responses. Neurophysiological experiments, including both extracellular recordings from the osphradial nerve and whole cell patch recordings from sensory neurons in the osphradium, demonstrate that osphradial chemoreceptors detect O_2, CO_2, amino acids, and more complex odor molecules. These sensory inputs regulate respiration (Wedemeyer and Schild 1995; Bell et al. 2007) and behavioral responses to kairomone detection (Karnik et al. 2012). There is also evidence that osphradial sensory cell discharges are affected by ionic concentrations, suggesting they may also be involved in osmoregulation (Kamardin, et al. 1999; Kamardin et al. 2001). Thus, chemoreception by the osphradium mediates homeostatic responses as well as true olfactory responses involving distance chemoreception. It remains unknown, however, whether the osphradium is involved in more traditional odor responses such as prey search or predator avoidance. Neuroanatomical observations differentiate two different sensory cell types based on size (Kamardin et al. 1999), but the various physiological responses have not been properly linked to the morphological categorizations. Given the diversity of responses recorded, further work matching the different sensory modalities to specific sensory cell types is needed to fully understand the olfactory capabilities of osphradial chemoreceptors in *Lymnaea* and their participation in odor-stimulated behavior.

Overall, *Lymnaea* presents one of the better opportunities for exploring the complete olfactory system in gastropods. The collected neuroanatomical and neurophysiological information on the cephalic sensory organs and osphradium is the most comprehensive of that for any species, and one chemoreceptor type has already been identified (or nearly so). An important part of this process also involves identifying mechanoreceptors to exclude from the pool of candidate chemoreceptors. Again, a number of studies on tactile responses and neurophysiology in *Lymnaea* (de Vlieger 1968; Janse 1976; Staras et al. 1999) provide an important foundation for further work assigning different modalities to the different cell types. At the moment, the unipolar nitrergic cells found in the lips and tentacles (Wyeth and Croll 2011) are the only candidates that match the chemosensory cells found in *Aplysia* (Cummins et al. 2009a.). If this is indeed the case, then there may be at least two populations of chemoreceptors in *Lymnaea*, both putatively nitrergic. Alternatively, if there is only one population of unipolar nitrergic cells (mistakenly described as bipolar by Elphick et al., 1995), then it is tempting to hypothesize that the chemosensory cells identified in *Aplysia* are nitrergic. Regardless, the information gathered thus far on the sensory neuroanatomy and function in *Lymnaea* provides a substantial foundation for further work on all sensory systems, including olfaction.

Terrestrial Gastropods

Achatina, *Limax*, and *Helix*

There is estimated to be approximately 35,000 extant species of terrestrial gastropods worldwide (van Bruggen 1995), most of which rely primarily on olfaction, as their senses of vision and hearing are less well developed. Four species in particular have largely contributed to our understanding of terrestrial gastropod olfaction: *Achatina fulica, Cantareus aspersus* (formally *Helix aspersa*), *Limax maximus,* and *Limax marginatus. Achatina fulica* is a giant snail native to East Africa but is now a widely invasive species, rated among the world's 100 most invasive species (Lowe et al. 2004). *Cantareus aspersus* is native to Western Europe but has been introduced, often unintentionally, to a variety of locations worldwide. It is now recognized as among the most widespread and abundant naturalized terrestrial molluscs (Barker 2002). Both species of *Limax* are terrestrial slugs from Europe. To date, no major differences in olfactory systems have been discovered among these species, and most authors have considered information gathered from one species equally applicable to all species. It should be noted, however, that our understanding of terrestrial gastropod olfaction is pieced together from studies of all four species, with many aspects of olfaction described in only one species. Furthermore, should differences among species arise, it will be important to consider current phylogenetic hypotheses (Figure 6) that *Limax* and *Cantareus* are more closely related than either is to the basally branching *Achatina* (contrary to the 'snail' versus 'slug' characteristics that superficially link *Achatina* and *Cantareus*).

Many studies demonstrate the importance of olfaction for both terrestrial slugs and snails, including aversion to noxious odors (Voss 2000), food preference (Farkas and Shorey 1976; Croll and Chase 1977; Sahley et al. 1981; Iglesias and Castillejo 1999), homing (Gelperin 1974; Lind 1989; Chelazzi 1990; Tomiyama 1992), and mate search (Chase et al. 1980). Moreover, following the discovery that *Limax* can learn to associate odors with aversive tastes (Gelperin, 1975), *Limax* (and to a lesser extent *Cantareus*) have been established as model systems for studying olfactory learning (Gelperin 1999; Balaban 2002; Watanabe et al. 2008; Matsuo et al. 2011).

The paired superior cephalic tentacles (also referred to as posterior, ommatophore, optic, or dorsal tentacles) and the similarly paired inferior (anterior, ventral) tentacles both function as olfactory organs (Cook 1985; Chase and Tolloczko 1993; Friedrich and Teyke 1998; Kimura et al. 1999). One study of *Cantareus* olfactory search behavior suggested that stereotypical movements of the tentacles are specifically adapted to clean the olfactory epithelia and exchange the adjacent air being sampled for odors (Lemaire and Chase 1998). The olfactory epithelia are found at the ventral surface of the tentacle terminal knob and protected by a brush border. Consistent with the importance of chemosensory detection in these animals, it is predicted that they dedicate a disproportionate number of neurons to olfaction.

Olfactory neural pathways involve sensory cells in the epithelia that form synapses in the tentacle ganglion, followed by projections to the procerebrum in the CNS. In *Achatina,* bipolar receptor neurons have dendrites that extend through the sensory pad epithelium to its surface and axons that project to three locations (Chase and Tolloczko 1993). Some axons (~10%) converge directly with the cerebral ganglia, whereas the remainder terminates within tentacle olfactory ganglion neuropils, at glomerulus-like structures in finger-like projections of the ganglion or in the main part of the ganglion. Interneurons originating from both locations in the tentacle ganglion also project to the CNS, with an additional pathway possible via interneuronal connections between the putative glomeruli and the tentacle ganglion. Thus, a total of four parallel pathways are potentially available for olfactory processing and transmission (Chase and Tolloczko 1993).

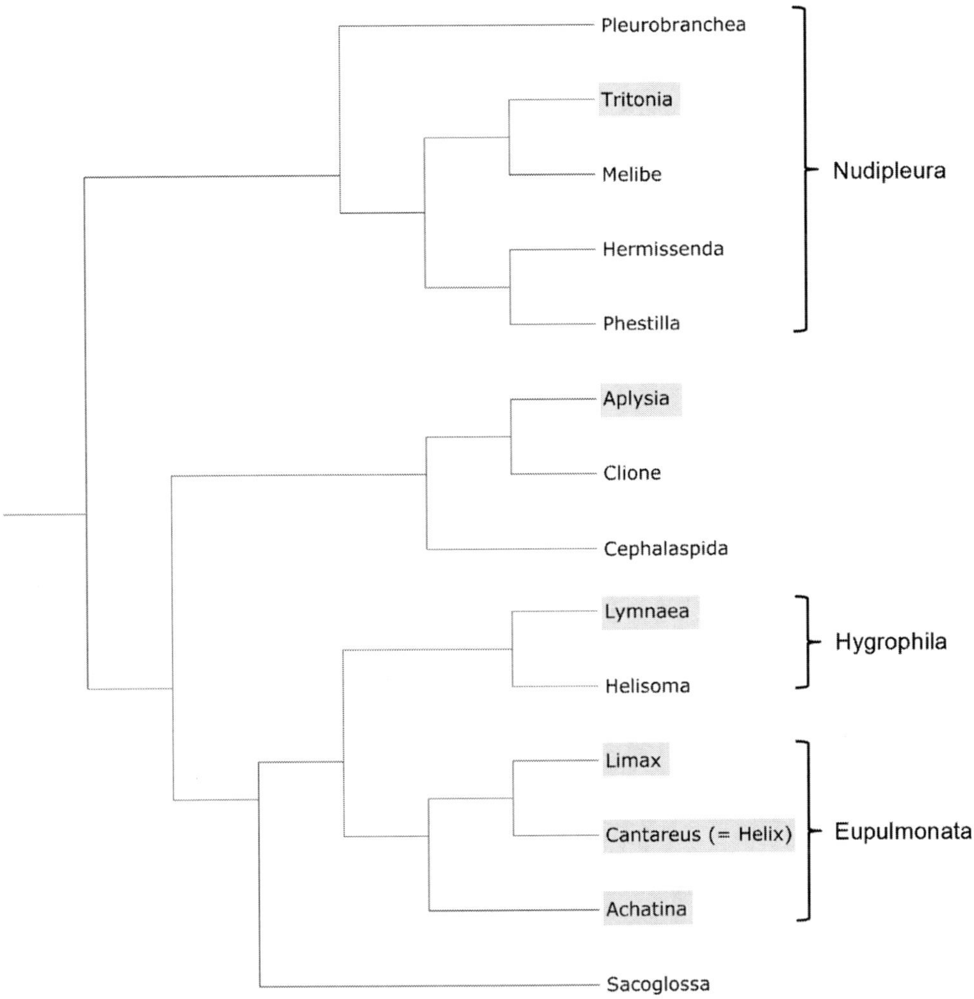

Figure 6. Phylogenetic relationships amongst the euthyneruan gastropods used as model systems in neuroscience, along with other major taxa. Highlighted genera are reviewed in this chapter. Tree constructed based on (Dinapoli and Klussmann-Kolb, 2010) and references therein.

Information about these different pathways and their various components is scattered across model systems. In *Limax,* using tentacle nerve backfills, at least three different categories of bipolar sensory neurons projecting to the cerebral ganglion were identified (Ito, et al. 2000). In addition, sensory neurons also project to the tentacle ganglion, which produces oscillating local field potentials that are modified by odor stimulation of the tentacle (Ito et al. 2000; Inokuma et al. 2002; Ito et al. 2003b; Ito et al. 2004; Ito et al. 2006). As yet, glomerulus-like structures in the tentacle ganglia have not been reported for *Limax* or *Cantareus* (but see unpublished observation note in Ito et al. 2000). The interneurons projecting to the CNS from the putative glomeruli and the main body of the tentacle ganglion are also found in *Achatina* and *Cantareus*, in which they specifically target the procerebrum (Ierusalimsky and Balaban 2010). Meanwhile, bipolar sensory cells that project directly from the sensory epithelium to the CNS were found to project only to the metacerebrum in

Cantareus (Ierusalimsky and Balaban, 2007); although the authors stated without substantiating evidence that sensory cells also project directly to the procerebrum (Chase 1985; Chase and Tolloczko 1993). The current hypothesis for *Cantareus* is that the direct-to-metacerebrum pathway mediates tentacle withdrawal in response to noxious odors (and mechanical stimuli, which also indicates multiple sensory neuron types). The polysynaptic pathway through the putative glomeruli and/or tentacle ganglion projects to the procerebrum and is thought to be involved in more complex odor processing.

Morphological and physiological evidence links both the tentacle ganglion and the procerebrum of the cerebral ganglia to olfaction (Croll and Chase 1980; Chase 1985). The procerebrum is a unique structure found in terrestrial gastropods, with tens of thousands of neurons collectively involved in olfactory learning, which has been primarily studied in several species of *Limax* (and extensively reviewed (Gelperin 1999; Watanabe et al. 2008; Matsuo et al. 2011) In brief, the procerebrum receives input via the polysynaptic pathway from the cephalic tentacles and sends output elsewhere in the CNS, ultimately leading to motor control of learned behavioral responses to previously encountered odors. Procerebral neurons, which include both bursting and non-bursting cells, generate waves of excitation that travel across the structure, creating regularly oscillating local field potentials. Although many intricacies of how oscillatory activity is produced by the procerebral network are understood, the exact role of procerebral neurons in olfactory processing is not yet clear. Certainly the procerebral lobe, as a whole, is necessary for learning to associate odors with aversive tastes (Kasai et al. 2006). Moreover, changes in field potential oscillations concomitant with associative olfactory learning have been recorded in both *Limax* and *Cantareus*. However, exactly how olfactory responses are modified by experience remains an open question.

Our understanding of the neurochemical content of the olfactory system in terrestrial gastropods is patchy, with a number of neurotransmitters identified as important. In *Limax,* bipolar sensory neurons with FMRFamide-like immunoreactivity project directly to cerebral ganglion (Suzuki et al. 1997) and may correspond to cells that project to the metacerebrum in *Cantareus* (Ierusalimsky and Balaban, 2010). In addition, in *Limax,* GABA, glutamate, and acetylcholine are implicated in the peripheral odor processing network and localized to the sensory epithelium and tentacle ganglion (Ito et al. 2003b; Ito et al. 2004). Histochemical and immunohistochemical evidence indicates that bipolar sensory neurons (as well as other neurons) in the tentacle ganglion and procerebrum generate and are responsive to nitric oxide. The pattern of labeling suggests that nitric oxide is involved throughout the polysynaptic pathway between the sensory epithelium and procerebrum, although the exact location of putative nitrergic interneuron projections to the CNS is not established (Fujie et al. 2002). Serotonin and dopamine have potential roles in regulating odor perception and tentacle movement in *Cantareus* (Roshchin and Balaban 2012). Cerebral cells send serotonergic fibers into the peripheral olfactory organ (Hernadi and Elekes 1999; Hernadi 2000), although no serotonin-containing cell bodies are present there (Hernadi and Elekes 1999). Other neuromodulators such as methionine-enkephalin, vasoactive intestinal peptide, and command neuron peptides (CNPs) have been identified in *Cantareus* tentacle sensory neurons and other components of peripheral olfactory pathways based on immunohistochemical localization (Marchand and Dubois 1986; Kaufmann et al. 1995; Ierusalimsky and Balaban 2007). In the CNS, a massive suite of classical neurotransmitters and neuromodulators have been identified in the procerebrum (see table in Gelperin (1999), as well as more recent findings for glutamate (Matsuo et al. 2011), GABA (Kobayashi et al. 2012), FMRFamide (Kobayashi et al. 2010), carbon dioxide (Gelperin et al. 2000), and nitric oxide (Fujie et al. 2002). This extensive list of putative neurotransmitters localized to certain cell types has the potential to increase our understanding of the olfactory system and other components of the nervous system. For example, CNPs may act cooperatively with FMRFamide in snail tentacle withdrawal behavior. However, proper synthesis will require double-labeling to confirm co-localization as well as corresponding physiology to help organize reconstruction of the olfactory system.

In summary, past exploration of olfaction in terrestrial gastropods sets the stage for further study of olfactory processing and learning. The accumulated knowledge regarding neurochemical content and experience-dependent changes in activity can be combined with better delineation of the afferent neural circuits as well as yet more study of procerebrum function. These twin approaches will afford a clearer understanding of how the olfactory system encodes olfactory afference and how that sensory input then creates the synaptic (or other) changes that underlie odor-based learning. In the long run, comparing the structure and function of the terrestrial gastropod olfactory system (and also aquatic gastropods) to other taxa will be important for delineating guiding principles in olfactory system design across all animals.

Future Priorities for Gastropod Olfaction

Olfaction enables gastropods to respond to and communicate with the environment and each other. In most gastropod species studied thus far, olfactory cues are critical in the regulation of food choice and intake, protection, and reproduction. Given the prominence of gastropod model systems in molecular and cellular neuroscience and neuroethology, further study of gastropod olfactory systems at multiple levels will contribute to a deeper understanding of the general principles of sensory detection and processing. The model systems discussed here have historically provided complementary information about the proximate mechanisms of olfaction, and we anticipate this pattern will continue. With the future in mind, a number of promising lines of further inquiry are apparent to us.

Gastropod chemoreceptor gene sequences need to be clearly established. Candidate chemoreceptors in *Aplysia* provide a foundation for identifying chemoreceptors in other gastropods. High-throughput next-generation sequencing now enables us to rapidly obtain copies of chemoreceptor-like genes from a broad selection of species. These data then provide opportunities for comparative analyses across species leading to clearer delineation of both the genes involved in chemoreception and the sensory cells that express them. Furthermore, experimental verification of receptor protein-odorant binding is necessary to confirm functionality of the identified olfactory receptor genes. Finally, through the process of elimination, identification of genes and cells associated with the olfactory system will help generate new hypotheses linking other sensory cell types to other sensory modalities (particularly mechanoreceptors, which are likely intermingled among chemosensory cells in the cephalic sensory organs in all species).

Pheromones need to be explored further in both aquatic and terrestrial gastropods. The documentation of pheromone-induced behavior, including attraction, arousal, alarm, and recognition, are plentiful in gastropod literature (for example Townsend 1974; Uhazy et al.

1978; Takeda and Tsuruoka 1979; Chase et al. 1980; Adamo and Chase 1990), yet few signaling molecules have been identified. In the experimental analysis of waterborne attraction pheromone in *Aplysia,* methods overcoming problems with sample contamination, breakdown, and concentration should expedite further discoveries of new pheromone communication systems. Their characterization will in turn lead to better understanding of the chemoreceptors and downstream neural circuits required for generating behavioral responses to the pheromones. Moreover, this information can have practical applications for gastropod aquaculture and pest control, where adjusting gastropod behavior is valuable.

By our assessment, two sets of questions about the peripheral olfactory nervous system particularly warrant further investigation. First, are unipolar nitrergic sensory cells found in the olfactory epithelia of all gastropods? Are they primary chemoreceptors? Do they have peripheral axosomatic synapses with second-order nitrergic cells? A mixture of data from *Aplysia*, *Limax,* and *Lymnaea* is consistent with these possibilities, but the complete circuit has not been described in any one species. Second, what is the distribution of glomeruli across gastropod olfactory systems? In *Achatina*, *Aplysia*, and *Phestilla* (Croll et al. 2003), glomerulus-like structures have been described based on histological observations. However, no modern, detailed comparison with the fine structure of either insect or vertebrate glomeruli (Bargmann, 2006) has been attempted. Critically, there has also been no confirmation that different glomeruli are involved in segregated responses to different odors. Therefore, several experiments are required to clarify the presence and role of glomeruli in gastropod olfaction. Glomerulus-like neuropils in gastropod sensory organs need to be tested for analogous functions to their counterparts in other phyla. Furthermore, *Tritonia* and *Lymnaea* need to be carefully re-examined for the presence of glomeruli (or alternate sites of synaptic integration in the olfactory system). The identification of olfactory neuron-specific genes and proteins suggested above may facilitate this effort by providing necessary tools for localizing axonal projections to glomeruli (although if the primary chemosensory cells are unipolar, then second-order projections will need to be traced). If glomeruli analogous in structure and function to those found in other olfactory systems are indeed absent in some or all gastropods, the functional consequences for gastropod olfactory systems will require exploration.

Future work on gastropod olfaction must also consider how animals acquire the odors detected by the olfactory system, which has been almost entirely neglected to date. Many taxa ‘refresh’ their olfactory epithelia; terrestrial vertebrates sniff the air, and aquatic arthropods flick their antennae through the water to exchange the odor-bearing medium on the surface of their sense organs (Koehl 2006; Zhao et al. 2006). These processes allow sampling of new odorants and are as integral to the proximate mechanisms of olfaction as are receptors and neural circuits (Mainland and Sobel 2006). To our knowledge, only one study has directly tackled this problem in a gastropod, reporting that *Cantareus* (Lemaire and Chase 1998) makes two separate tentacle movements that are possibly involved in facilitating access of odor molecules to the olfactory epithelium. In aquatic gastropods, however, this sampling process is presumably based on flows generated by cilia. We know that motile cilia are abundant on sense organs (Emery 1992), but our knowledge ends there. Little to no information exists on the role of ciliary flow in olfaction or the neural regulation of ciliary activity, which may control olfactory sampling (as occurs in other taxa). Given the integration of cilia and cilated cells in olfactory epithelia, we suggest future work should consider these two aspects of olfaction together. Studies of the proximate mechanisms of olfaction should include surveying patterns of ciliation and ciliary activity and experimentally exploring the motor control of cilia beat rates that regulate medium exchange across the epithelia surface.

In the past, despite known differences among gastropod taxa, there has been an overarching assumption of similarity in both form and function of proximate olfactory mechanisms. Consequently, the available data represent a widespread patchwork of information from many different species, including the major model systems covered in this review and other species. Yet it is possible that olfactory systems (and indeed any other aspect of the nervous system) have undergone substantial changes during the different evolutions of gastropod species. For example, the osphradium is present in two of the major taxa studied (*Aplysia* and the freshwater Hygrophila pulmonates, including *Lymnaea*), whereas it is absent in at least two major taxa (in the nudibranchs and Eupulmonata, including the terrestrial *Cantareus, Limax,* and *Achatina*). This difference clearly impacts any olfactory processing based on afference from the osphradium, and it could be linked to substantial compensatory changes in the cephalic sensory organs of species lacking osphradia. On one hand, it is worth seeking to unite the disparate information on neuroanatomy and function currently scattered across species in an effort to build a more comprehensive view of the gastropod olfactory system. Thus, a goal should be to fully understand neuroanatomy, neurochemistry, and neurophysiology in a single species. However, there is value in investigating different model systems for both taxonomic breadth (and thus properly testing the assumption that all gastropods have similar olfactory systems) and their amenability to different types of experiments. To reconcile these contrasting approaches, it will be necessary to delve into the olfactory details of a few species (scientifically tractable or economically important, such as *Lymnaea*, *Aplysia,* and *Biomphalaria*) while continuing to sample widely across gastropod phylogeny.

In this context, it is important to bear in mind the recently revised hypotheses regarding the evolutionary relationships among gastropods (Figure 6). In particular, in the Euthyneura (which is the primary taxon studied with regard to olfactory mechanisms due to the amenability of their nervous systems for experimentation), nudibranchs branch basally in the latest phylogenies. Although the authors urge caution in accepting this as the true evolutionary history of the group (Klussmann-Kolb et al. 2008; Dinapoli and Klussmann-Kolb 2010), it is clear that the nudibranchs differ substantially from the other euthyneuran taxa, and these differences may extend to their olfactory systems. In addition, it is worth noting the repeated support for *Helix* and *Limax* being more closely related than either is to *Achatina* (Wade et al. 2006). These relationships need to be considered when attempting to collate information about the olfactory system from multiple species, and also when predicting the presence of olfactory system components in one species based on evidence gathered in other species.

Regardless of the evolutionary relationships, one common theme is apparent across all gastropod olfactory systems studied thus far: the complex peripheral nervous system likely plays an important role in olfaction. The basic architecture of olfactory systems begins with peripheral sensory neurons concentrated in sense organs. Some of these cells form synapses with other neurons that project centrally (and perhaps also peripherally), and others are likely to project directly to the CNS. Hoyle and Willows (Hoyle and Willows 1973), and several other researchers (for example Lukowiak and Jacklet 1972; Janse 1976; Fredman and Jahan-Parwar 1980), report evidence of peripheral networks contributing to behavioral responses in several gastropods, with some responses requiring absolutely no involvement of the CNS. Yet we know very little about how the peripheral nervous system functions, and its components have not been fully described. Recent work in *Lymnaea* (Wyeth and Croll) and *Aplysia* (Wyeth and Croll, in preparation) highlights the massive complexity of the peripheral nervous system, and future work must consider peripheral sensory cells and potential peripheral processing contributing to olfaction in gastropods.

Conclusion

In conclusion, many of the proximate mechanisms of gastropod olfaction have been unraveled, but many new opportunities for extending our knowledge have arisen. Further exploration is needed to unify understanding across all levels of organization, starting at the sampling of odors, to the function of receptor genes and sensory cells, and ultimately the peripheral and central circuits that process olfactory inputs to produce adaptive behavior. These pursuits will best be done with explicit awareness of the evolutionary relationships among the species chosen for study. In this manner, the somewhat piecemeal set of current information will be unified into a more comprehensive understanding of olfaction in a group of animals for which the sense of smell is paramount.

References

Adamo, S. A. and Chase, R. (1990). The "love dart" of the snail *Helix aspersa* injects a pheromone that decreases courtship duration. *Reprod. Biol.* 255(1): 80-87.

Agersborg, H. P. K. (1925). The sensory receptors and the structure of the oral tentacles of the nudibranchiate mollusk *Hermissenda crassicornis. Acta Zoologica* 6: 167–182.

Audesirk, G. and Audesirk, T. (1980a). Complex mechanoreceptors in *Tritonia diomedea*. I. Responses to mechanical and chemcial stimuli. *Journal of Comparative Physiology* 141: 101-109.

Audesirk, G. and Audesirk, T. (1980b). Complex mechanoreceptors in *Tritonia diomedea*. II. Neuronal correlates of a change in behavioral responsiveness. *Journal of Comparative Physiology* 141: 111-122.

Audesirk, T. E. (1975). Chemoreception in *Aplysia californica*. I. Behavioral localization of distance chemoreceptors used in food-finding. *Behav Biol* 15(1): 45-55.

Balaban, P. M. (2002). Cellular mechanisms of behavioral plasticity in terrestrial snail. *Neuroscience and Biobehavioral Reviews* 26(5): 597-630.

Baratte, S. and Bonnaud, L. (2009). Evidence of early nervous differentiation and early catecholaminergic sensory system during *Sepia officinalis* embryogenesis. *Journal of Comparative Neurology* 517(4): 539-549.

Bargmann, C. I. (2006). Comparative chemosensation from receptors to ecology. *Nature* 444(7117): 295-301.

Barker, G. M. (2002). *Molluscs as crop pests.* Wallingford, UK: CABI Publishing.

Bell, H. J., Inoue, T., Shum, K., Luk, C. andSyed, N. I. (2007). Peripheral oxygen-sensing cells directly modulate the output of an identified respiratory central pattern generating neuron. *European Journal of Neuroscience* 25(12): 3537-3550.

Benjamin, P. (2008). *Lymnaea. Scholarpedia* 3(1).

Benton, R., Vannice, K. S., Gomez-Diaz, C. and Vosshall, L. B. (2009). Variant ionotropic glutamate receptors as chemosensory receptors in Drosophila. *Cell* 136(1): 149-162.

Bicker, G., Davis, W. J., Matera, E. M., Kovac, M. P. and Stormogipson, D. J. (1982). Chemoreception and mechanoreception in the gastropod mollusk *Pleurobranchea californica*. I. Extracellular analysis of afferent pathways. *J Comp Physiol* 149: 221-234.

Birkeland, C. (1974). Interactions between a sea pen and seven of its predators. *Ecol. Monogr.* 44: 211–232.

Boudko, D. Y., Switzer-Dunlap, M. and Hadfield, M. G. (1999). Cellular and subcellular structure of anterior sensory pathways in *Phestilla sibogae* (Gastropoda, Nudibranchia). *Journal of Comparative Neurology* 403: 39-52.

Bovbjerg, R. V. (1968). Responses to Food in Lymnaeid Snails. *Physiological Zoology* 41(4): 412-423.

Cain, S. D., Clark, M. and Kelly, D. (2010). The Functional Morphology and Transmitter Distribution an Olfactory Organ of *Tritonia diomedea. Integrative and Comparative Biology* 50: E212-E212.

Chase, R. (1985). Responses to odors mapped in snail tentacles and brain by 14C-2-deoxyglucose autoradiography. *J. Neurosci.* 5: 2930–2939.

Chase, R., Croll, R. P. and Zeichner, L. L. (1980). Aggregation in snails, Achatina fulica. *Behav Neural Biol* 30(2): 218-230.

Chase, R. and Tolloczko, B. (1993). Tracing neural pathways in snail olfaction: from the tip of the tentacles to the brain and beyond. *Microsc. Res. Tech.* 24: 214–230.

Chelazzi, G. (1990). Eco-ethological aspects of homing behaviour in molluscs. *Ethology Ecology and Evolution* 2(1): 11-26.

Cook, A. (1985). Tentacular function in trail following by the pulmonate slug *Limax pseudoflavus* evans. *Journal of Molluscan Studies* 51(3): 240-247.

Cooke, I. R. C., Edwards, S. L. and Anderson, C. R. (1994). The distribution of NADPH-diaphorase activity and immunoreactivity to nitric oxide synthase in the nervous system of the pulmonate mollusc *Helix aspersa. Cell Tissue Res.* 277: 565–572.

Croll, R. P. (2001). Catecholamine-containing cells in the central nervous system and periphery of Aplysia californica. *J Comp Neurol* 441(2): 91-105.

Croll, R. P. (2003). Complexities of a simple system: new lessons, old challenges and peripheral questions for the gill withdrawal reflex of Aplysia. *Brain Res Brain Res Rev* 43(3): 266-274.

Croll, R. P., Boudko, D. Y. and Hadfield, M. G. (2001). Histochemical survey of transmitters in the central ganglia of the gastropod mollusc Phestilla sibogae. *Cell Tissue Res* 305(3): 417-432.

Croll, R. P., Boudko, D. Y., Pires, A. and Hadfield, M. G. (2003). Transmitter contents of cells and fibers in the cephalic sensory organs of the gastropod mollusc Phestilla sibogae. *Cell Tissue Res* 314(3): 437-448.

Croll, R. P. and Chase, R. (1977). A long-term memory for food odors in the land snail, Achatina fulica. *Behav Biol* 19(2): 261-268.

Croll, R. P. and Chase, R. (1980). Plasticity of olfactory orientation to food in the snail *Achatina fulica. J Comp Physiol A* 136: 267-277.

Croll, R. P., Voronezhskaya, E. E., Hiripi, L. and Elekes, K. (1999). Development of catecholaminergic neurons in the pond snail, Lymnaea stagnalis: II. Postembryonic development of central and peripheral cells. *J Comp Neurol* 404(3): 297-309.

Cummins, S. F., Erpenbeck, D., Zou, Z., Claudianos, C., Moroz, L. L., Nagle, G. T. and Degnan, B. M. (2009a). Candidate chemoreceptor subfamilies differentially expressed in the chemosensory organs of the mollusc Aplysia. *BMC Biol* 7: 28.

Cummins, S. F., Leblanc, L., Degnan, B. M. and Nagle, G. T. (2009b). Molecular identification of candidate chemoreceptor genes and signal transduction components in the sensory epithelium of Aplysia. *J Exp Biol* 212(Pt 13): 2037-2044.

Cummins, S. F., Nichols, A. E., Amare, A., Hummon, A. B., Sweedler, J. V. and Nagle, G. T. (2004). Characterization of *Aplysia* enticin and temptin, two novel water-borne protein pheromones that act in concert with attractin to stimulate mate attraction. *J Biol Chem* 279(24): 25614-25622.

Cummins, S. F., Schein, C. H., Xu, Y., Braun, W. and Nagle, G. T. (2005). Molluscan attractins, a family of water-borne protein pheromones with interspecific attractiveness. *Peptides* 26(1): 121-129.

de Vlieger, T. A. (1968). An experimental study of the tactile system of *Lymnaea stagnalis* (L.). *Neth. J. Zool.* 18: 105-154.

Dinapoli, A. and Klussmann-Kolb, A. (2010). The long way to diversity-phylogeny and evolution of the Heterobranchia (Mollusca: Gastropoda). *Mol Phylogenet Evol* 55(1): 60-76.

Ellis, R. (2003). *Aquagenesis: The origin and evolution of life in the sea.* New York: Penguin.

Elphick, M. R., Kemenes, G., Staras, K. and O'Shea, M. (1995). Behavioral role for nitric-oxide in chemosensory activation of feeding in a mollusk. *Journal of Neuroscience* 15(11): 7653-7664.

Emery, D. G. (1992). Fine structure of olfactory epithelia of gastropod molluscs. *Microsc Res Tech* 22: 307-324.

Emery, D. G. and Audesirk, T. E. (1978). Sensory cells in *Aplysia*. *J Neurobiol* 9(2): 173-179.

Farkas, S. R. and Shorey, H. H. (1976). Anemotaxis and odour-trail following by the terrestrial snail Helix aspersa. *Animal Behaviour* 24(3): 686-689.

Field, L. H. and Macmillan, D. L. (1973). An electrophysiological and behavioural study of sensory responses in *Tritonia* (Gastropoda, Nudibranchia). *Marine Behaviour and Physiology* 2: 171-185.

Fink, P., Elert, E. v. and Jüttner, F. (2006). Volatile foraging kairomones in the littoral zone: attraction of an herbivorous freshwater gastropod to algal odors. *Journal of Chemical Ecology* 32(9): 1867-1881.

Fredman, S. M. and Jahan-Parwar, B. (1980). Processing of chemosensory and mechanosensory information in identifiable *Aplysia* neurons. *Comparative Biochemistry and Physiology* 66A: 25-34.

Friedrich, A. and Teyke, T. (1998). Identification of stimuli and input pathways mediating food-attraction conditioning in the snail, *Helix*. *J. Comp. Physiol. A* 183: 247–254.

Frost, W. N., Hoppe, T. A., Wang, J. and Tian, L. M. (2001). Swim initiation neurons in *Tritonia diomedea*. *American Zoologist* 41(4): 952-961.

Fujie, S., Aonuma, H., Ito, I., Gelperin, A. andIto, E. (2002). The nitric oxide/cyclic GMP pathway in the olfactory processing system of the terrestrial slug Limax marginatus. *Zoological Science (VSP International Science Publishers)* 19(1).

Gallagher, C. A., Rohozinski, D., Stanton, K., Murray, J. A. and Cain, S. D. (2011). Description of a novel chemosensory organ in the nudibranch *Tritonia diomedea*. *Integrative and Comparative Biology* 51: E191-E191.

Gelperin, A. (1974). Olfactory basis of homing behavior in the giant slug *Limax maximus*. *PNAS* 71: 966-970.

Gelperin, A. (1975). Rapid food-aversion learning by a terrestrial mollusk. *Science* 189(4202): 567-570.

Gelperin, A. (1999). Oscillatory dynamics and information processing in olfactory systems. *Journal of Experimental Biology* 202(14): 1855-1864.

Gelperin, A., Flores, J., Raccuia-Behling, F. and Cooke, I. R. C. (2000). Nitric oxide and carbon monoxide modulate oscillations of olfactory interneurons in a terrestrial mollusk. *Journal of Neurophysiology* 83(1): 116-127.

Getting, P. A. (1976). Afferent neurons mediating escape swimming of the marine mollusc, Tritonia. *Journal of Comparative Physiology A: Neuroethology, Sensory, Neural, and Behavioral Physiology* 110(3): 271-286.

Gobbeler, K. and Klussmann-Kolb, A. (2007). A comparative ultrastructural investigation of the cephalic sensory organs in Opisthobranchia (Mollusca, Gastropoda). *Tissue Cell* 39(6): 399-414.

Goldschmeding, J. T. and Jager, J. C. (1973). Feeding responses to sucrose in the pond snail *Lymnaea stagnalis* after nerve section and tentacle amputation. *Neth. J. Zool.* 23: 118-124.

Hatakeyama, D., Aonuma, H., Ito, E. and Elekes, K. (2007). Localization of glutamate-like immunoreactive neurons in the central and peripheral nervous system of the adult and developing pond snail, *Lymnaea stagnalis. Biol Bull* 213: 172–186.

Hegedus, E., Kaslin, J., Hiripi, L., Kiss, T., Panula, P. and Elekes, K. (2004). Histaminergic neurons in the central and peripheral nervous system of gastropods (*Helix*, *Lymnaea*): an immunocytochemical, biochemical, and electrophysiological approach. *J Comp Neurol* 475: 391–405.

Hernadi, L. (2000). Topographic organization of efferent neurons with different neurochemical characters in the cerebral ganglia of the snail *Helix pomatia. Microsc Res Tech* 49(6): 521–533.

Hernadi, L. and Elekes, K. (1999). Topographic organization of serotonergic and dopaminergic neurons in the cerebral ganglia and their peripheral projection patterns in the head areas of the snail *Helix pomatia*. *J Comp Neurol* 411(2): 274–287.

Hill, E. S., Moore-Kochlacs, C., Vasireddi, S. K., Sejnowski, T. J. and Frost, W. N. (2010). Validation of independent component analysis for rapid spike sorting of optical recording data. *J. Neurophysiol.* 104(6): 3721-3731.

Hoyle, G. and Willows, A. O. D. (1973). Neuronal basis of behaviour in *Tritoma*. II. Relationship of muscular contraction to nerve impulse pattern. *J. Neurobiol.* 4: 239-254.

Ierusalimsky, V. N. and Balaban, P. M. (2007). Primary sensory neurons containing command neuron peptide constitute a morphologically distinct class of sensory neurons in the terrestrial snail. *Cell and Tissue Research* 330(1): 169-177.

Ierusalimsky, V. N. and Balaban, P. M. (2010). Two morphological sub-systems within the olfactory organs of a terrestrial snail. *Brain Res* 1326: 68-74.

Iglesias, J. and Castillejo, J. (1999). Field observations of the land snail *Helix aspersa* Müller. *J Moll Stud* 65: 411-423.

Inokuma, Y., Inoue, T., Watanabe, S. and Kirino, Y. (2002). Two types of network oscillations and their odor responses in the primary olfactory center of a terrestrial mollusk. *Journal of Neurophysiology* 87(6): 3160-3164.

Ito, I., Kimura, T., Watanabe, S., Kirino, Y. and Ito, E. (2004). Modulation of two oscillatory networks in the peripheral olfactory system by gamma-aminobutyric acid, glutamate, and acetylcholine in the terrestrial slug *Limax marginatus*. *Journal of Neurobiology* 59(3): 304-318.

Ito, I., Nakamura, H., Kimura, T., Suzuki, H., Sekiguchi, T., Kawabata, K. and Ito, E. (2000). Neuronal components of the superior and inferior tentacles in the terrestrial slug, *Limax marginatus*. *Neuroscience Research* 37(3): 191-200.

Ito, I., Watanabe, S., Kimura, T., Kirino, Y. and Ito, E. (2003a). Distributions of gamma-aminobutyric acid immunoreactive and acetylcholinesterase-containing cells in the primary olfactory system in the terrestrial slug *Limax marginatus*. *Zoological Science* 20(11): 1327-1335.

Ito, I., Watanabe, S., Kimura, T., Kirino, Y. and Ito, E. (2003b). Distributions of Œ≥-Aminobutyric Acid Immunoreactive and Acetylcholinesterase-Containing Cells in the Primary Olfactory System in the Terrestrial Slug Limax marginatus. *Zoological Science* 20(11): 1337-1346.

Ito, I., Watanabe, S. and Kirino, Y. (2006). Mapping of odor-related neuronal activity using a fluorescent derivative of glucose. *Neuroscience Letters* 398(3): 224-229.

Jager, J. C. (1971). A quantitative study of a chemoresponse to sugars in *Lymnaea stagnalis* (L.). *Neth. J. Zool.* 21: 1-59.

Jahan-Parwar, B. (1972). Behavioral and electrophysiological studies on chemoreceptors in *Aplysia*. *Am. Zool.* 12: 525-537.

Janse, C. (1976).The tactile system of *Lymnaea stagnalis*. In *Neurobiology of invertebrates, Gastropoda brain*, 473-485 Budapest: Akademiai Kiado.

Kamardin, N., Szücs, A. and Rózsa, K. S. (1999). Distinct responses of osphradial neurons to chemical stimuli and neurotransmitters in *Lymnaea stagnalis*. *Cellular and Molecular Neurobiology* 19(2): 235-247.

Kamardin, N. N., Shalanki, Y., Sh.-Rozha, K. and Nozdrachev, A. D. (2001). Studies of chemoreceptor perception in mollusks. *Neuroscience and Behavioral Physiology* 31(2): 227-235.

Karnik, V., Braun, M., Dalesman, S. and Lukowiak, K. (2012). Sensory input from the osphradium modulates the response to memory-enhancing stressors in *Lymnaea stagnalis*. *The Journal of Experimental Biology* 215(3): 536-542.

Kasai, Y., Watanabe, S., Kirino, Y. and Matsuo, R. (2006). The procerebrum is necessary for odor-aversion learning in the terrestrial slug *Limax valentianus*. *Learning and Memory* 13(4): 482-488.

Kaufmann, W., Kerschbaum, H. H., Hauser-Kronberger, C., Hacker, G. W. and Hermann, A. (1995). Distribution and seasonal variation of vasoactive intestinal (VIP)-like peptides in the nervous system of *Helix pomatia*. *Brain Res* 695: 125–136.

Kehoe, E. F., Webber, M. P. and Wyeth, R. C. (2012). Catecholaminergic peripheral neuroanatomy in the nudibranch *Tritonia diomedea. Bulletin - Canadian Society of Zoologists.*

Kemenes, G., Elliott, C. J. H. and Benjamin, P. R. (1986). Chemical and tactile inputs to the *Lymnaea* feeding system - effects on behavior and neural circuitry. *Journal of Experimental Biology* 122: 113-137.

Keunzi, F. M. and Carew, T. J. (1991). Identification of neuronal pathways mediating phototactic modulation of head-waving in *Aplysia californica. Behav Neural Biol* 55(3): 338-355.

Kimura, T., Iwama, A. and Sekiguchi, T. (1999). Contributions of superior and inferior tentacles to learned food-avoidance behavior in *Limax marginatus. Zoological Science* 16(4): 595-602.

Klussmann-Kolb, A. (2004). Phylogeny of the Aplysiidae (Gastropoda, Opisthobranchia) with new aspects of the evolution of the seahares. *Zoologica Scripta* 33: 439-462.

Klussmann-Kolb, A., Dinapoli, A., Kuhn, K., Streit, B. and Albrecht, C. (2008). From sea to land and beyond--new insights into the evolution of euthyneuran Gastropoda (Mollusca). *BMC Evol Biol* 8: 57.

Kobayashi, S., Hattori, M., Elekes, K. r., Ito, E. and Matsuo, R. (2010). FMRFamide regulates oscillatory activity of the olfactory center in the slug. *European Journal of Neuroscience* 32(7): 1180-1192.

Kobayashi, S., Matsuo, R., Sadamoto, H., Watanabe, S. and Ito, E. (2012). Excitatory effects of GABA on procerebrum neurons in a slug. *Journal of Neurophysiology* 108(4): 989-998.

Koehl, M. A. R. (2006). The fluid mechanics of arthropod sniffing in turbulent odor plumes. *Chemical Senses* 31(2): 93-105.

Kosaka, K., Toida, K., Aika, Y. and Kosaka, T. (1998). How simple is the organization of the olfactory glomerulus?: the heterogeneity of so-called periglomerular cells. *Neuroscience Research* 30: 101-110.

Kriegstein, A. R. (1977). Development of the nervous system of *Aplysia californica. Proc Natl Acad Sci U S A* 74(1): 375-378.

Kupfermann, I. and Weiss, K. (1976). Water regulation by a presumptive hormone contained in identified neurosecretory cell R15 of *Aplysia. J Gen Physiol* 67: 113-123.

Lemaire, M. and Chase, R. (1998). Twitching and quivering of the tentacles during snail olfactory orientation. *Journal of Comparative Physiology A* 182(1): 81-87.

Leonard, J. L. and Lukowiak, K. (1986). The behavior of *Aplysia californica* Cooper (Gastropoda; Opisthobranchia). I. Ethogram. *Behaviour* 98: 320–360.

Levy, M., Blumberg, M. and Susswein, A. J. (1997). The rhinophores sense pheromones regulating multiple behaviors in *Aplysia fasciata. Neuroscience Letters* 225: 113-116.

Lind, H. (1989). Homing to hibernating sites in *Helix pomatia* involving detailed long-term memory. *Ethology Ecology and Evolution* 81: 221-234.

Lombardo, F., Maramaldo, R., Fratello, B. and Sonetti, D. (1991). Sequential analysis of chemoreception in *Planorbarius corneus* (L.). *Comparative Biochemistry and Physiology Part A: Physiology* 99(4): 627-632.

Lowe, S., Browne, M., Boudjelas, S. and De Poorter, M. (2004). *100 of the world's worst invasive alien species: a selection from the global invasive species database.* Invasive Species Specialist Group Auckland, New Zealand.

Lukowiak, K. andJacklet, J. W. (1972). Habituation and dishabituation: interactions between peripheral and central nervous systems in *Aplysia*. *Science* 178(67): 1306-1308.

Madsen, H. (1992). A Comparative Study on the Food-Locating Ability of *Helisoma duryi, Biomphalaria camerunensis* and *Bulinus truncatus* (Pulmonata: Planorbidae). *Journal of Applied Ecology* 29(1): 70-78.

Mainland, J. and Sobel, N. (2006). The sniff is part of the olfactory percept. *Chemical Senses* 31(2): 181-196.

Marchand, C. R. and Dubois, M. P. (1986). Immunocytochemical detection of substances related to methionine enkephalin in the tentacles and foot of the snail (*Helix aspersa*). *C R Seances Soc Biol Fil* 180: 184–189.

Martin, J. P., Beyerlein, A., Dacks, A. M., Reisenman, C. E., Riffell, J. A., Lei, H. and Hildebrand, J. G. (2011). The neurobiology of insect olfaction: Sensory processing in a comparative context. *Progress in Neurobiology* 95: 427-447.

Matsuo, R., Kobayashi, S., Watanabe, S., Namiki, S., Iinuma, S., Sakamoto, H., Hirose, K. and Ito, E. (2009). Glutamatergic neurotransmission in the procerebrum (Olfactory center) of a terrestrial mollusk. *Journal of Neuroscience Research* 87(13): 3011-3023.

Matsuo, R., Kobayashi, S., Yamagishi, M. and Ito, E. (2011). Two pairs of tentacles and a pair of procerebra: optimized functions and redundant structures in the sensory and central organs involved in olfactory learning of terrestrial pulmonates. *The Journal of Experimental Biology* 214(6): 879-886.

McCullagh, G.B. (2013). The rhinophores provide sufficient sensory input for odour-based navigation by the nudibranch mollusc, *Tritonia diomedea*. M.Sc. Thesis, St. Francis Xavier University, Antigonish, NS, Canada.

Megalou, E. V., Brandon, C. J. and Frost, W. N. (2009). Evidence that the swim afferent neurons of *Tritonia diomedea* are glutamatergic. *Biological Bulletin* 216(2): 103-112.

Moroz, L. L. (2006). Localization of putative nitrergic neurons in peripheral chemosensory areas and the central nervous system of *Aplysia californica*. *J Comp Neurol* 495(1): 10-20.

Moroz, L. L., Bulloch, A. G. M., Lukowiak, K. and Syed, N. (1994a). Putative NO-synthesizing neurons of *Lymnaea in vivo* and *in vitro*. *Neth. J. Zool.* 44: 535–549.

Moroz, L. L. and Gillette, R. (1996). NADPH-diaphorase localization in the CNS and peripheral tissues of the predatory sea-slug Pleurobranchaea californica. *J Comp Neurol* 367(4): 607-622.

Moroz, L. L., Norekian, T. P., Pirtle, T. J., Robertson, K. J. and Satterlie, R. A. (2000). Distribution of NADPH-diaphorase reactivity and effects of nitric oxide on feeding and locomotory circuitry in the pteropod mollusc, Clione limacina. *J Comp Neurol* 427(2): 274-284.

Moroz, L. L., Park, J. H. and Winlow, W. (1993). Nitric oxide activates buccal motor patterns in Lymnaea stagnalis. *Neuroreport* 4(6): 643-646.

Moroz, L. L., Winlow, W., Turner, R. W., Bulloch, A. G., Lukowiak, K. and Syed, N. I. (1994b). Nitric oxide synthase-immunoreactive cells in the CNS and periphery of Lymnaea. *Neuroreport* 5(10): 1277-1280.

Murray, J. A., Estepp, J. and D., C. S. (2006). Advances in the neural bases of orientation and navigation. *Integrative and Comparative Biology* 46(6): 871-879.

Murray, J. A., Hewes, R. S. and Willows, A. O. (1992). Water-flow sensitive pedal neurons in *Tritonia*: role in rheotaxis. *Journal of comparative physiology* 171(3): 373-385.

Nakamura, H., Ito, I., Kojima, S., Fujito, Y., Suzuki, H. and Ito, E. (1999a). Histological characterization of lip and tentacle nerves in *Lymnaea stagnalis. Neuroscience Research* 33(2): 127-136.

Nakamura, H., Kojima, S., Kobayashi, S., Ito, I., Fujito, Y., Suzuki, H. and Ito, E. (1999b). Physiological characterization of lip and tentacle nerves in *Lymnaea stagnalis. Neuroscience Research* 33(4): 291-298.

Nelson, G. M. and Audesirk, T. E. (1986). Identification of central neurons innervating peripheral chemoreceptive structures in *Lymnaea stagnalis. Comparative Biochemistry and Physiology A-Physiology* 83(1): 113-120.

Newcomb, J. M. and Watson, W. H. I. (2001). Identifiable nitrergic neurons in the central nervous system of the nudibranch *Melibe leonina* localized with NADPH-diaphorase histochemistry and nitric oxide synthase immunoreactivity. *J Comp Neurol* 437: 70-78.

Painter, S. D., Cummins, S. F., Nichols, A. E., Akalal, D. B., Schein, C. H., Braun, W., Smith, J. S., Susswein, A. J., Levy, M., de Boer, P. A., ter Maat, A., Miller, M. W., Scanlan, C., Milberg, R. M., Sweedler, J. V. and Nagle, G. T. (2004). Structural and functional analysis of *Aplysia* attractins, a family of water-borne protein pheromones with interspecific attractiveness. *Proc Natl Acad Sci U S A* 101(18): 6929-6933.

Pinching, A. J. and Powell, T. P. S. (1971). The neuropil of the glomeruli of the olfactory bulb. *J. Cell Sci.* 9: 347-377.

Preston, R. J. and Lee, R. M. (1973). Feeding behavior in *Aplysia calfornica:* Role of chemical and tactile stimuli. *J. Comp. Physiol. Psychol.* 82: 368-381.

Redondo, R. L. and Murray, J. A. (2005). Pedal neuron 3 serves a significant role in effecting turning during crawling by the marine slug Tritonia diomedea. *J Comp Physiol A Neuroethol Sens Neural Behav Physiol.* 191(5): 435-444.

Robertson, H. M., Warr, C. G. and Carlson, J. R. (2003). Molecular evolution of the insect chemoreceptor gene superfamily in *Drosophila melanogaster. Proc Natl Acad Sci U S A* 100 Suppl 2: 14537-14542.

Roshchin, M. and Balaban, P. M. (2012). Neural control of olfaction and tentacle movements by serotonin and dopamine in terrestrial snail. *J Comp Physiol A* 198: 145-158.

Sahley, C., Gelperin, A. and Rudy, J. W. (1981). One-trial associative learning modifies food odor preferences of a terrestrial mollusc. *Proceedings of the National Academy of Sciences* 78(1): 640-642.

Serfozo, Z., Elekes, K. and Varga, V. (1998). NADPH-diaphorase activity in the nervous system of the embryonic and juvenile pond snail, *Lymnaea stagnalis. Cell Tissue Res* 292: 579–586.

Staras, K., Kemenes, G. and Benjamin, P. R. (1999). Electrophysiological and behavioral analysis of lip touch as a component of the food stimulus in the snail *Lymnaea. Journal of Neurophysiology* 81(3): 1261-1273.

Suzuki, H., Kimura, T., Sekiguchi, T. and Mizukami, A. (1997). FMRF amide-like-immunoreactive primary sensory neurons in the olfactory system of the terrestrial mollusc, *Limax marginatus. Cell and Tissue Research* 289(2): 339-345.

Takeda, N. and Tsuruoka, H. (1979). A sex pheromone secreting gland in the terrestrial snail, *Euhadra peliomphala. J Exp Biol Zool* 207: 17-26.

Tomiyama, K. (1992). Homing behaviour of the giant African snail, *Achatina fulica* (Ferussac) (Gastropoda; Pulmonata). *Journal of Ethology* 10(2): 139-146.

Townsend, C. R. (1973). The food-finding orientation mechanism of *Biomphalaria glabrata* (Say). *Animal Behaviour* 21: 544-548.

Townsend, C. R. (1974). Mucus trail following by the snail *Biomphalaria glabrata* (Say). *Animal Behaviour* 22: 549-556.

Uhazy, L., Tanaka, R. D. and MacInnis, A. J. (1978). *Schistosoma manoni*: identification of chemicals that attract or trap its snail vector, *Biomphalaria glabrata. Science* 201: 924-926.

van Bruggen, A. C. (1995).Biodiversity of Mollusca: Time for a new approach. In *Biodiversity and Conservation of the Mollusca*, 1–19 (Ed W. S. van Bruggen AC, Kemperman TCM). Oegstgeest-Leiden: Backhuys.

Voss, M. (2000). Neurophysiological and behavioral responses to olfactory stimuli in the snail *Helix pomatia* L. *Physiological Response* 49: 463-469.

Wade, C. M., Mordan, P. B. and Naggs, F. (2006). Evolutionary relationships among the Pulmonate land snails and slugs (Pulmonata, Stylommatophora). *Biological Journal of the Linnean Society* 87(4): 593-610.

Wagele, H. and Willan, R. C. (2000). Phylogeny of the Nudibranchia. *Zoological Journal of the Linnean Society* 130(1): 83-181.

Watanabe, S., Kirino, Y. and Gelperin, A. (2008). Neural and molecular mechanisms of microcognition in *Limax. Learning and Memory* 15(9): 633-642.

Wedemeyer, H. and Schild, D. (1995). Chemosensitivity of the osphradium of the pond snail *Lymnaea stagnalis. Journal of Experimental Biology* 198(8): 1743-1754.

Weiss, K. R., Bayley, H., Lloyd, P. E., Tenenbaum, R., Kolks, M. A., Buck, L., Cropper, E. C., Rosen, S. C. and I., K. (1989). Purification and sequencing of neuropeptides contained in neuron R15 of *Aplysia californica. Proc Natl Acad Sci USA* 86: 2913–2917.

Wertz, A., Rossler, W., Obermayer, M. and Bickmeyer, U. (2006). Functional neuroanatomy of the rhinophore of *Aplysia punctata. Front Zool* 3: 6.

Wertz, A., Rössler, W., Obermayer, M. and Bickmeyer, U. (2007). Functional neuroanatomy of the rhinophore of *Archidoris pseudoargus*. 61(2): 135-142.

Willows, A. O., Dorsett, D. A. and Hoyle, G. (1973). The neuronal basis of behavior in *Tritonia*. I. Functional organization of the central nervous system. *J. Neurobiol.* 4(3): 207-237.

Wollesen, T., Wanninger, A. and Klussmann-Kolb, A. (2007). Neurogenesis of cephalic sensory organs of Aplysia californica. *Cell Tissue Res* 330(2): 361-379.

Wyeth, R. C. and Croll, R. P. (2011). Peripheral sensory cells in the cephalic sensory organs of Lymnaea stagnalis. *J Comp Neurol* 519(10): 1894-1913.

Wyeth, R. C., Kehoe, E. F. and McCullagh, G. B. (2012).Odour-based navigation behaviour in *Tritonia*: sensory systems and strategies. In *10th International Congress of Neuroethology*.

Wyeth, R. C. and Willows, A. O. D. (2006a). Field behavior of the nudibranch mollusc *Tritonia diomedea. Biol Bull* 210(2): 81-96.

Wyeth, R. C. and Willows, A. O. D. (2006b). Field behavior of the nudibranch mollusc *Tritonia diomedea. Biological Bulletin* 210(2): 81-96.

Wyeth, R. C. and Willows, A. O. D. (2006c). Odours detected by rhinophores mediate orientation to flow in the nudibranch mollusc, Tritonia diomedea. *Journal of Experimental Biology* 209(8): 1441-1453.

Wyeth, R. C., Woodward, O. M. and Willows, A. O. D. (2006). Orientation and navigation relative to water flow, prey, conspecifics, and predators by the nudibranch mollusc *Tritonia diomedea*. *Biological Bulletin* 210(2): 97-108.

Zaitseva, O. V. (1994). 2. Sensory systems of the snail: structural organization of the sensory systems of the snail. *Neuroscience and Behavioral Physiology* 24(1): 47-57.

Zaitseva, O. V. and Bocharova, L. S. (1981). Sensory cells in the head skin of pond snails. *Cell Tissue Res* 220: 797-807.

Zhao, K., Dalton, P., Yang, G. C. and Scherer, P. W. (2006). Numerical modeling of turbulent and laminar airflow and odorant transport during sniffing in the human and rat nose. *Chemical Senses* 31(2): 107-118.

Zylstra, U. (1972). Distribution and ultrastructure of epidermal sensory cells in the freshwater snails *Lymnaea stagnalis* and *Biomphalaria pfeifferi*. *Neth. J. Zool.* 22(283-298).

In: Neuroecology and Neuroethology in Molluscs
Editors: Anna Di Cosmo and William Winlow
ISBN: 978-1-62948-970-4

Chapter 4

Communications between Predator and Prey

Matthew Nusnbaum
Department of Biology
Georgia State University, GA, US

Abstract

The study of complex behavioral and neurological phenomena requires an understanding of the context in which the behavior takes place and the relevant pressures under which they evolved. Predator-prey dynamics play an important role in the survival and fecundity of species and the traits and behaviors shaped by these dynamics provide compelling questions for these studies. Adaptations that improve predatory success and those that help prey species avoid or escape attack each have strong influence on individual survival and success. Molluscs are important consumers and resources in many ecosystems and their adaptations as predators and prey are well studied. This chapter provides examples of cues and elicited behaviors in diverse molluscs as prey and as predators. By studying the signals through which molluscs identify and avoid threats and locate and consume food, researchers can connect the dots between stimuli, neural pathways, behavior, and ecology.

Introduction

The broad goal of neuroethology is to link behavior and biological context to underlying neural control. In order to study these connections, we must have an understanding of the relevant environmental cues and ecological context as well as behavioral responses in this context. To study concepts that are both complex and intricate, we must know a great deal about the system in which subjects interact. Predator-prey dynamics lend themselves to this study because of the high degree of importance of food search and survival behaviors. Strong selective pressure leads to stereotypical behaviors and robust physiology. In this chapter I summarize work that has been done to elucidate the role of communication between predators and prey in shaping these relationships. The molluscs represent excellent model systems for the study of predator prey dynamics in neuroethological contexts. The phylum Mollusca is amongst the most diverse animal taxa, comprising over 80,000 species. Molluscs occupy diverse niches in a wide variety of habitats and represent important species in terms of understanding ecology, behavior, physiology, neuroscience and conservation biology. The molluscs include members that are benthic or sessile as well as active pelagic swimmers. They include members that represent prey for diverse taxa as well as active and ambush predators. In treating the subject of predator-prey communication, I will address the role of molluscs as prey and as predators.

Predation is a significant mortality factor in many organisms. A predation event can be identified as an encounter in which the predator searches, encounters, attacks, captures and, finally, ingests the prey (Endler 1986). Interrupting this cycle as soon as possible is adaptive for prey species and this strong selective pressure has favored the evolution of many elaborate defense adaptations, including behavioral, morphological, and chemical adaptations. For example, a prey species may adjust its behavior so that the probability of being encountered by a predator decreases. Alternatively, prey may have morphological defense structures that decrease capture success. Many of these adaptations may be costly, impacting fitness in a variety of ways. The development of defensive compounds or morphological structures may divert energy away from growth or reproduction. A change of habitat may impact foraging rate or chances of encountering a suitable mate. Therefore, the costs and benefits of anti-predator defenses must be balanced. Behaviorally, this can be accomplished by assessing predation risk and adjusting activity accordingly. For example, many marine and freshwater snails will react to chemical stimuli from predators by exhibiting escape behaviors that include increased locomotion or even crawling out of the water (Snyder and Snyder 1971, Alexander and Covich 1991, Dalesman et al. 2007).

Predators require continuous successful predation events to maintain metabolic activities and must therefore overcome the defensive adaptations of their prey. These predator adaptations can include sensory systems to detect and identify prey as well as morphological adaptations to subjugate prey. A number of studies have shown that predators and omnivores in aquatic systems, including molluscs, crustaceans, echinoderms and fish, are attracted to wounded or dead and decaying prey organisms which release low-molecular metabolites including amino acids, nucleotides and organic acids (for examples see Carr 1988). While prey benefit from interrupting the predatory encounter as soon as possible through defenses, predators wish to reach the conclusion quickly through counter defenses. The interactions between predators and prey are an evolutionary arms race in which there is selection on predators to become ever more efficient and effective at detecting and capturing prey. Predators that find prey unaware will usually dominate and successfully capture them. Prey that can identify a predator during approach will usually evade detection through crypsis, area avoidance, hiding, or initiating early flight (Lima and Dill 1990). The challenge for predators is to locate and identify prey before prey can locate, identify and respond to the approaching threat. These sets of goals can be achieved by gathering information about an organism's environment and the inhabitants of that environment. Not all information that is gathered is relevant to survival and the recipient must be able to filter the signals from the noise.

In this chapter, communication will be broadly defined as any signal being released by a signaler that provides relevant information to a receiver. For clarity, the catchall term "cues" will be used to describe signals that are being detected regardless of the mode or nature of the communication. Under this definition signals passively released by prey species that are used by predators for tracking as well as active and passive defenses used by prey to avoid or deter predators can all be considered communication cues because they give information to the receiver that can then be acted on. All extant organisms possess chemical senses to allow them to detect resources and dangers in their environment. In aquatic environments where visual and auditory cues can be less effective, chemical cues are important signals between and among species (Burks and Lodge, 2002). A great deal of study has been done to date into the roles of chemical senses in predator-prey interactions; in the interest of space and the context of this book we will limit discussions to the role of communication in molluscs as predators and prey (for other groups see: Greenstone and Dickens 2004, Derby and Zimmer 2011).

Molluscs As Prey

Molluscs, being abundant and soft bodied, represent an important food source for many different predators. Predators can impose important selection pressures on their prey and in molluscs this has resulted in the evolution of an impressive diversity of morphological and behavioral defenses that include a wide variety of chemical defenses, shell thickening and sculpturing, as well as avoidance, escape and some aggressive behaviors (Appleton and Palmer 1988, Dix and Hamilton 1993, Harper and Skelton 1993, McClintock et al. 1994, Kicklighter et al. 2005). These various adaptations can be expressed in a predator-specific way, requiring prey not only to detect the presence of a credible threat but also to identify the nature of the predator and perform the appropriate response (Rundle and Bronmark 2001). Being able to respond to cues that indicate the presence of a threat may allow individuals to alter their physiology or behavior and reduce chances of a threatening encounter. In a relatively simple experiment, octopuses displayed body patterning associated with alarm to water from a tank in which a Moray eel had been living (MacGinitie and MacGinitie, 1968). This experiment does not demonstrate that the octopus' behaviors would have protected them from attack by an eel, but it does show that the octopus could detect the potential threat and reliably alter its behavior in response.

Phenotypic plasticity is an organism's ability to express a range of different morphological, physiological, life history, or behavioral phenotypes in response to environmental conditions. Numerous studies indicate that plastic phenotypes may play a more important role in the co-evolution between predators and molluscs than genetically fixed phenotypes (Leonard et al. 1999, Tollrian and Harvell, 1999, Ernande and Dieckmann, 2004). For example, when exposed to predator's odor *Mytilus edulis* increases byssal thread production and shell thickness, and grows a larger adductor muscle (Côté, 1995, Reimer and Tedengren, 1996, Leonard et al., 1999). These changes may be adaptive because increased byssus helps mussels attach to the substrate, a thick shell better resists crushing, and stronger adductor muscles help mussels keep their shells closed. These types of adaptations can serve to help prey species avoid predator encounters or make individuals less attractive targets by being more difficult to overpower.

Predatory decapod crustaceans are prevalent consumers of molluscs and represent conspicuous dangers (Schweitzer and Feldmann 2010). Filter feeding bivalves emit attractive chemicals into the water column through their feeding process, and crustaceans can follow these aquatic trails to locate prey (Weissburg and Zimmer-Faust, 1993, Weissburg et al. 2002). These crustaceans may kill and eat the bivalves, but they may also snip the siphons or other protruding soft tissue and thereby reduce bivalve fitness through sub-lethal damage. Chemical cues emanating from crabs and lobsters have been shown to produce responses in a wide range of molluscan prey including reduced feeding behavior, aggregation, fleeing and seeking refuges (Irlandi and Peterson 1991, Côté and Jelnikar 1999, Nakaoka 2000, Turner et al. 2000, Jacobsen and Stabell 2004).

If a prey species possesses a variety of responses to the presence of predators, they would be best served by being able to determine not only that a predator is nearby but also what the identity of that predator may be so that they can initiate the appropriate response. Defenses that are successful against one predator may a) be effective against another predator, b) have no effect on defense against another predator, or c) be detrimental in protecting against another predator. In fact, the pulmonate pond snail *Lymnaea stagnalis* responds to both fish and crayfish predator cues and will respond by either hiding in rock crevices or moving above the waterline (Rundle and Bronmark 2001, Dalesman et al. 2006). Emerging out of the water would not protect the snail from crayfish and would not be an appropriate defense against this predator. Behaviors can be immediate, such as hiding and seeking shelter, or long lasting, such as habitat avoidance or morphological adaptations. Physid snails can alter foraging activity and habitat use in response to predator or alarm cues (Alexander and Covich 1991) while many species of bivalves and gastropods will change the shape, thickness or size of their shells in response to the presence of predators (Bertness and Cunningham 1981, Appleton and Palmer 1988, Harper and Skelton 1993).

Chemical cues from prey species can include attractive compounds that identify prey as present or vulnerable, but they can also include signals that reduce the selection and ingestion of that prey item. These protective cues include suppressants that reduce predator search or acquisition as well as deterrents that protect prey from attack. This may function by disrupting the attack through distraction, distastefulness, toxicity, irritation or other immediately unpleasant effects. Numerous studies have shown that chemical defenses protect prey from predation at multiple points in the encounter (Eisner and Meinwald 1966, Paul 1992, Pawlik 1993, Kelley et al. 2003, Nusnbaum and Derby 2010). A chemically defended animal avoids being eaten by using its chemical weaponry, which, depending on the mechanism, may allow them to avoid damage and injury during the encounter. While many molluscs are protected by shells, some, including opisthbranch gastropods and coleoid cephalopods have an impressive defensive behavior of releasing ink and other chemical weaponry when attacked. Defensive cues do not need to be actively released to be effective and there are many examples of passive defenses. Mucus from gastropods embeds released compounds and can limit their dispersal, improving chances that they reach their targets. This mucus can contain chemical deterrents that cause avoidance by potential predators (Branch 1981, Rice 1985, Avila et al. 1991). Secondary metabolites embedded in skin tissue may be diet-derived or created *de novo* and many of these have been demonstrated to be effective deterrents (Faulkner and Ghiselin 1983, Ginsburg and Paul 2001, Cimino and Gavagnin 2006). Beyond deterrence the biological effects of many of these compounds have not been tested, but some of them have been shown to be toxic and therefore honest signals (Thompson 1960, Marin et al 1999). Some molluscan chemical defenses include secretion of acids, which may be directly deterrent or enhance the effectiveness of other deterrent compounds (Gillette et al. 1999, Shabani et al. 2007). Passive defenses serve to provide a prey species with continuous protection, but require continuous production and cannot be selectively engaged during periods of heightened risk.

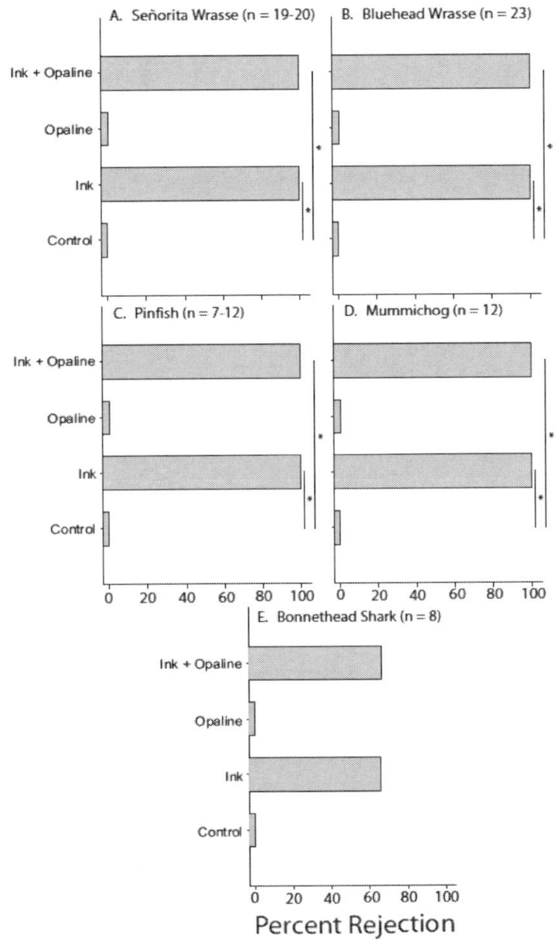

Figure 1. Responses of five fish species in ingestion assay using ink and opaline. Responses in A–D represent the percentage of fish that rejected shrimp-flavored alginate pellets or freeze-dried shrimp to which the indicated substance had been added, where *n* = number of individual fish on which each substance was tested. Responses in E represent the percentage of trials in which a single group of 20 sharks rejected freeze-dried shrimp to which the indicated substance had been added, where *n* = number of trials in which each substance was tested on the group of sharks. Rejection of shrimp-flavored pellets or shrimp containing the indicated substance was statistically compared to rejection of shrimp-flavored pellets or shrimp without an additive (which all fish ate before and after the experimental pellet or shrimp), using Cochran's *Q* test and *post hoc* one-tailed McNemar's tests. Ink or ink + opaline produced significant rejection ($P = 0.0001$), as indicated by asterisks, in A–D: (A) señorita wrasses, $Q = 51.08$, df = 3, $P = 0.0001$; (B) bluehead wrasses, $Q = 61.23$, df = 3, $P = 0.0001$); (C) pinfish, $Q = 30$, df = 3, $P = 0.0001$); (D) mummichogs, $Q = 33.90$, df = 3, $P = 0.0001$. For bonnethead sharks (E), there was an overall difference in the responses to the substances ($Q = 9$, df = 3, $P = 0.029$); however, none of the test substances was significantly different from the seawater control, although there was a strong but non-significant trend ($P = 0.06$) for shrimp treated with ink or ink + opaline to be rejected more than the control. For all five species of fishes A–E, opaline did not cause rejection ($P = 0.05$). Redrawn from Nusnbaum and Derby (2010a).

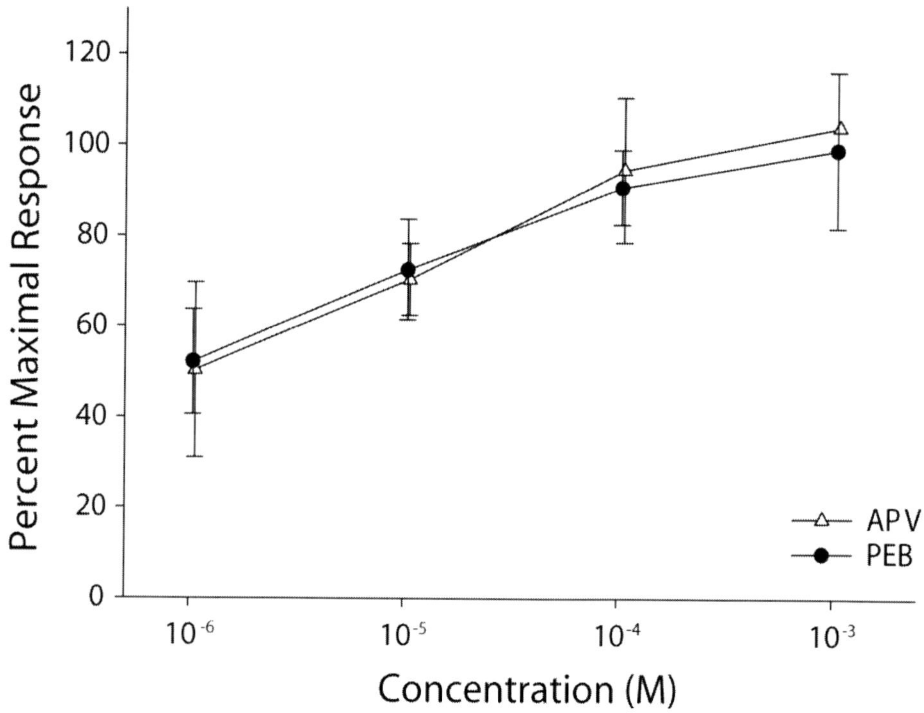

Figure 2. Electrophysiological concentration-response curve for APV and PEB. Integrated multi-unit recordings were taken from bundles of the facial-trigeminal nerve complex while presenting the maxillary barbel with different concentrations of APV and PEB. The integrated signal amplitudes (± SD) are displayed relative to the maximal response for APV. Response amplitude increased with concentration for APV and PEB, and the stimuli are not differentially stimulatory at similar concentrations. Concentration affected response magnitude for APV and PEB. Redrawn from Nusnbaum et al. (2012).

Active defenses are reserved for encounters in which other solutions such as avoidance or hiding have been ineffective and an attack has become imminent. After high intensity attack sea hares will release an ink secretion that is composed of the products of two glands. The ink gland produces a deep purple ink and the opaline gland releases a sticky, cloudy secretion. The purple color of ink is derived from phycoerythrobilin in the red algal diet of these grazers (Rüdiger 1967, Chapman and Fox 1969, Nolen et al. 1995). Combined, the product is a sticky, deep purple secretion that functions as a distracter (visual and chemical) as well as a deterrent against a variety of predators including sea anemones, crustaceans and fishes (Kicklighter et al. 2005, Derby et al. 2007, Nusnbaum and Derby 2010). Work with these extracts showed the product of the ink gland to be deterrent to a variety of fish species (Figure 1) (Nusnbaum and Derby 2010a). Further study with this product identified aplysioviolin (APV) and phycoerythrobilin (PEB) as deterrent components (Nusnbaum and Derby 2010b). These identified components could then be used to test the neural responses to behaviorally relevant deterrent compounds (Figures 2 & 3) (Nusnbaum and Derby 2012). Using relevant compounds to understand behavioral responses and neural coding underscores the neuroethological approach.

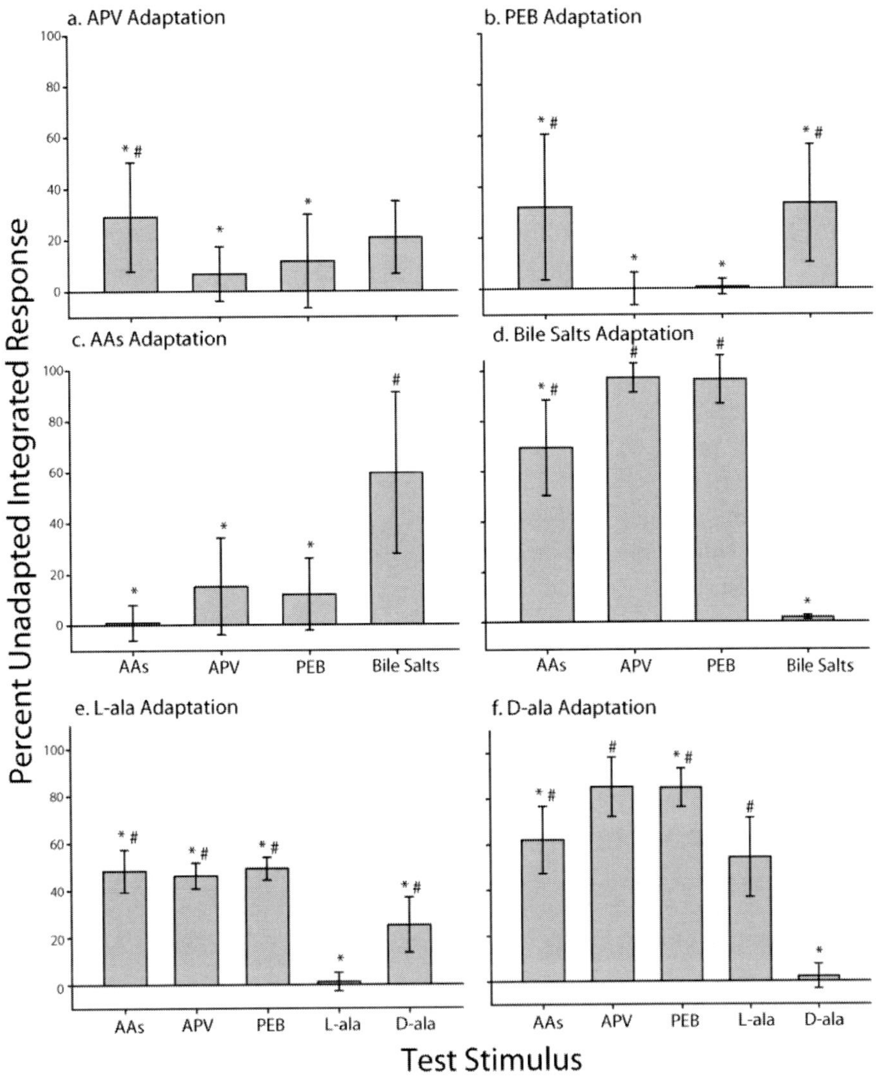

Figure 3. Cross-adaptation effects of different stimuli presented to the maxillary barbel. Results are depicted as the percent amplitude of the unadapted response, and values are mean ± SD. * - Adapted response is significantly different from 100% unadapted response. # - Cross-adapted response is significantly different from control (self-adaptation = the test response to the adapting stimulus). All adapting stimuli showed significant cross-adaptation of at least one test stimulus (ANOVA, APV: $n = 10$, $F = 3.161$, $df = 3$, $p < 0.05$; PEB: $n = 9$, $F = 6.402$, $p < 0.05$; AAs: $n = 10$, $F = 13.77$, $p < 0.05$; bile salts: $n = 4$, $F = 52.35$, $p < 0.05$; L-ala: $n = 4$, $F = 30.48$, $p < 0.05$; D-ala: $n = 4$, $F = 29.32$, $p < 0.05$). Statistical values for each figure are presented below. (a) Adaptation with APV reduced responses to APV, PEB, and AAs and showed a trend towards reduction of responses to bile salts. Adapted responses were reduced to control levels for APV and PEB, but not AAs. (b) Adaptation with PEB reduced responses to APV, PEB, AAs, and bile salts. Adapted responses were reduced to control levels for APV and PEB, but not AAs or bile salts. (c) Adaptation with AAs reduced responses to APV, PEB, and AAs. Adapted responses were reduced to control levels for APV, PEB, and AAs. (d) Adaptation with bile salts was effective in self-adaptation and showed a low level of adaptation of AAs. Adapted responses were only reduced to control levels for bile salts. (e) Adaptation with L-ala reduced responses to APV, PEB, AAs, L-ala, and D-ala. Adapted responses were only reduced to control levels for L-ala. (f) Adaptation with D-ala reduced responses to AAs, L-ala, and D-ala. Adapted responses were only reduced to control levels for D-ala. Redrawn from Nusnbaum et al. (2012).

Active defenses can protect molluscs from imminent threat by deterring predators or distracting them for long enough for the organism to escape, but they can also function to help a predator learn to avoid defended prey. Mummichogs and striped blennies learned to avoid nudibranchs through exposure to the sesquiterpene polygodial (Long and Hay 2006). Cephalopods are highly active pelagic organisms that are prey for diverse predators ranging from sea birds to marine mammals, sharks, and teleosts (Aronson 1991, Hanlon and Messenger 1996, Wood et al. 2003). They are famous for their numerous behavioral and physiological defenses which include color change that can function through crypsis or deimatism; escape movements using adapted water jets; or chemical release such as toxins or inks (Hanlon and Messenger 1996, King and Adamo 2006, Hanlon 2007, Bush et al. 2009). Cephalopods use ink as a last resort defense and only employ it after prolonged harassment, but it can be an effective part of their protean escape behaviors. Inking involves the release of a mass of chemicals produced in the ink sac and ejected through the funnel as the cephalopod flees. Observation would suggest that cephalopod ink is a visual stimulus that hides the animal as it escapes; though it may also serve as a distractor that attracts the attention of the attacking predator, or in some other way that may startle or distract the predator (Hanlon and Messenger, 1996, Bush and Robison, 2007). However, Wood et al. (2010) studying ink from *Sepioteuthis sepioidea* showed that the squid's ink could function through visual and chemical mechanisms during the capture and consummatory phases of predation by fish. Adding squid ink to otherwise palatable food increased the predators' handling time, indicating that components of the ink were unpalatable. There are a number of other studies that have identified potential components of cephalopod inks that may signal unpalatability to predators and therefore deter feeding. Tyrosinase, a cytotoxin present in the ink of several cephalopods, may irritate chemosensory receptors and deter predators (Prota et al. 1981; Russo et al. 2003). Ink from some octopuses contains tetrodotoxin although this compound's deterrent effects have not been experimentally demonstrated in octopus predators (Williams and Caldwell, 2009; Williams, 2010).

Molluscs As Predators

From the perspective of a predator, prey search and acquisition is affected by a wide variety of signals. Attractants initiate search for prey over distance; stimulants increase local phases of feeding including acquisition and acceptance; and deterrents and defenses reduce acquisition and acceptance (Derby and Carr 1987, Zimmer-Faust 1993, Nusnbaum and Derby 2010). Therefore, a single act of predation may require a predator to integrate a variety of complex, and sometimes contradictory, chemosensory information. *Tritonia diomedea*, a nudibranch that feeds on soft corals, will orient upstream when it senses food but will swim or crawl downstream in response to predator odors (Wyeth et al. 2006). Successful chemosensory based predation depends on an individual's ability to identify attractive chemical stimuli in the environment and locate the source of prey odors faster than competitors. Slow-moving predatory whelks can actually navigate in odor plumes that might otherwise be difficult for more active crustacean and fish predators because of dynamic properties of turbulent environment (Ferner and Weissburg 2005). While the gastropod molluscs do include a number of predatory members, the hunting behavior of the exclusively predatory cephalopods has been studied to a much greater degree. Conversely, the gastropod chemosensory systems and chemoreceptive behavior have been relatively well established when compared with hunting in cephalopods, which is generally ascribed to be primarily, if not exclusively, controlled by visual and mechanical cues.

Gastropods that hunt other molluscan prey can sometimes find that prey by following chemosensory signals in the mucus trails that are laid down for locomotion. These trails have been shown in many cases to function as aggregation cues in conspecifics and mating frequently occurs in these aggregations (Chase et al. 1978, Paul and Pennings 1991, Bryan and Slattery 1996). In the case of the mollusk-hunters, this cue can function as a reliable signal that a prey item is nearby. Simply following the trail can yield a reward that would not be as easily achieved through chemically mediated rheotaxis. Some opisthobranch and pulmonate carnivores such as *Navanax* and *Euglandina* use this method to hunt their molluscan prey. *Euglandina* is a terrestrial slug and snail predator and once it has contacted a slime trail it may follow it for as long as 20 cm, maintaining contact using its chemonsensitive lips. Experiments indicate that *Euglandina* as well as another snail predator, *Haplotrema concavum*, can even differentiate the species whose trail they are following (Cook 1985, Pearce and Gaertner 1996).

Signals between predator and prey do not need to be innate. Predators can learn to associate specific cues with the food reward of a prey item. In a study by Avila (1998) the nudibranch *Hermissenda crassicornis* was cultured in a lab, allowing researchers to manipulate its diet and cue exposure. *Hermissenda crassicornis* is a generalist predator that feeds on cnidarians, tunicates, small crustaceans and other nudibranchs. The animals in the experiment were fed a single food source after their metamorphosis. When tested for chemotaxis in a Y-maze the animals oriented towards this food source, *Tubularia crocea*, and another hydroid, *Pennaria*, but not to a tunicate, mussel or anemone. Following these tests individuals were fed exclusively on a new, non-oriented, food source: mussel, tunicate, or anemone. The newly conditioned animals were shown to prefer the new food source in the Y-maze trials, sometimes to the exclusion of the original favorites. Here, the nudibranch was demonstrated to not only associate the chemosensory stimulus with a food reward, but also to take on a new feeding preference using that same modality. The large group of cone snails, well known for their toxic barbs and ambush predation, would be of great interest for the cues that stimulate their attack. Unfortunately, a lot of emphasis has been given to the components of the diverse conotoxins and not as much attention has been given to the behaviors of these fascinating gastropods.

Cephalopods are fast-moving hunters and their predatory behaviors include active search and subjugation. A great deal of attention has been paid to the excellent vision seen in cephalopod molluscs. However, many cephalopods are either nocturnal or live at depths where little light penetrates, and distance and contact chemoreception may be quite important to these species. The cuttlefish, *Sepia officinalis*, is a very active predator able to catch comparatively large living prey such as shrimps, fishes and crabs. Crabs are favorite prey and when they are identified, cuttlefish will approach a crab while raising their first and second pair of arms and modifying their body patterning (Hanlon and Messenger 1996). The raised arms may direct attention away from the striking tentacles and the body patterning appears to distract or mesmerize the prey. Interestingly, this predatory tactic is complex and young cuttlefish often fail on their first attempt and get pinched by the crab (Boal et al. 2000). The cuttlefish can learn the most successful rear attack method within a few trials and avoid getting pinched after the first attempt. Boal et al. hypothesized that exposure to crab odor enhances the naïve cuttlefishes' attention to the crabs as prey items and enhances their ability to learn to overcome the prize.

Distance chemoreception in cephalopods was first noted in experiments with octopuses that became active when fish extract was added to tank water (Wells 1963, Messenger 1967). Electrophysiological recordings from olfactory pits in octopuses, cuttlefishes and squids demonstrates that they are lined with chemoreceptors that are sensitive to chemicals as low as 10^{-5} M (Budelmann et al. 1997). The specific cues that initiate alertness or search behavior in cephalopods have not yet been identified, but this research clearly shows that they can respond to relevant chemosensory cues behaviorally and have the capacity to detect chemical stimuli at behaviorally relevant concentrations.

Conclusion

Clearly, predation has promoted a wide array of prey defenses, and the question arises as to whether there have been reciprocal adaptations in predators. Have predators and prey engaged in an evolutionary arms race such that reciprocal selection has promoted a continuing increase of predator offense and prey defense? For instance, the drilling abilities of predatory gastropods and the shell thickness of their bivalve prey have increased through geologic time (Vermeij, 1994). Similarly, aquatic snails have increased their armor, while the correlated response in predatory crabs has been the evolution of larger claws for crushing the better-defended snails. In both of these instances, predators may have evolved greater weaponry in response to improved prey defense or predator adaptations may have evolved in response to other predators or competitors.

A small but growing field of neuroecology has in the last decade embraced the need to coordinate studies of the physiology of nervous systems with the behaviors they produce and the effects of intra-and inter-specific organismal interactions. These studies identify relevant cues and study animal behavior and physiology in response to these signals. Few investigations to date have identified the complete structures of ecologically important natural products, and even fewer studies have reported results from experiments in which chemical stimuli are presented in ecologically relevant contexts. While studies of responses to novel, alien compounds have provided us with a great deal of insight into the function of chemosensory systems, the ecological relevance of these published works is questionable.

This approach has answered many questions about the transduction mechanisms involved in behavioral responses to deterrent compounds, but it does not allow the researcher to ask questions about the evolutionary or ecological relationships between the producers of these compounds and the behaviors that detection elicits in the consumers. The integration of insights gained from many studies across many levels of biological organization, from chemical to ecological, ultimately accomplishes the goals of the neuroecological approach. Molluscs, which frequently use chemical signals to avoid predators as well as protect themselves from predation, provide an excellent resource for studying communication between predators and prey.

This communication is so important to individual survival that the behaviors and neural pathways involved will be quite robust and can make an excellent substrate for the neuroecological study of complex systems.

References

Agrawal, A.A., Laforsch, C., Tollrian, R. 1999. Transgenerational induction of defenses in animals and plants. Nature. 401, 60-63.

Alexander, J. E., Covich, J. A. P. 1991. Predator avoidance by the freshwater snail *Physella virgata* in response to the crayfish *Procambarus simulans*. Oecologia 87: 435–442.

Appleton, R.D., Palmer, A.R. 1988. Water borne stimuli released by predatory crabs and damaged prey induce more predator resistant shells in a marine gastropod. *PNAS*. 85, 4387-4391.

Aronson, R.B., 1991. Ecology, paleobiology and evolutionary constraint in the octopus. *Bull. Mar. Sci*. 49, 245–255.

Appleton, R.D., Palmer, A.R., 1988.Water-borne stimuli released by predatory crabs and damaged prey induce more predator-resistant shells in a marine gastropod. *Proc. Natl. Acad. Sci. USA* 85, 4387–4391.

Avila, C. 1998. Chemotaxis in the nudibranch *Hermissenda crassicornis*: does ingestive conditioning influence its behavior in a Y-maze? *J. Moll. Stud*. 64, 215-222.

Avila, C., G. Cimino, A. Fontana, A. F. M. Gavagnin, J. Ortea, and E. Trivellone. 1991. Defensive strategy of two *Hypselodoris* from Italian and Spanish coast. *J. Chem. Ecol*. 17: 625–636.

Bertness, M.D., Cunningham, C., 1981. Crab shell-crushing predation and gastropod architectural defense. *J. Exp. Mar. Biol. Ecol*. 50, 213–230.

Boal, J.G., Wittenberg, K.M., Hanlon, R.T. 2000. Observational learning does not explain improvement in predation tactics by cuttlefish (Mollusca: Cephalopoda). *Behav. Proces*. 52, 141-153.

Branch, G. M. 1981. The biology of limpets: physical factors, energy flow and ecological interactions. *Oceanogr. Mar. Biol. Annu. Rev*. 19, 235–380.

Budelmann, B.U., Schipp, R., von Boletzky, S. 1997. Cephalopoda. In: F.W. Harrison, A. Kohn eds. Microscopic Anatomy of Invertebrates. vol. 6A. Wiley-Liss, New York.

Burks, R.L., Lodge, D.M. 2002. Cued in: advances and opportunities in freshwater chemical ecology. *J. Chem. Ecol*. 28. 1901-1917.

Bush, S.L., Robison, B.H. 2007. Ink utilization by mesopelagic squid. *Mar. Biol*. 152, 485-494.

Bush, S.L., Robison, B.H., Caldwell, R.L., 2009. Behaving in the dark: locomotor, chromatic, postural, and bioluminescent behaviors of the deep-sea squid *Octopoteuthis deletron* Young 1972. Biol. Bull. 216, 7–22.

Carr, W. E. S. 1988. The molecular nature of chemical stimuli in the aquatic environment. – In: Atema, J., Fay, R. R., Popper, A. N. and Tavolga, W. N. (eds), Sensory biology of aquatic animals. *Springer-Verlag*, Berlin, pp. 3–27.

Chapman D.J., Fox D.L. 1969. Bile pigment metabolism in the sea hare *Aplysia*. *J Exp Mar Biol Ecol* 4, 71-78.

Chase, R., Pryer, K., Baker, R., Madison, D. 1978. Responses to conspecific chemical stimuli in the terrestrial snail *Achatina fulica*. *Behav. Biol*. 22, 302-315.

Cimino, G. Gavagnin, M. 2006. Molluscs: From chemo-ecological study to biotechnological application, Vol. 43. *Springer,* 387 pp.

Cook, A. 1985. Functional aspects of trail following by the carnivorous snail *Euglandina rosea*. *Malacologia*. 26, 173-181.

Côté, I.M., 1995. Effect of predatory crab effluent on byssus production in mussels. *J. Exp. Mar. Biol. Ecol*. 188, 233–241.

Côté, I.M., Jelnikar, E. 1999. Predator-induced clumping behaviour in mussels (Mytilusedulis Linnaeus). JEMBE. 235, 201-211.

Dalesman, S., Rundle, S. D., Coleman, R. A. and Cotton, P. A. 2006. Cue association and antipredator behaviour in a pulmonate snail, *Lymnaea stagnalis*. *Anim. Behav*. 71, 789-797.

Dalesman, S, SD Rundle, and PA Cotton. 2007. Predator regime influences innate anti-predator behavior in the freshwater gastropod Lymnaea stagnalis. *Freshwater Biology*, 52:2134-2140.

Derby CD, Zimmer RK 2007. Biological Bulletin Virtual Symposium: The Neuroecology of Chemical Defense. *Biol Bull* 213: 205-207.

Dix, T.L., Hamilton, P.V., 1993. Chemically mediated escape behavior in the marsh periwinkle *Littorina irrorata*. *J. Exp. Mar. Biol. Ecol*. 166, 135–149.

Eisner, T. and Meinwald, J. 1966. Defensive secretions of Arthropods. *Science*. 153, 1341-1350.

Endler J.A. 1986. Defense against predators. In: Feder ME, Lander GV (eds) Predator Prey Relationships: Perspectives and Approaches from the Study of Low Vertebrates Chicago: Chicago University Press, pp 100-134.

Ernande, B. and Dieckmann, U. 2004. The evolution of phenotypic plasticity in spatially structured environments: implications of intraspecific competition, plasticity costs and environmental characteristics. *J. Evol. Biol*. 17, 613-628.

Faulkner, D.J., Ghiselin, M.T. 1983. Chemical defense and evolutionary ecology of dorid nudibranchs and some other opisthobranch gastropods. *Mar. Ecol. Prog. Ser*. 13, 295-301.

Ferner, M.C., Weissburg, M.J. 2005. Slow-moving predatory gastropods track prey odors in fast and turbulent flow. *J. Exp. Biol*. 208, 809-819.

Ginsburg D.W., Paul V.J. 2001. Chemical defenses in the sea hare *Aplysia parvula*: importance of diet and sequestration of algal secondary metabolites. *Mar Ecol Prog Ser* 215, 261-274.

Greenstone, M.H., Dickens, J.C. 2005. The production and appropriation of chemical signals amount plants, herbivores and predators.. *Ecology of Predator-Prey Interactions*., p. 139-165.

Hanlon, R.T., 2007. Cephalopod dynamic camouflage. *Curr. Biol*. 17, 400–404.

Hanlon, R.T., Messenger, J.B., 1996. Cephalopod Behaviour. Cambridge University Press, Cambridge, U.K.

Harper, E.M., Skelton, P.W., 1993. A defensive value of the thickened periostracum in the *Mytiloidea*. *Veliger* 36, 36–42

Jacobsen, H. P. and Stabell, O. B. 2004. Antipredator behaviour mediated by chemical cues: the role of conspecific alarm signaling and predator labeling in the avoidance response of a marine gastropod. *Oikos* 104, 43- 50.

Kelley S.R., Jensen P.R., Henkel T.P., Fenical W., Pawlik J.R. 2003. Effects of Caribbean sponge extracts on bacterial attachment. *Aquat. Micro Ecol*. 31,175–182.

Kicklighter C.E., Derby C.D. 2006. Multiple components in ink of the sea hare Aplysia californica are aversive to the sea anemone Anthopleura sola. *J. Exp. Mar. Biol. Ecol.* 334:256–268

Kicklighter C.E., Shabani S., Johnson P.M., Derby C.D. 2005. Sea hares use novel antipredatory chemical defenses. *Curr. Biol.* 15:549-554

King, A.J., Adamo, S.A. 2006. The ventilatory, cardiac, and behavioural responses of resting cuttlefish (Sepia officinalis Linnaeus) to sudden visual stimuli. *J. Exp. Biol.* 209, 1101–1111.

Leonard, G., Bertness, M.D., and Yund, P.O. 1999. Crab predation, waterborne cues, and inducible defenses in the blue mussel *Mytilus edulis*. *Ecology,* 80,1-14.

Lima, SL, and LM Dill. 1990. Behavioral decisions made under the risk of predation: a review and prospectus. *Canadian Journal of Zoology*, 68:619-640.

Long J.D., Hay M.E. 2006. Fishes learn aversions to a nudibranch's chemical defense. *Mar Ecol Prog Ser* 307, 199-208.

MacGinitie, G.E., MacGinitie, N., 1968. Natural History of Marine Animals. McGraw-Hill, New York, 2nd edition, 523 pp.

Marin, A., Alvarez, L.A., Cimino, G., Spinella, A. 1999. Chemical defence in cephalaspidean gastropods: origin, anatomical location, and ecological role. *J. Mollusc. Stud.* 65. 121-135.

McClintock, J.B., Baker, B.J., Hamann, M.T., Yoshida, W., Slattery, M., Heine, J.N., Bryan, P.J., Jayatilake, G.S., Moon, B.H., 1994. Homarine as a feeding deterrent in common shallow-water antarctic lamellarian gastropod *Marseniopsis mollis*: a rare example of chemical defense in a marine prosobranch. *J. Chem. Ecol.* 20, 2539–2549.

Messenger, J.B. 1967. The effects on locomotion of lesions to the visuo-motor system in *Octopus*. *Proc. Roy. Soc. London.* 167, 252-281.

Nakaoka, M. 2000. Nonlethal effects of predators on prey populations: predator-mediated change in bivalve growth. *Ecology* 81:1031–1045.

Nolen T.G., Johnson P.M., Kicklighter C.E., Capo T. 1995. Ink secretion by the marine snail *Aplysia californica* enhances its ability to escape from a natural predator. *J Comp Physiol* A 176, 239–254.

Nusnbaum, M. Aggio, J.F., Derby, C.D. 2012. Taste-mediated behavioral and electrophysiological responses by the predatory fish *Ariopsis felis* to deterrent pigments from *Aplysia californica* ink. *J. Comp. Physiol.* A 198, 283-294.

Nusnbaum M., Derby C.D. 2010a. Effects of sea hare ink secretion and its escapin-generated components on a variety of predatory fishes. *Biol. Bull* 218, 282–292.

Nusnbaum M., Derby C.D. 2010b. Ink secretion protects sea hares by acting on the olfactory and nonolfactory chemical senses of a predatory fish. *Anim. Behav.* 79, 1067-1076.

Orr M.V., El-Bekai M., Lui M., Watson K., Lukowiak K. 2007. Predator detection in *Lymnaea stagnalis*. *J Exp Biol* 210:4150–4158.

Orr, M. V. and Lukowiak K. 2008. Electrophysiological and behavioral evidence demonstrating that predator detection alters adaptive behaviors in the snail *Lymnaea*. *J. Neurosci.* 28, 2726-2734.

Paul V.J., Arthur K.E., Ritson-Williams R., Ross C., Sharp K. 2007. Chemical defenses: from compounds to communities. *Biol. Bull* 213, 226–251.

Paul V.J., Pennings S.C. 1991. Diet-derived chemical defenses in the sea hare *Stylocheilus longicauda*. *J. Exp. Mar. Biol. Ecol.* 151, 227-243.

Pearce, T.A., Gaertner, A. 1996. Optimal foraging and mucus-trail following in the carnivorous land snail *Haplotrema concavum*. *Malacol. Rev*. 29, 85-99.

Prota, G., Ortonne, J.P., Voulot, C., Khatchadourian, C., Nardi, G., Palumbo, A. 1981. Occurrence and properties of tyrosinase in the ejected ink of cephalopods. *Comp. Biochem. Physiol. B*. 68, 415-419.

Rice, S. H. 1985. An antipredatory chemical defense of the marine pulmonate gastropod Trimusculus reticulatus (Sowerby). *J. Exp. Mar. Biol. Ecol*. 93: 83–89.

Reimer, O and Tedengren, M. 1996. Phenotypical improvement of morphological defences in the mussel *Mytilus edulis* induced by exposure to the predator *Asterias rubens*. *Oikos* 75, 383-390.

Rüdiger W. 1994. Phycobiliproteins and phycobilins. In: Round FE, Chapman DJ (eds) Progress in Phycological Research. Vol. 10 pp. 97-135. Bristol: Biopress

Rundle, S. D. and Bronmark, C. 2001. Inter- and intraspecific trait compensation of defence mechanisms in freshwater snails. *Proc. R. Soc. Lond. B Biol. Sci*. 268, 1463-1468.

Russo, G.L., De Nisco, E., Fiore, G., Donato, P.D., d'Ischia M., Palumbo, A. 2003. Toxicity of melanin-free ink of *Sepia officianalis* to transformed cell lines: identification of the active factor as tyrosinase. *Biochem. Biophys. Res. Comm*. 308, 293-299.

Schweitzer, C.E. and Feldmann, R.M. 2010. The Decapoda (Crustacea) as predators on Mollusca through geologic time. *Palaios*. 25, 167-182.

Shabani S., Yaldiz S., Vu L., Derby C.D. 2007. Acidity enhances the effectiveness of active chemical defensive secretions of sea hares, *Aplysia californica*, against spiny lobsters, *Panulirus interruptus*. *J Comp Physiol* A 193, 1195-1204.

Snyder, N. F. R. and Snyder, H. A. 1971. Pheromone-mediated behaviour of Fasciolaria tulipa. –*Anim. Behav*. 19, 257–268.

Thompson, T.E. 1960. Defensive acid-secretion in marine gastropods. *J. Exp. Mar. Biol.* Assoc. 39, 115-123.

Tollrian, R. and Harvell, C.D. 1998. The ecology and evolution of inducible defenses. Princeton University Press. 1st edition 383 pp.

Turner, A. M., Bernot, R. J. and Boes, C. M. 2000. Chemical cues modify species interactions: the ecological consequences of predator avoidance by freshwater snails. Oikos 88, 148-158.

Vermeij, G. J. 1994. The evolutionary interaction among species: Selection, escalation, and coevolution. Ann. Rev. Ecol. System. 25, 219–236.

Walderon, M.D., Nolt, K.J., Haas, R.E., Prosser, K.N., Holm, J.B., Nagle, G.T., Boal, J.G. 2011. Distance chemoreception and the detection of conspecifics in *Octopus bimaculoides*. *J. Mollusc. Stud*. 77, 309-311.

Weissburg, M. J., and Zimmer-Faust, R. J. 1993. Life and death in moving fluids: hydrodynamic effects on chemosensory-mediated predation. *Ecology* 74, 1428-1443.

Wells, J. 1963. Taste by touch: some preliminary experiments with octopus. *J. Exp. Biol*. 40, 187-193.

Williams, B.L. 2010. Behavioral and chemical ecology of marine organisms with respect to tetrodotoxin. *Mar. Drugs*, 8, 381-398.

Williams, B.L., Caldwell, R.L. 2009. Intra-organismal distribution of tetrodotoxin in two species of blue-ringed octopus. *Toxicon*. 54, 345-353.

Wood, J.B., Byrne, R.B., Monks, N. 2003. Understanding the past by looking at the present; using CephBase to examine behavior and selection in coleoid cephalopods. Berliner Geowissenschaftliche *Abhandlungen* 3, 261–266.

Wyeth, R.C., Woodward, O.M., Dennis Willows, A.O. 2006. Orientation and navigation relative to water flow, prey, conspecifics, and predators by the nudibranch mollusk *Tritonia diomedea*. *Biol. Bull.* 210, 97-108.

In: Neuroecology and Neuroethology in Molluscs
Editors: Anna Di Cosmo and William Winlow
ISBN: 978-1-62948-970-4

Chapter 5

Ecological Consequences of Chemically-Mediated Prey Perception in Molluscan Communities

Miranda Wilson and Marc Weissburg
School of Biology, Georgia Institute of Technology, Atlanta, GA, US

Abstract

We take a sensory ecological approach to examining the role of chemical information in regulating ecological interactions in molluscan communities. Current and past studies suggest that molluscs can use chemical cues to determine both the type of predator and general information about it's proximity, thus allowing potential prey to evaluate the level of potential threat, and mount effective defenses. Depending on the predator, defenses range from changes in activity patterns, to evasion and escape, to morphological and life-historical responses that reduce vulnerability. These anti-predator strategies produce a variety of cascading effects by altering demography, patterns of resource and habitat utilization and energy allocation of potential mollusc prey. The ensuing indirect effects on other species include changes in resources consumed by molluscs, altered interactions among different predators that consume particular molluscan species, and changes in the strength and outcome of competition between molluscs and other species in the community. Because the fluid environment affects how chemical cues are dispersed, fluid flow alters the information available to potential prey and provides a strong context for interactions between predators and prey. There is good evidence that this environmental context alters the form and strength by which predators exert their effects on prey by affecting the ability of prey to obtain information about predator threat.

Introduction

The survival of many organisms depends on their ability to detect and respond to potential predators in a manner that enables them to avoid injury or death. Consequently, the detection of predators can invoke a variety of defensive behaviors (evasion, escape), modulate the level and type of morphological defenses, and alter life history traits (e.g., Vermeij and Covich 1978, Menge and Lubchenco 1981, Chivers and Smith 1998, Kats and Dill 1998). Such changes in individuals can propagate to affect other species with which they interact. The obvious canonical example is when reductions in activity of a prey species increase the abundance of that prey's resources, but the variety of interactions any one species participates in creates a cornucopia of potential ecological effects, often termed non-consumptive or trait-mediated interactions (Peacor and Werner 2001). Ecologists then, must often understand how potential prey obtain information about predators and the resulting response in order to gain insight into interactions occurring within assemblages of organisms.

The role of information transfer in modulating ecological interactions can be broken down into 4 broad areas: the properties of the cue that contains information about the predator or its activities; the processes that alter the cue as it propagates from sender to receiver; the response of the receiver; and how the response affects resulting interactions. Many animals receive information regarding predators in the form of chemical messages, which have a number of convenient properties when both specificity and forewarning are important criteria. First – the chemosensory systems of animals have the capacity to encode an effectively infinite number of chemical blends (Jacquin-Jolly and Merlin 2004, Bruce et al. 2005), allowing chemosensory detection to be both specific and flexible. Second – the release of chemicals into the environment, where they can be transported to potential prey, is inevitable as all organisms must eliminate metabolic by-products. Thus, chemical cues can be ubiquitous informants of predator presence for animals of every taxon (Kats and Dill 1998), and comprise roughly two-thirds of current studies (reviewed in Preisser et al. 2005). Our goals for this chapter are to review how the processes listed above account for the strength and variety of chemically-mediated ecological effects of predators in molluscan communities, examine the nature of these effects, and where possible, suggest important knowledge gaps that may lead to future insights. Our focus is on molluscs as prey for other creatures. However, molluscs can be important intermediate predators (e.g., grazing snails), and so our treatment often includes how odor-mediated cues from creatures that prey on molluscs affect the organisms that molluscs themselves consume.

Important Cue Characteristics

Prey responding to predators incur costs of reduced activity or energy reallocation, implying the intensity of antipredator response should be consistent with the perceived level of risk (Risk Allocation Hypothesis; Lima and Bednekoff 1999). Using chemical cues as a reliable assessment of risk allow prey to respond to predators before the predator actually contacts the prey. Thus, the chemical make-up of predator odors or odor concentration should inform prey of their level of risk by containing information about predator identity, proximity, size, etc. Investigations largely focus on how prey determine the risk associated with different types of predators, presumably since predators that do not commonly consume a given prey species (because the predator is not common or does not eat that prey) represent little threat. Molluscs are clearly able to discern different predators using chemical cues and show heightened responses to predators that represent a greater threat (Barbeau and Schiebling

1994, Turner et al. 2000). There is a general hierarchy of response intensity exhibited by prey that also is related to presumptive risk; prey respond most strongly to cues from injured conspecifics, followed by metabolites released by predators consuming conspecific prey. Metabolites released by predators that have consumed heterospecific prey generally, but not always, produce weaker responses (Turner 2008). Super-imposed on this hierarchy is the overall response sensitivity, which can be modulated by historical patterns of predation intensity (Rochette et al. 1998, Smee and Weissburg 2008).

Prey often respond to cues released by injured conspecifics and sometimes to cues released by threatened conspecifics (Kats and Dill 1998), termed damage-released and disturbance-released cues, respectively (Chivers and Smith 1998). As noted, cues from injured conspecifics generally produce extremely strong responses (Stenzler and Atema 1977, Leonard et al. 1999, Rahman et al. 2000, Trussell and Nicklin 2002, Large and Smee 2010; Figure 1), no doubt because damage to conspecifics is the most reliable indicator of imminent and mortal threat. For example, *Nassarius* sp. responds more strongly to crushed conspecifics than closely related heterospecifics, but it is unknown how these antipredator responses compare to responses from prey exposed to fed predators (Stenzler and Atema 1977). Results from a variety of other studies suggest that the intensity of antipredator responses to cues from injured conspecifics is at least of a magnitude similar to that of cues from predators fed conspecifics (Smee and Weissburg 2008, Large and Smee 2010; Figure 1). Damage to heterospecifics sometimes produces responses, particularly if species are ecologically related as well. There is a strong positive relationship between phylogenetic relatedness and the strength of antipredator responses in *Lymnaea stagnalis* when responding to metabolites released from other species found sympatrically, but the relationship was weaker for other species found allopatrically (Dalesman et al. 2007). Presumably, damage to a either a related species or a species found in the same environment might indicate a threat, and consequently both taxonomic and ecological similarity are important. Clarifying the extent to which these two types of relatatedness affect response strength remains an important goal.

Disturbance-released alarm cues have been shown in many non-molluscan systems including fish, amphibians, and crayfish (reviewed in Chivers and Smith 1998), but are rare in molluscs. One such rare study determined that ink and opaline (a combination of uracil, uridine, and cytidine) function as alarm cues from *Aplysia* sp. and illicit antipredator responses in conspecifics (Kicklighter et al. 2007).

Metabolites released by fed predators commonly produce strong responses from prey individuals, but predator diet often is important as well (Figure 1). Turner (1998) finds that snails respond more weakly to metabolites from crayfish predators that are fed non-snail prey. Interestingly, taxonomic relatedness seems more important here than ecological relatedness. The response of *H. trivolvis* to the predators fed the congener *H. anceps* is stronger than that of predators fed *Physa acuta*, even though the ecological overlap is greater for *H. trivolvis* and *P. acuta*. This again illustrates the importance of ecological vs. taxonomic relatedness requires further resolution.

Appleton and Palmer (1988) found that the growth of apertural teeth in the gastropod *Nucella* sp. was greater when exposed to odors from a *Cancer* sp. crab fed conspecifics than when the crab was fed fish. Since injured conspecific odor was a component of the cue when crabs were fed snails but not fish, response differences also reflect the potential effect of the cue from injured conspecifics, which is a common complication in experiments of this type (see also Trussel and Smith, 2000). Jacobsen and Stabell (2004) similarly observed that

Tegula sp. exhibit a climbing response to predators fed conspecifics, but do not respond to predators fed heterospecifics. It is unclear in many of these types of experiments whether the prey does not recognize a predator fed alternative prey or if they fail to respond appropriately. Sometimes the combination of cues from predators fed conspecifics and injured conspecifics is necessary to elicit a response from prey (Rochette et al. 1998, Leonard et al. 1999, Smee and Weissburg 2008). The combination may represent a fairly reliable cue by conveying information that a predator is nearby and that it has an affinity for conspecifics.

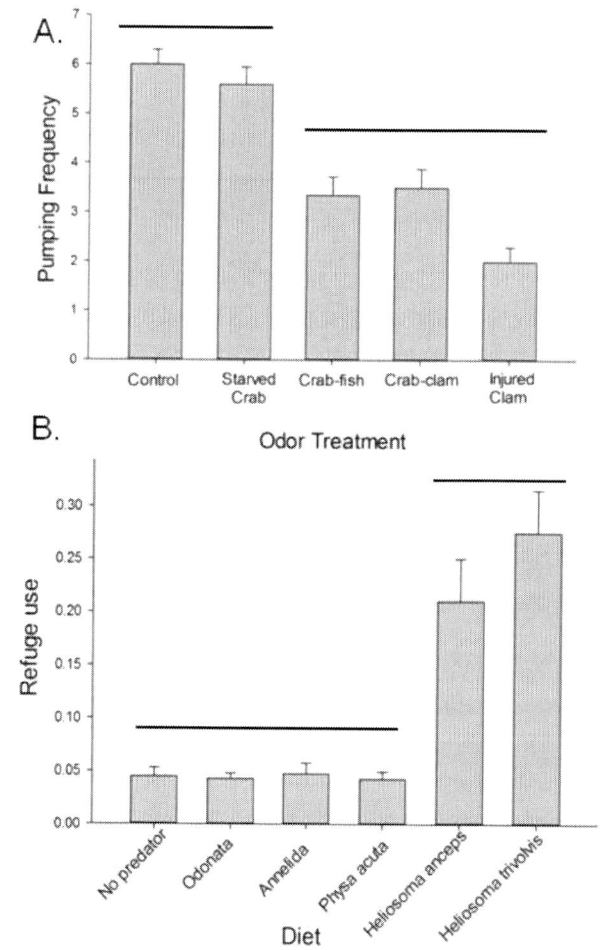

Redrawn with permission from Turner (2008).

Figure 1. The effects of diet on responses to predators. A. Behavioral responses of the clam *Mercenaria mercenaria* to predator and conspecific odors. The y-axis (with std. err) is the average number of times clams were observed pumping in a 30 minute interval (maximum = 6). The x-axis is odor treatement, which consists of no odor, odor of a crab starved 48 hours, odor of a crab fed fish, odor of a crab fed clams, and the odor of injured conspecifics. Lines join treatment groups not significantly different using a Tukey-Kramer *post hoc* test. Redrawn with permission from Smee and Weissburg, 2006a. B. Responses of the snail *Heliosoma trivolvis* to crayfish fed different diets. Refuge use is the proportion of snails that moved away from caged crayfish to the water's surface. Lines join treatment groups not significantly different using a Tukey-Kramer post hoc test.

Predators feeding on conspecifics may be more likely to produce a reliable cue for assessing risk when that predator is a specialist. Generalists by definition have varied diets, suggesting that responding to odors from predators fed only conspecifics would result in an inaccurate assessment of risk by prey. The hard clam *Mercenaria mercenaria* responds to blue crabs (*Callinectes sapidus*) regardless of their diet (Smee and Weissburg 2006a; Figure 1). Blue crabs are very generalist predators in nature, feeding on a variety of different bivalve species (Hines et al. 1990). Thus, the recent dining habits of a blue crab are not indicative of what else may be on the menu, and it is in the best interest of prey to be able to detect them and respond appropriately. The same trend has been documented for *Nucella* sp., which responds to green crabs (*Carcinas menas*) regardless of their diet (Large and Smee 2010). This of course, contrasts with the results of Palmer 1998; crayfish are a generalist predator too, yet diet strongly determined prey responses in the crayfish-snail system. It is clear we currently lack a firm understanding of why diet-specificity occurs.

Very little is known about the specific chemicals that molluscs use to assess predatory risk, either in metabolites released by predators or damaged prey. It is unknown if the chemical signals used to assess predation risk are single compounds as often occurs with defensive chemicals (Hay and Fenical 1996) or as mixtures, which occur for substances that elicit food search. It has been suggested that basic peptides make good chemical signals in aquatic environments because they are not readily taken up by bacteria and are persistent (Rittschof 1990, Decho et al. 1998), but there are no cases where a water-borne predation cue has been structurally identified. Lacking such information, we have no insights into how the generality vs. specificity of chemical cues is determined, which limits our understanding of the evolutionary development of antipredator behavior.

Identifying the specific molecules, and whether they are found singly or in mixtures, is needed to formulate hypotheses as to the impact of multiple predator species on antipredator behavior, the intensity or length of antipredator responses, and the impact that other chemicals or changes in water chemistry may have on the ability of prey to assess risk. For instance, an important issue is whether cues from injured conspecifics are additive when combined with metabolites released by predators, since this will determine the degree of morphological or behavioral response to different situations. If the cues are the same, then one would expect the response to be a function of the amount of cue released by each source, and the combined cues should produce additivity relative to each tested alone. Note that experiments testing additivity need to expose prey to cues from the predator with and without the presence of metabolites from injured conspecifics if predator diet itself has an effect, although this is not always the experimental design. Experiments often give predators an opportunity to attack and injure prey, and it is this water that is used as a cue. This design conflates diet-specific differences in cue quality or concentration with the combined effect of predator and predator+conspecific odor.

There also is considerable uncertainty regarding whether chemical cues from two different predators produce additive or intermediate responses compared with the response of each individual predator, although this has rarely been examined specifically for molluscs (see Hoverman and Relyea 2007). A priori, responses cannot be additive if the chemical cues are the same and response intensity to single predator is a function of concentration. This is because the combination of two different predators will produce an intermediate concentration relative to the concentrations produced by each predator alone, assuming equal biomass.

Predator characteristics that alter odor concentration also will affect the intensity or duration of prey responses, as suggested above. In other systems prey respond more when exposed to greater volumes of predator conditioned water or predator conditioned water produced with more predator biomass (Harvell 1990, Kats and Dill 1998, Hill and Weissburg, 2012). Such studies are uncommon in any system, and have not been performed with molluscs. Consequently, it is unknown whether molluscs have a graded or threshold response to the concentration of predator odors, which has obvious importance in predicting the impacts of predators or the effect of predator biomass. Starved predators rarely illicit reactions from prey (Figure 1), likely because they are not producing recognizable cues or they are producing cues in small amounts. This has been seen for many mollusc species that otherwise respond to fed predators (*M. mercenaria* – Smee and Weissburg 2006a, *Nucella* sp. – Large and Smee 2010, *Tridacna squamosa* – Neo and Todd 2011). The incidence of starved predators in natural environments has not been assessed in this context and may be an essential component of future studies examining risky habitats for prey species.

The Role of Chemical Cue Transport

Although chemical signal source properties (e.g., source biomass or dietary history) determine the initial concentration of the chemical cue, transmission through the environment subsequently affects the strength of the cue reaching the receiver. Any predator will have a "sphere of influence" comprising the area where chemical signal intensity is above the detection limits, and which is determined by prey sensitivity, but also the properties of the predator governing cue generation and the environmental properties governing cue transmission (Figure 2). A small sphere of influence would suggest risk perception only by prey that are close to the predator, and the attendant direct and cascading non-consumptive effects vary spatially and temporally as a function of predator presence. In contrast, predators with a large sphere of influence will affect many prey over a broader area and create a more homogenous "landscape of fear". When measured, predators are detected at distances up to 1-2 m (Turner and Montgomery 2003, Smee and Weissburg 2006a, Dickey and McCarthy 2007), but this is an extremely crude estimate given the lack of comprehensive measurements across either environments or taxa.

In contrast to our lack of knowledge regarding chemical signal source properties such as concentration-response functions, or molecular identity, we have good general knowledge for how transmission of chemical plumes sets limits on the ability of animals to detect and respond to predator chemical cues. In many environments relevant to molluscan species, chemicals are released into flow, where turbulent mixing subsequently alters the intensity and distribution of these chemicals as they are transported downstream. Rivers and inter- and subtidal environments are typical cases. Even in nominally still waters, such as ponds and small lakes, most transport will occur by flow, as a result of wind or thermal mixing, or animal movements. Because diffusion is simply too slow a process (Dusenbery 1992), the majority of chemical signals will be entrained in plumes transported by flows with some degree of turbulence.

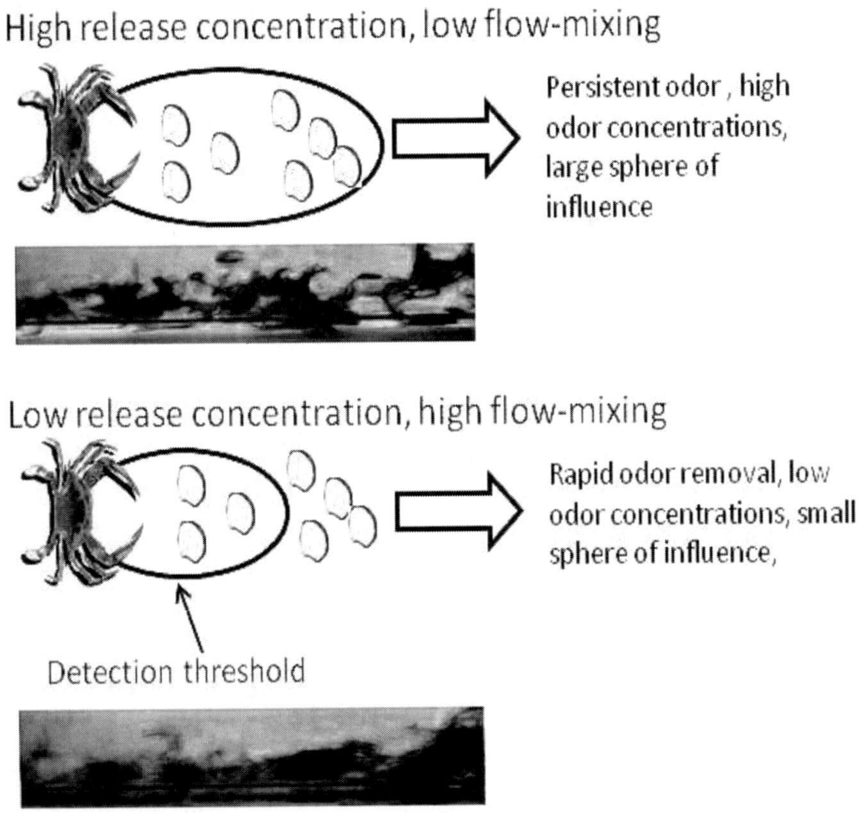

Figure 2. Effects of odor source properties and flow environment on predator sphere of influence. The photographs of odor plumes show the character of odor signals at low and high flows.

Turbulent plumes initially are composed of concentrated odor filaments entrained in unscented water, producing strong, but intermittent signals. As these filaments are transported downstream, turbulent mixing spreads them over a larger area, creating a more diffuse but homogeneous plume. Greater mixing increases both the rate of spread and dilution. As mixing increases, a strong signal may be initially spread out over a larger area to potentially affect more prey, but will decay more rapidly with distance downstream of the source. Generally speaking, the greater dilution of chemicals in turbulent conditions will reduce the range of predator-prey information exchange.

A variety of studies have shown that the flow/mixing regime affects the ability of prey to acquire information from chemical signals (Powers and Kittinger 2002, Smee and Weissburg 2006b, Smee et al. 2008). Increasing flow velocity reduces the perceptual range of hard clams to their crustacean and whelk predators (i.e., reduces the predator sphere of influence). These effects are predator specific; whelks are detected at a greater range, although it is unclear whether this effect is due to different release properties (e.g., concentration or production rate) or prey sensitivity to different chemicals (Smee and Weissburg 2006b). These effects on the sphere of influence have important implications in the field. By reducing the sphere of influence, mixing decreases the ability of clams to perceive predators and respond by retracting their siphon and decreasing their own metabolite release in an effort to become cryptic. The lessened deployment of antipredator strategies increases clam mortality to ambient consumers in the field. In an another example, the effects of pumpkinseed sunfish chemical cues on *Physa* sp. habitat use and growth rate are an exponentially decreasing function with distance to 1-2 m from the source (Turner and Montgomery 2003). This seems logical since diffusive and turbulent transport processes generally result in exponential decreases in time-averaged concentration with distance from the source (Rahman and Webster 2005). Patterns of habitat use or growth in these experiments may reflect the time-averaged concentration, since the effects of the chemical cues are integrated spatially and temporally as a result of numerous animals responding over intervals of hours to days. Thus, not only do transport processes set limits on the number of prey that can detect chemical cues, but they produce spatial patterning in response intensity.

Prey Responses to Predators

Prey detecting predators respond behaviorally, morphologically, and with changes in life history. Although behavioral changes (such as reductions in activity) can secondarily affect morphology (e.g., body mass), morphology also can be altered in animals that are not given food, indicating a physiological component to effects such as shell thickness or ornamentation (Appleton and Palmer, 1988). Changes in a focal prey species also indirectly influence other species (predators, competitors, resources) as a result of shared interactions. There is a rich body of literature on how molluscs respond to predator cues.

One of the primary behavioral responses of prey when they detect predator odors is escape. Escape can take on various forms including rapid escape, movements that result in decreased access to prey, and hiding. Each of these behaviors result in decreased contact with predators and can result in decreased predation.

Rapid escape from predators can be seen in highly mobile molluscs that can move comparatively or more quickly than their potential predators. Scallops initiate swimming by rapidly opening and closing their shell when encountered by predators (Ordize and Garafalo 1980, Peterson et al. 1982, Legault and Himmelman 1993) or predator odors (Kats and Dill 1998, Powers and Kittinger 2002). The scallop, *Placopecten magellanicus*, can differentiate between predator species, demonstrating a rapid escape response only when confronted with slowly moving predators that are unlikely to capture quickly moving prey (Barbeau and Scheibling 1994).

Movements that reduce access to prey can include burial, climbing structure, or clumping behaviors. Infaunal bivalves and some gastropods increase their burial depth when exposed to predator odor cues (Stenzler and Atema 1977, Doering 1982, Peterson 1982, Flynn and Smee 2010), which translates into enhanced prey survival in laboratory and natural habitats (Blundon and Kennedy 1982, Haddon et al. 1987, Roberts et al. 1989, Flynn and Smee 2010). The refuge provided by increasing burial depth could be limited by predator digging capabilities or predator size.

Molluscs that live in habitats with structure can seek refuge within or on the structure to reduce their contact with potential predators. Odor cues from predatory blue crabs induce *Littoraria* sp. to climb *Spartina alterniflora* shoots in salt marshes (Lewis and Eby 2002, Siliman and Bertness 2002). The density of the *S. alterniflora* shoots can further mediate contact between predator and prey (Lewis and Eby 2002), with higher densities resulting in reduced contact. *Nucella* sp. snails reduce their movement and increase their use of refuge habitat (cracks in rocks) when exposed to predator odors (Vadas et al. 1994, Trussell et al. 2003, Large and Smee 2010). The freshwater snail *Physella gyrina* increases its use of covered habitat in the presence of odors from pumpkinseed sunfish, but not odors from crayfish (Turner et al. 2000). This difference in refuge use likely stems from variation in the feeding ability of the two predators, as crayfish are more likely to gain access to refuge habitat.

Another antipredator strategy that prey use to reduce predation is to change their density or distribution to make it more difficult for predators to exploit patches (Parrish and Edelstein-Kedshet 1999). Many bivalve species form aggregations in the presence of predator odor cues (mussels – Cote and Jelnikar 1999, Kobak et al. 2010). Predation by blue crabs was 86% lower on mussels in clumps compared to solitary mussels (Brown et al. 2011). Predation by drilling, as opposed to crushing, also decreased when mussel prey were found in aggregations compared to solitary individuals (Casey and Chattopadhyay 2008). It is likely harder for crushing and drilling predators to handle or distinguish individual prey when prey are found in aggregations.

In contrast to running away or seeking physical refuge, molluscs sometimes hide from their predators by ceasing to exude odors that predators can use to locate them. Many bivalve species that rely on pumping for collecting food and exchanging oxygen and waste reduce their pumping, or "clam up" in response to predator odors. This prey behavior is likely an attempt to hide or reduce the prevalence of their own secondary metabolites in the environment, as these cues are used by predators to locate prey individuals (e.g., Weissburg and Zimmer-Faust 1993). The hard clam *M. mercenaria* reduces its' feeding and pumping rates in response to odor cues from predatory crabs (Irlandi and Peterson 1991, Smee and Weissburg 2006a) and sea stars (Doering 1982). Molluscs are also known to reduce their movement and feeding behaviors in the presence of predator odors (Trussell et al. 2003, Large and Smee 2010). As any movement or feeding could release prey odors into the environment, a reduction in these behaviors would presumably reduce prey apparency to potential predators, and the survival benefit of these behaviors has occasionally has been demonstrated in the field (Smee and Weissburg 2006b, Flynn and Smee 2010). More interestingly, bivalve behavioral repertoires may be more sophisticated than simple on-off functions. *M. mercenaria* alter the velocity and duration of excurrent flow when exposed to blue crab predators (Delavan and Webster, In Press). The net result is to create a more well-mixed and less discrete signal that may be harder for blue crab predators to detect.

Rapid escape, seeking refuge or hiding could be beneficial to prey that are unpredictably exposed to predators, but an inducible morphological defense may be more cost effective for prey that are continually exposed to predators (Harvell 1990). Inducible defenses in response to predator odors have been demonstrated across a wide range of taxa including: rotifers (Gilbert 1966), cladocereans (Dodson 1989), fish (Bronmark and Miner 1992), amphibians (Relyea 2004), and molluscs (Appleton and Palmer 1988; Figure 3). Here, shell morphology can exhibit remarkable phenotypic plasticity in response to predator odor cues, exhibiting an array of inducible morphological defenses that increase predator handling time or make it harder for predators to access the soft tissues.

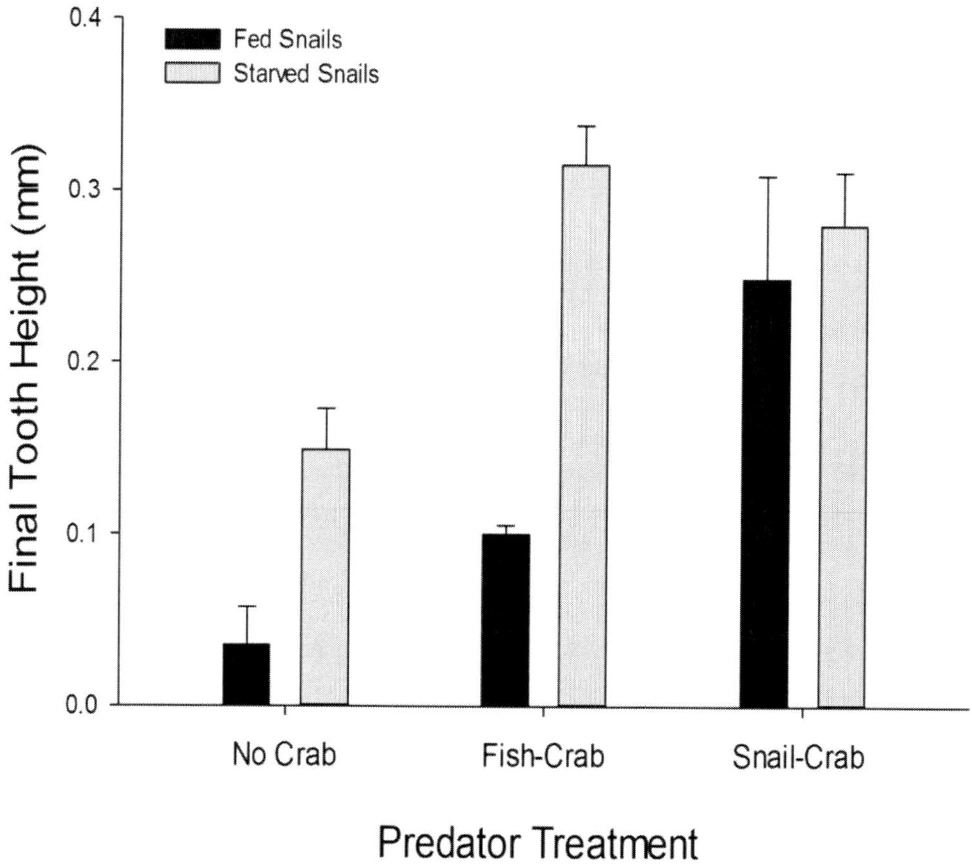

Redrawn with permission from Appleton and Palmer (1988).

Figure 3. Final length (with std err.) of the apertural tooth in large, thin shelled *Nucella lamellose* after 76 days of exposure to no odor, the odor of crabs fed fish, and the odor of crabs fed snails. Note that snails were exposed to the odor of both the food and the predator, so that snails in the Snail-Crab treatment were exposed to cues from damaged conspecifics. Fed snails were fed twice every 6 days with live intact barnacles, *Balanus glandula*. Note that predators induce morphological changes even in starved snails, indicating a direct physiological response. Note also this figure also illustrates responses to prey fed different diets.

Appleton and Palmer (1988) examined the impact of predatory *Cancer productus* odors on the morphology of *Nucella* sp. Snails (Figure 3). Exposure to predator odors resulted in the development of aperatural teeth along the shell, which may make the *Nucella* sp. shell harder for predatory crabs to crack and gain access to soft parts. Similar changes in morphology in response to predator odors have been documented in a variety of other bivalves and gastropods (mussel lip increased in thickness– Smith and Jennings 2000, *Littorina obtusata* increased in shell thickness – Trussell and Smith 2000, *Littoraria irrorata* increased in aperatural lip thickness – Moody and Aronson 2012, *Tridacna squamosa* increased in shell strength – Neo and Todd 2011). The extent of morphological changes observed in the above studies was frequently dependent on predator identity or predator diet. Despite the abundance of studies on inducible morphological defenses, relatively few of these studies include confirmation that those changes actually confer protection from predation or reduced predation rates, although some studies may infer this by measuring proxies such as changes in the force required to crush the shell (e.g., Trussell and Smith, 2000)

Another common morphological change observed in molluscs that attach themselves to hard substrates is the development or enhancement of attachment mechanisms when exposed to predator odors. Predators frequently attempt to dislodge prey from substrates in an effort to remove prey to safer areas or to enhance their ability to manipulate prey. Mussels respond to predator odors by increasing byssal thread production (Leonard et al. 1999, Cheung et al. 2006, Caro et al. 2008) and attachment strength (Kobak et al. 2010). Increasing the strength of attachment increases the predator handling cost. Greater mussel attachment strength has been shown to correlate to reduced predation and is more common in areas of greater predator abundance (Caro et al. 2008), indicating that increased attachment strength can be an effective antipredator defense. Ishida and Iwasaki (2003) showed contradictory impacts of predatory odors on byssal thread production (i.e., they observed reduced byssal thread production). The attachment of byssal threads by mussels to the substrate requires the exposure of a soft foot that may provide an enhanced visual and chemical cue to potential predators. Therefore, Ishida and Iwasaki (2003) suggest their findings are the result of mussels trying to decrease their apparency to potential predators. The tradeoff between byssal attachment and prey apparency may therefore be mediated by mixing conditions that determine the sphere of predator influence relative to the perception of prey by predators.

Ecological Consequences of Prey Responses

A) Impacts on Focal and Cascading Interactions

Antipredator responses have potentially strong ecological effects on both the focal species and those species with which it interacts, for a number of reasons. First, behavioral defenses often preclude prey from contributing to other essential biological processes such as growth or reproduction. Changing patterns of resource allocation, either because of reductions in foraging or increased allocations to morphological defenses, may affect competitive abilities. Moreover, prey reactions to predators affect decisions on when and where to forage, what level of vigilance to maintain, what types of food to eat, and how best to consume food based on the level of risk they perceive (Lima and Dill 1990), thus altering interactions between prey species and it's resources and predators. Finally, a defense against one predator can enhance the ability of other predators to exploit prey populations (so called antagonistic defenses; Kishida et al. 2009). Chemically-mediated antipredator responses of molluscs display all of these effects.

Antipredator behaviors such as hiding, seeking refuge habitat, or reducing contact with potential predators usually result in reductions in food acquisition. For example, *M. mercenaria* reduce their pumping behavior when exposed to predator odor cues (Smee and Weissburg 2006a), suggesting that their food intake is also reduced, as pumping is the primary behavior used to acquire food. This translates into reduced growth in the presence of predator odors over longer temporal periods (1 month; Nakaoka 2000). A variety of studies show reduced growth in snail prey as a consequence of exposure to predator odors (e.g., Appleton and Palmer, 1988, Turner 2008, Trussell et al. 2006).

Reduced growth can have important biological and ecological consequences such as increasing the time needed to reach size refuges from predators. Many mollusc species can gain refuge from gape or crush-limited predators by growing to a larger size. Blue crab predators cannot consume bivalve prey (*M. mercenaria*, *Crassostrea virginica*, *Geukensia demissa*) that are greater than approximately 35 mm in shell length (Arnold 1984, Eggleston 1990). This may have resulted in a drastically greater abundance of large bivalve prey in natural systems where blue crabs are the primary predators (Smee and Weissburg 2008, Wilson 2011).

Another important consequence of reduced prey growth is an increase in the time to reach maturity and/or a reduction in the resources (time/energy) allocated to reproduction. There have been limited studies examining the influence of predator odors on the investment of energy to reproductive organs in molluscs (Lima and Dill 1990, Guerra et al. 2012). Scallops exposed to predators that were prevented from consuming prey allocate more energy to muscle growth as opposed to gonad growth compared to scallops that were not exposed to predators (Guerra et al. 2012). More directly, reductions in mating activity in response to predators could have important consequences for population demographics. *Nucella* sp. stop spawning in the presence of predatory odors (Rawlings 1994), likely in an effort to reduce their apparency to predators, or in anticipation of other antipredator behaviors such as seeking refuge. *Littorina* sp. is significantly less likely to form mating pairs in the presence of odors from predators feeding on conspecifics (Koch et al. 2007).

The ability to detect and respond to predator cues at the larval stage may influence the demography of prey populations. Larvae respond to a variety of settlement chemical cues from conspecifics, resources, predators, and habitat (Pawlik 1992). Chemical cues from resources generally represent habitats where resources are plentiful and result in greater larval settlement (Hadfield and Koehl 2004). In contrast, the presence of predatory odors results in lower larval settlement rates, as these habitats could represent potentially high predation risk areas (Pawlik 1992). To our knowledge there are no studies examining the settlement behavior of molluscs in relation to predator odor cues, although it has been investigated in a variety of other taxa (blue crabs – Forward et al. 2003; lobsters – Boudreau et al. 1993; frogs – Gonzalo et al. 2006; polychaetes – Dahms and Qian 2005). Future studies should examine the how the potential role of predator odors in mediating molluscan larval settlement could impact the density and distribution of adult populations.

The density and distribution of prey populations also can be mediated by antipredator behavior of adults. As discussed previously, one antipredator behavior common to bivalves is to form aggregations (see Prey Responses to Predators). It is thought that this provides protection from predators by increasing handling time and reducing the risk to any one individual (Brown et al. 2011). However, bivalve density affects a number of other processes such as filtration rate (Fulford et al. 2007), benthic boundary layer depletion (Butman et al. 1994), and refuge value for other organisms (Tolley and Voltey 2005), suggesting the consequences of changes in aggregation density may propagate to other community members.

Aside from autecological consequences, the responses of prey to predator odors also can indirectly mediate interactions between prey species and their competitors. Predator-dependent alterations in competitive ability occur when prey species respond differently to the same predator, either through different antipredator behaviors or a different response magnitude of a similar behavior. Such changes may reduce the superiority, or even change the identity of the competitive dominant. By example, *Littorina littorea* generally is a superior competitor to *Gibbula umbilicalis* in the absence of predation risk, but *L. littorea* is more susceptible to green crab predators and responds more strongly to predator odors than *G. umbilicalis* (via a reduction in feeding behavior; Mowles et al. 2011). The presence of green crab odor reverses the competitive relationship between *L. littorea* and *G. umbilicalis* because of the different magnitudes of their antipredator responses (Mowles et al. 2011). These processes add context-dependence to competitive interactions, altering competitively-mediated patterns in prey species abundance in areas of varying predation risk, and possibly the likelihood of competitive exclusion among prey that respond differently to predators.

Apparent competition, the indirect negative interaction between two prey species mediated by a common predator (Holt 1977, Schmitt 1987), also could be impacted by the ability of prey to respond to predators. Schmitt (1987) showed that the presence of preferred bivalve prey increased the density of predators and increased predation rates on both preferred bivalve prey and non-preferred gastropod prey. Schmitt (1987) suggests that the ability of non-preferred prey to "resist predation" will determine the strength of the interaction, although to our knowledge the role of antipredator behavior in shaping apparent competitive interactions is lacking despite its potential importance in numerous systems. For instance, oyster reefs in intertidal estuaries are known to negatively impact associated clams by attracting and enhancing the foraging success of potential predators in the area (Wilson and Weissburg, 2013; Figure 4). The scent associated with oyster reefs brings predators into the vicinity of clam prey as predators move upstream toward the reef. Not only does the scent of oyster reef function as an attractant, but because predators attracted by oyster reef chemical cues would largely approach clam prey from the downstream direction as they transit towards the reef, predator chemical cues would not be available to modulate clam behavior. The close spatial association of two prey species means that the failure of one species to respond to a predation threat produces danger to both, or a shared doom effect (Wahl and Hay 1995). Additional studies are needed to determine the role of odor-mediated predator detection and antipredator responses in the indirect interactions between prey species.

Reductions in foraging time or efficiency, or changes in habitat utilization may indirectly affect the resources of a given prey species (e.g., a non-consumptive or trait-mediated trophic cascade; Schmitz et al. 1997, Peacor and Werner 2001, Heiathus and Dill 2002). A plethora of studies document trait-mediated trophic cascades in which mollusc species that exhibit antipredator behavior release consumer pressure on lower trophic levels (Turner 1997, Turner et al. 2000, Trussell et al. 2003, 2004, Siliman and Bertness 2002). Turner (1997) document that predator-induced habit shifts and reductions in activity of snail grazers change the abundance of stream periphyton. The effect of these trait-mediated cascades can be quite significant when compared to the effects of direct consumption. *Nucella* sp. consume less algae when exposed to predatory green crab odor cues (Trussell et al. 2003), and the strength of this behaviorally-mediated trophic cascade was equal to that of a similar density-mediated trophic cascade (i.e., where predation on *Nucella* sp. occurred; Trussell et al. 2004). Similarly dramatic behaviorally-mediated trophic cascades have been observed in salt marsh systems. Lewis and Eby (2002) showed that predator odor cues from blue crabs cause *Littoraria* sp. to climb *Spartina alterniflora* shoots. Left unchecked, *Littoraria* sp. can consume of all the *S. alterniflora* within 8 months, which transforms the salt marsh into a mud flat (Siliman and Bertness 2002). Given that basal producers in these systems often provide structure for other creatures, these chemically-mediated trophic cascades may have consequences for numerous other species.

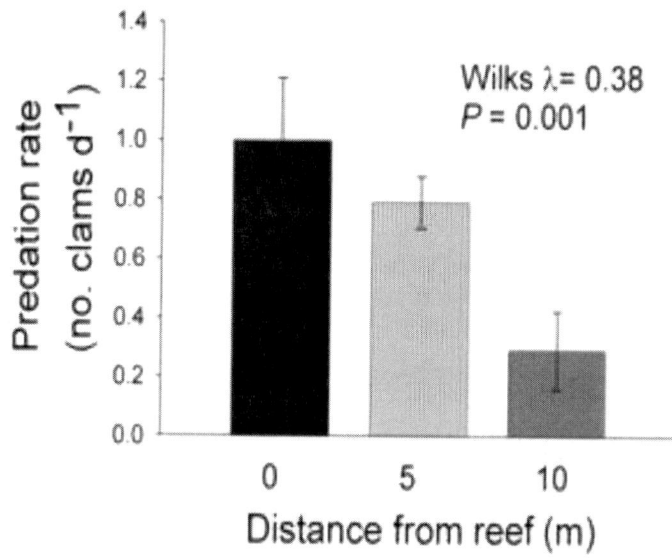

Redrawn from Wilson and Weissburg (2013).

Figure 4. Influence of naturally occurring oyster reefs on predation by blue crabs and whelks on hard clam prey. Data represents predation rates (no. clams per day ± SE) on clams within patches (each patch = 15 clams; 60 clams per square meter) placed at 0, 5, and 10 m from oyster reefs (N = 8) in the field. Clam patches were placed parallel to the main channel and at the low tide level. Each replicate oyster reef was tested with a single distance treatment during each three day period in order to prevent predator aggregation, and all distances (0, 5, and 10 m) at each reef were tested over the same 9 day period. Clams were left in the field for 3 days before the number of missing and consumed clams was assessed. Significance was assessed using a multivariate analysis of variance.

Mollusc prey species also can be impacted by behaviorally-mediated trophic cascades when they represent the basal species examined. Grabowski (2004) examined a trophic cascade on oyster reefs with toadfish, mud crabs, and the eastern oyster as the predator, prey, and basal resource, respectively. Antipredator behavior of mud crabs in the presence of toadfish odor cues released oysters from predation in mesocosm (Grabowski 2004) and field experiments (Hill and Weissburg, 2012; Figure 5). The antipredator behavior of mud crabs has also been shown to release juvenile hard clams from predation in the same oyster reef system (Grabowski and Kimbro 2005).

Research suggests that the strength of trait-mediated direct and indirect trophic interactions can vary based on other factors that influence how strongly and often prey respond to predator signals. These include factors that affect the perception of predator chemical signals by prey, but also those that change the risk-reward functions by which prey “determine” the appropriate investment in antipredator strategies (Figure 6). This occurs, for instance, when environments vary in available refuges that affect the ability of predators to access prey. Refuges make the environment less “risky” than environments without refuge, even when predators are present.

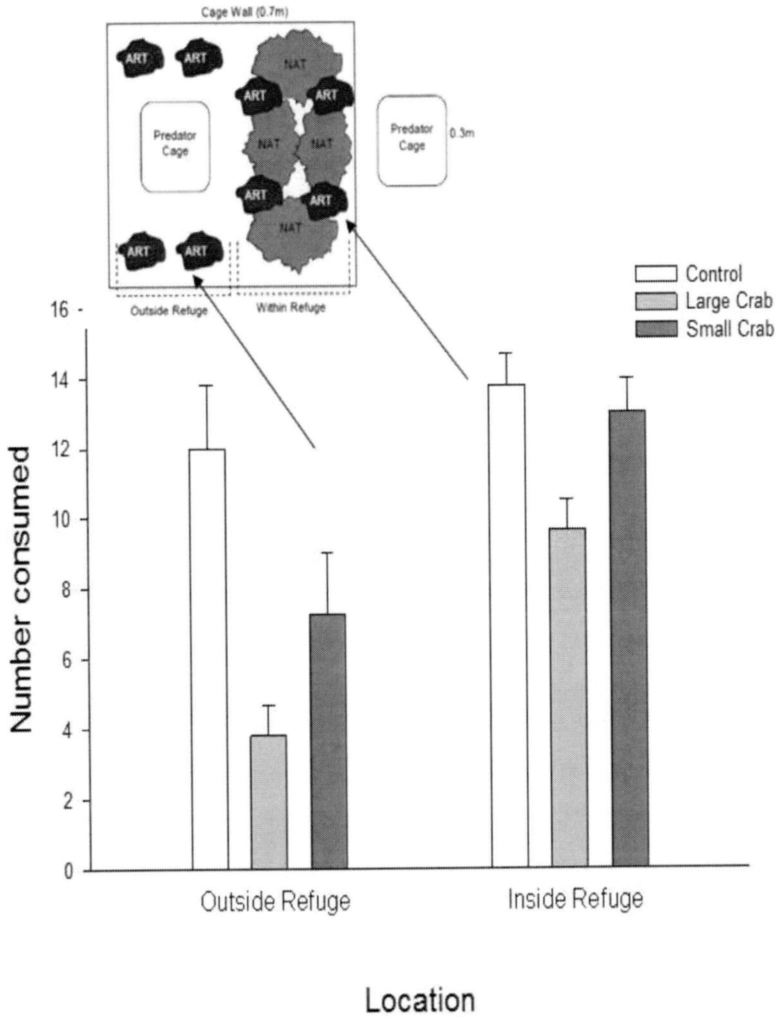

Redrawn from Hill and Weissburg (2012).

Figure 5. Effects of blue crab odor cues on survival of oysters exposed to mud crab predation in the field. Data shows number of oysters consumed (out of 16) inside and outside of refugia as a function of predator type (a single crab 10-14 cm carapace width, or a single crab 8-10 cm carapace width). Data shows mean (with std.err) of 3 replicate treatments of each type. Inset shows the caging design, where mud crabs and oyster clusters were confined within cages, and blue crab predators confined to smaller predator cages to prevent any direct interactions. Oyster prey were affixed to valves of bleached oyster shells (artificial cluster-ART) located outside of, and within refugia consisting of natural bleached oyster clusters (NAT). Caged blue crabs were fed mud crabs prior to the start of the 48 hour experiments. The odor of large, but not small blue crabs significantly decreased oyster consumption and increased the relative proportion of oysters consumed inside the refuge. Small blue crabs likely produced insufficient quantities of metabolites to affect mud crab behavior.

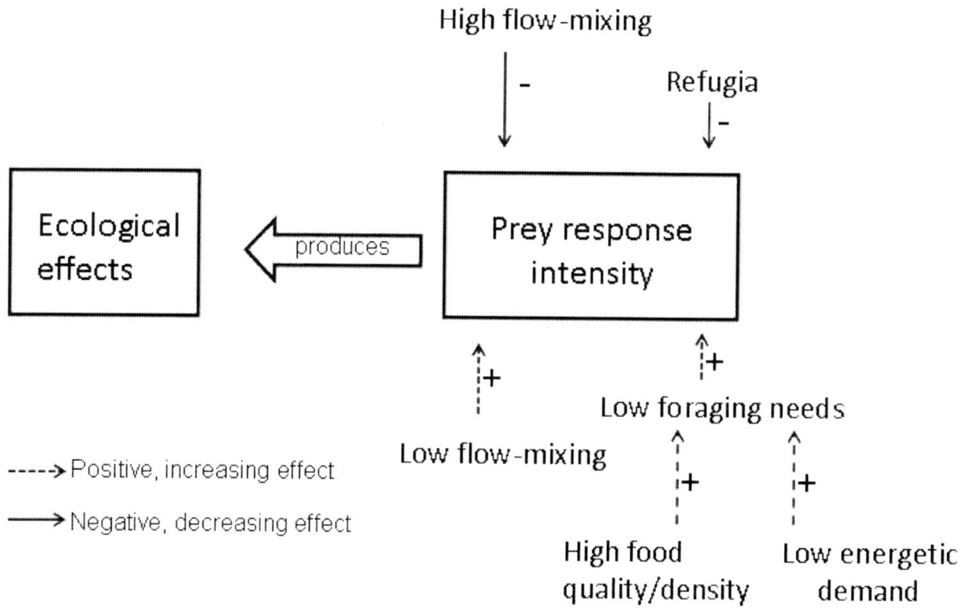

Figure 6. Summary of potential effects on prey responses to predator chemical signals, showing how various factors may modulate prey response intensity and mediate the level of ecological effects. Dashed arrows denote positive, or increasing effects whereas solid arrows denote negative or decreasing effects. For instance, low foraging needs increase prey responses since low foraging needs reduce the cost of investing in behavioral or morphological defenses. High food quality acts positively and further reduces foraging needs.

Spatial heterogeneity in refuge availability may produce patterns in activity of molluscan prey that impose structure on other species. Turner et al. (2000) showed that the decrease in stream periphyton discussed above occurs within refuge habitats that are predator specific. Refugia differ for crayfish and fish predators, and so snails shift to different habitats in a predator specific manner, which produces a decline in periphyton abundance in the refuge habitat. Foraging *Nucella* sp. exposed to predatory green crab odors consumed significantly more barnacle prey near refuge habitats than further away (Matassa and Trussell 2011), resulting in a gradient of barnacle abundance directly related to the distance from refuge habitat. In contrast, patterns of prey consumption by *Nucella* sp. that were culled to represent purely consumptive effects did not produce spatial patterning, presumably because of the absence of antipredator behavior in remaining individuals (Matassa and Trussell 2011). Interactions such as these, where antipredator behavior can mediate the spatial distribution of basal resources, could further influence community processes such as succession in patches, facilitation of colonization, and resilience.

Similarly, resource identity also can mediate the strength of behaviorally-mediated trophic cascades by modifying the relative risk while foraging, such as when resources also act as prey refugia. Predatory odors caused more intense antipredator behaviors when *Nucella lapillus* foraged for barnacles than when they foraged for mussels, suggesting that the additional relief (i.e., refuge) provided by mussels lowers the risk perceived by *Nucella* (Trussell et al. 2008). In this case, increased levels of perceived risk resulted in strong behaviorally-mediated trophic cascades when barnacles were the basal resource but not when mussels were the basal resource. By mediating the level of risk, resources could modulate foraging behavior and by extension the abundance and distribution of resources.

Influences of the sensory environment on chemically-mediated interactions have been investigated in the context of the potential strength of consumptive vs. non-consumptive (i.e., trait-mediated) effects on molluscan prey. Gradients in the perceptual environment may function to modulate the strength of chemically-mediated interactions much the same way that physical gradients (such as wave energy) alter the strength of direct consumption (i.e., consumer and producer stress models; Menge and Olson 1990). This may be especially true for interactions between molluscan prey and their predators, since both participants in the predator-prey duet employ chemical sensing. Very low levels of fluid mixing favor detection of predators by prey, since the pressure to avoid being eaten is stronger than that of finding food (i.e., the life-dinner principle; Dawkins and Krebs 1979). Such environments produce strong trait-mediated chemical effects. As fluid mixing increases, predators may become more effective due to the decline in perceptual ability of prey, even though the predators themselves are no longer in an environment that maximizes their own perceptual ability. The impact of predators in these environments shifts to direct consumption. At higher levels of fluid mixing, the sensory abilities of both predators and prey are impacted such that the overall effect of predators is low. In support, predation intensity in experimental plots of bivalves is a hump shaped function with respect to variables measuring flow/mixing (Smee and Weissburg 2006b, Smee et al. 2010). This combined with the well known decreases in perceptual range of both predators and prey suggest that the lowest levels of fluid mixing favor prey detection of predators that produce trait-mediated interactions, intermediate levels favor predators and produce more direct consumption, and higher levels that compromise perceptual abilities of both predators and prey result in diminished predator effects of any kind. Note that examining the role of fluid mixing in these scenarios requires joint examination of predators and prey, since it is the relative advantage of one species over the other that is explanatory, and not the absolute level of sensory performance of each species. Note further that the cascading and indirect effects of predators will be strongest in those situations where prey perception of predators is greatest (e.g., "low" flows), although we have yet to extend this analysis to additional ecological interactions.

B) Larger Scale Ecological and Evolutionary Effects

The process of chemically-mediated predator detection, when summed over a number of prey species over time, has consequences beyond even the indirect effects discussed above. Community properties such as diversity or resistance may be affected by the presence and intensity of chemically-mediated antipredator behavior. Interactions between predators and prey over time may produce an arms race and alter basal levels of behavioral or morphological defenses against predators.

Antipredator strategies (i.e., rapid escape, hiding, seeking refuge, or morphological change) are dependent on the methods that predators use to find and consume prey. As noted, this specificity may produce situations where defenses against one predator are counterproductive with respect to another. Prey thus may adopt particular morphological or behavioral defenses that correspond to the dominant local predators. Covich (2010) suggests that the diversity of freshwater gastropod shell morphologies and antipredator defenses is a result of the variety of predator feeding strategies.

The fact that prey respond specifically to predators, and uniquely identify threatening predators by particular odor properties, can have important consequences for invasion resistance. The ability of mollusc prey to detect and respond to non-native predators will determine their susceptibility to invasive predator species, mediating the ability of invasive species to expand their ranges. Many prey species do not recognize non-native predators (Freeman and Byers 2006, Sih et al. 2010), presumably because the odors produced by predators contain foreign chemicals or unrecognizable blends. Sometimes prey respond to non-native predators but fail to do so in a manner that reduces their risk of predation (Banks and Dickman 2007). An invasive predator that does not evoke defensive responses enjoys a competitive advantage and may then be able to insert itself in the community, with potentially destructive consequences. The evolution and development of morphological antipredator defenses in some bivalves may require approximately 15 years of contact with non-native predators (Freeman and Byers 2006), but this could be longer than the ability of some prey species to withstand predation pressure, resulting in extinction.

Non-native prey species also are capable of invading communities if they can recognize and respond to native predators. For example, non-native oyster drills display appropriate antipredator behaviors when exposed to odors from native predators and from a combination of native predator and injured conspecific odors (Grason and Miner 2012). The introduction of predators is a common management practice to help control prey species (Fagan et al. 2002). If prey species are able to recognize and respond appropriately to predators it could reduce predation events and thus the efficiency of these types of management practices. In herbivore-plant interactions the introduction of non-native herbivores can result in an "invasional meltdown" because native plant species lack the defenses needed to resist herbivory (Parker et al. 2006), suggesting that the ability of exotic molluscan grazers to recognize predators may have far reaching consequences. Plasticity in the ability of native and non-native prey species to respond to the scent of novel predators may be an important feature in invasion resistance, but this hypothesis has not been examined systematically.

Although plasticity in the sensitivity of prey to odors has not been well investigated, it is clear that the history of exposure to predators can alter prey responsiveness; prey generally respond to predators more strongly if they are from populations that suffer from higher predation risk. For example, *M. mercenaria* from Georgia, USA (an area of high predation pressure) responded more to predators fed conspecifics and injured conspecifics than *M. mercenaria* from Maine, USA (an area of low predation pressure; Smee and Weissburg 2008). Siphon withdrawal (i.e., clamming up) was exhibited more often and to a greater degree in southern populations. Consistent with the risk allocation hypothesis, predation pressure seems to correlate with antipredator behavior. The intensity and incidence of antipredator responses can also be related to historical contact with predators over short time periods. Increased exposure to predatory asteroid odors increased the incidence of antipredator behaviors in whelks (Rochette et al. 1998). Although models suggest that prey in perpetually high risk environments may exhibit antipredator behavior, the need for food can outweigh the risk of predation in these types of environments (Lima and Bednekoff 1999) and result in less intense antipredator behavior in high risk environments compared with low risk environments, if food is limiting.

The intensity or specificity of morphological defenses also can be correlated with differences in predation pressure or risk (Trussell and Smith 2000, Moody and Aronson 2012; Figure 7), consistent with the risk allocation hypothesis. For example, *Littorina* sp. from geographically separated populations that were transplanted and raised in the presence of green crab predatory odors developed thicker shells regardless of their original geographic locale (Trussell and Smith 2000; Figure 7). This suggests that the development of morphological defenses is plastic and that morphology, at least partially, reflects perceived risk as opposed to inherent genetic variation between geographically separated populations (Trussell 1996). However, the level of plasticity (e.g., responses relative to control) is greater in the Northern (low predation) site even though the absolute level of morphological defense is less than in the southern (high predation) site. This suggests additional tradeoffs associated with shell architecture impose an upper limit to the degree morphological change designed to reduce predation (Trussell and Nicklin, 2002).

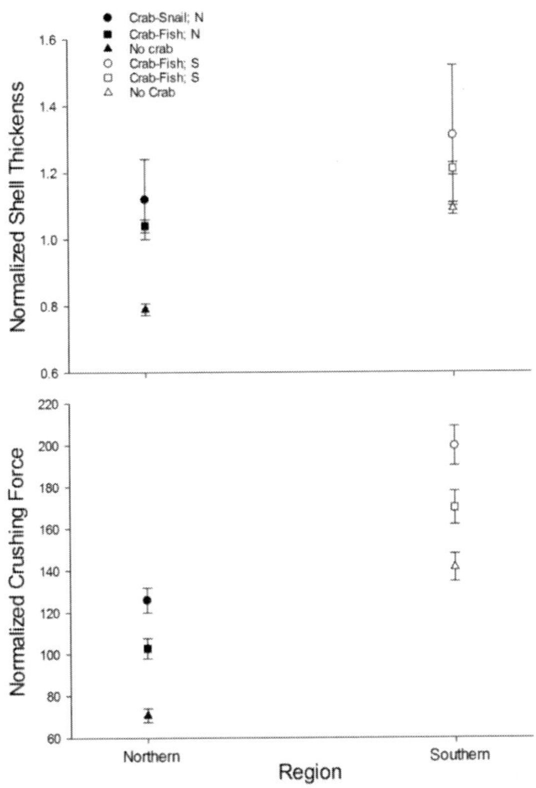

Redrawn with permission from Trussell and Nicklin (2000).

Figure 7. Response of *Littorina obtusa* to chemical cues from predators and injured prey. Figure shows mean increase in shell thickness and force required to crush the shell (with std. error), where responses have been normalized to shell length. Northern refers to populations from the Gulf of Maine whereas Southern refers to populations from Massachusetts, representing low and high historical predation rates. Snails were exposed to cues from *Carcinus maenus* that had been given fish, or snails (Crab-Fish, Crab-Snail), so note that the Crab-Snail treatment includes the odor of injured conspecifics. Snails form northern regions displayed lower absolute responses, but greater plasticity, that is, a greater difference between control and odor treatment.

Finally, as we have pointed out, chemically-mediated trophic cascades often include species such as oysters or plants that provide ecosystem services. Clearly the suppression of snail grazing of *Spartina* sp. has very important consequences for estuarine systems (Siliman and Bertness 2002). However, molluscan species themselves may be critical ecosystem components, and alterations in their feeding time due to antipredator behavior also may have system wide consequences, although this possibility remains largely speculative. Oyster populations along the Atlantic coast provide a good system in which to start examining odor-mediated impacts on ecosystem services, because oyster feeding activities enhance water quality, remove toxic phytoplankters and improve nutrient cycling (Newell 2004, Coen et al. 2007). Reduced feeding would obviously impact all these functions. Exploring the connection between odor-mediated antipredator behavior and ecosystem services could provide insight into better management and remediation strategies in systems where bivalves provide key services.

Conclusion

Chemical information is of great significance for molluscan communities, and alters interactions between molluscan prey and their predators, as well as having indirect effects on other species in the community. The range and breadth of these effects makes it important to understand how chemical cues govern these interactions, particularly since molluscs often are key members of communities in which they reside. Although we have made progress towards this goal, there are still significant gaps in our understanding.

First-although we know molluscs use qualitative and quantitative aspects of chemical cues to discern the degree of threat imposed by a predator, we still don't know much about how this is accomplished. In particular, the lack of identified chemical deterrents prevents us from understanding how combined signals from multiple predators may affect potential prey. Without knowledge of concentrations emitted by the predator, properties such as the perceptual range, and hence, the number of prey organisms that may be able to discern the presence of a given predator, will be difficult to establish.

Second, although we know that the fluid environment can control how easily prey can detect predators, the consequences of gradients in fluid motion for patterns of community organization remain unclear. Much like the strong physical (wave-induced) disturbances that affect regulatory forces in hard substrate communities (Menge and Lubchenko, 1981), changes in perceptual abilities in different flow environments may alter the strength and significance of predatory vs. competitive interactions.

Finally, we have yet to comprehensively examine both short and long term plasticity in molluscan responses to chemical cues. The apparent variability in sensitivity to chemical cues between populations and species suggests that predation intensity as experienced over a variety of temporal and spatial scales changes the response of molluscan prey to predator scent. Not only does this affect the consequences of chemically-mediated prey perception in different populations, but it has consequences when the community is subject to invasion by predators or prey with a different evolutionary history.

Despite the level of uncertainty in some key areas, molluscan communities will continue to be important for examining chemical ecology. The range of flow environments, the relative ease of understanding chemical signal transport in aquatic habitats, the tractability of lab and manipulative field experiments, and the frequent key status of molluscan species in their community all conspire to make molluscs model organisms for understanding the impact of chemical cues.

References

Arnold, WS. 1984. The effect of prey size, predator size, and sediment composition on the rate of predation of the blue crab, *Callinectes sapidus* Rathbun, on the hard clam *Mercenaria mercenaria* (Linne). *Journal of Experimental Marine Biology and Ecology*, 80: 207-219.

Appleton, R, and AR Palmer. 1988. Water-bourne stimuli released by predatory crabs and damaged prey induce more predator-resistant shells in a marine gastropod. *PNAS*, 85:4387-4391.

Banks, PB, and CR Dickman. 2007. Alien predation and the effect of multiple levels of prey naïveté. *Trends in Ecology and Evolution*, 22: 229-230.

Barbeau, MA, and RE Scheibling. 1994. Procedural effects of prey tethering experiments: predation of juvenile scallops by crabs and sea stars. *Marine Ecology Progress Series*, 111:305-310.

Blundon, J, and VS Kennedy. 1982. Refuges for infaunal bivalves from blue crab, *Callinectes sapidus*, predation in Chesapeake Bay. *Journal of Experimental Marine Biology*, 65: 67-81.

Boudreau, B, E Bourget, and Y Simard. 1993. Effect of age, injury, and predator odors on settlement and shelter selection by lobster *Homarus americanus* postlarvae. *Marine Ecology Progress Series*, 93: 119-129.

Bronmark, C, and JG Miner. 1992. Predator-induced phenotypical change in body morphology in crucian carp. *Science*, 258: 1348-1350.

Brown, KM, B Aronhime, and X Wang. 2011. Predatory blue crabs induce byssal thread production in hooked mussels. *Invertebrate Biology*, 30:43-48.

Bruce, T, LJ Wadhams, and CM Woodcock. 2005. Insect host location: a volatile situation. *Trends in Plant Science*, 10: 269-274.

Butman, C, M Frechette, WR Geyes, and VR Starczak. 1994. Flume experiments on food supply to the blue mussel Mylilus edulis as a function of boundary-layer flow. *Limnology and Oceanography*, 39: 1755-1768.

Caro, AU, J Escobar, F Bozinovic, SA Navarrete, and JC Castilla. 2008. Phenotypic variability in byssus thread production of intertidal mussels induced by predators with different feeding strategies. *Marine Ecology Progress Series*, 372:127-134.

Casey, MM, and D Chattopadhyay. 2008. Clumping behavior as a strategy against drilling predation: implications for the fossil record. *Journal of Experimental Marine Biology and Ecology*, 367:174-179.

Cheung, SG, KC Luk, and PKS Shin. 2006. Predator-labeling effect on byssus production in marine mussels *Perna viridis* (L.) and *Brachidontes variabilis* (Krauss). *Journal of Chemical Ecology*, 32:1501-512.

Chivers, DP, and RJF Smith. 1998. Chemical alarm signaling in aquatic predator-prey systems: a review and prospectus. *Ecoscience*, 5:338-352.

Coen, L, RD Brumbaugh, D Bushek, R Grizzle, MW Luckenbach, MH Posey, SP Powers, and SG Tolley. 2007. Ecosystem services related to oyster restoration. *Marine Ecology Progress Series*. 341: 303-307.

Cote, IM, and E Jelnikar. 1999. Predator-induced clumping behavior in mussels (*Mytilus edulis* Linnaeus). *Journal of Experimental Marine Biology and Ecology*, 235: 201-211.

Covich, AP. 2010. Winning the biodiversity arms race among freshwater gastropods: competition and coexistence through shell variability and predator avoidance. *Hydrobiologia*, 653:191-215.

Dahms, H, and P Qian. 2005. Exposure of biofilms to meiofaunal copepods affects the larval settlement of *Hydroides elegans* (Polychaeta). *Marine Ecology Progress Series*, 297: 203-214.

Dalesman, S, SD Rundle, and PA Cotton. 2007. Predator regime influences innate anti-predator behavior in the freshwater gastropod Lymnaea stagnalis. *Freshwater Biology*, 52:2134-2140.

Dawkins, R, and JR Krebs. 1979. Arms races between and within species. *Proc. R. Soc. Lond.* B, 205: 489-511.

Decho, A, KA Browne, and RK Zimmer-Faust. 1998. Chemical cues: why basic peptides are signal molecules in marine environments. *Limnology and Oceanography*, 43:1410-1417.

Delavan, SK, and DR Webster. In Press. Predator and flow influence on bivalve clam excurrent jet characteristics. *Journal of Experimental Marine Biology and Ecology*.

Dickey, BF, and TM McCarthy. 2007. Predator-prey interactions between crayfish (*Orconectes juvenilis*) and snails (*Physa gyrina*) are affects by spatial scale and chemical cues. *Invertebrate Biology*, 126: 57-66.

Dodson, SI. 1989. The ecological role of chemical stimuli for the zooplankton: predator-induced morphology in Daphnia. *Oecologia*, 78:361-367.

Doering, P. 1982. Reduction of sea star predation by the burrowing response of the hard clam *Mercenaria mercenaria* (Mollusca: Bivalvia). *Estuaries*, 5:310-315.

Dusenbery, D. 1992. Sensory ecology: how organisms acquire and respond to information. WH Freeman and Company, New York.

Eggleston, D. 1990. Foraging behavior of the blue crab, *Callinectes sapidus* on juvenile oysters, *Crassostrea virginica*: effects of prey density and size. *Bulletin of Marine Science*, 46: 62-82.

Fagan, WF, MA Lewis, MG Neubert, and P van den Dreissche. 2002. Invasion theory and biological control. *Ecology Letters*, 5: 148-157.

Flynn, AM, and DL Smee. 2010. Behavioral plasticity of the soft-shelled clam *Mya arenaria* in the presence of predators increases survival in the field. *Journal of Experimental Marine Biology and Ecology*, 383: 32-38.

Forward, RB Jr., RA Tankersley, KA Smith, and JM Welch. 2003. Effects of chemical cues on orientation of blue crab, *Callinectes sapidus*, megalopae in flow: implications for location of nursery areas. *Marine Biology*, 142: 747-756.

Freeman, AS, and JE Byers. 2006. Divergent induced responses to and invasive predator in marine mussel populations. *Science*, 313: 831-833.

Fulford, R, DL Breitburg, RIE Newell, WM Kemp, and M Luckenbach. 2007. Effects of oyster population restoration strategies on phytoplankton biomass in Chesapeake Bay: a flexible modeling approach. *Marine Ecology Progress Series*, 336: 43-61.

Gilbert, JJ. 1966. Rotifer ecology and embryological induction. *Science*, 151:1234-1237.

Grason, EW, and BG Miner. 2012. Behavioral plasticity in an invaded system: non-native whelks recognize risk from native crabs. *Oecologia*, 169: 105-115.

Gonzalo, A, C Cabido, P Galan, P Lopez, and J Martin. 2006. Predator, but not conspecific, chemical cues influence pond selection by recently metamorphosed Iberian green frogs, *Rana perezi*. *Canadian Journal of Zoology*, 84: 1295-1299.

Grabowski, J. 2004. Habitat complexity disrupts predator-prey interactions but not the trophic cascade on oyster reefs. *Ecology*, 85: 995-1004.

Grabowski, J, and DL Kimbro. 2005. Predator avoidance behavior extends trohpic cascades to refuge habitats. *Ecology*, 86: 1312-1319.

Guerra, C, AN Madea-Martinez, A Hernandez-Llamas, MT Sicard-Gonzalez, S Koenigstein, D Abele, and EER Phillip. 2012. The influence of temperature and the presence of predators on growth, survival and energy allocation for reproduction in the Catarina Scallop *Argopecten ventricosus*. *Aquaculture Research*, 3:756-766.

Haddon, M, RG Wear, and HA Packer. 1987. Depth and density of burial by the bivalve *Paphies ventrocosa* as refuges from predation by the crab *Ovalipes catharus*. *Marine Biology*, 94: 25-30.

Hadfield, M, and MAR Koehl. 2004. Rapid behavioral responses of an invertebrate larva to dissolved settlement cue. *Biological Bulletin*, 207: 28-43.

Harvell, C. 1990. The ecology and evolution of inducible defenses. *Quarterly Review of Biology*, 65: 323-340.

Hay, ME, and W Fenical. 1996. Chemical ecology and marine biodiversity: insights and products from the sea. *Oceanography*, 9: 10-20.

Heiathus, M, and LM Dill. 2002. Food availability and tiger shark predation risk influence bottlenose dolphin habitat use. *Ecology*, 83: 480-491.

Hill, JM, and MJ Weissburg. 2012. The influence of predator body size on the magnitude of NCEs: experiments between blue crabs, mud crabs, and their shared oyster prey. *Oecologia*, DOI 10.1007/s00442-012-2488-4

Hines, A, AM Haddon, and LE Weichert. 1990. Guild structure and foraging impact of blue crabs and epibenthic fish in a subestuary of Chesapeake Bay. *Marine Ecology Progress Series,* 67:105-126.

Holt, R. 1977. Predation, apparent competition, and the structure of prey communities. *Theoretical Population Biology*, 12: 197- 229.

Hoverman, JT, and RA Relyea. 2007. Rules of Engagement: how to defend against combinations of predators. *Oecologia*, 154: 551-560.

Ishida, S, and K Iwasaki. 2003. Reduced byssal thread production and movement by the intertidal mussel Hormomya mutabilis in response to effluent from predators. *Journal of Ethology*, 21:117-122.

Irlandi, E, and CH Peterson. 1991. Modification of animal habitat by large plants: mechanisms by which seagrasses influence clam growth. *Oecologia*, 87: 307-318.

Jacobsen, H, and OB Stabell. 2004. Antipredator behaviour mediated by chemical cues: the role of conspecific alarm signaling and predator labeling in the avoidance response of a marine gastropod. *OIKOS*, 104:43-50.

Jacquin-Jolly, E, and C Merlin. 2004. Insect olfactory receptors: contributions of molecular biology to chemical ecology. *Journal of Chemical Ecology*, 30: 2359-2397.

Kats, LB, and LM Dill. 1998. The scent of death: chemosensory assessment of predation risk by prey animals. *Ecoscience*, 5:361-394.

Kicklighter, CE, MW Germann, M Kamio, and CD Derby. 2007. Molecular identification of alarm cues in the defensive secretions of the sea hare *Aplysia californica. Animal Behavior,* 74: 1481-1492.

Kishida, O, GC Trussell, K Nishimura. 2009. Top-down effects on antagonistic inducible defense and offense. *Ecology*, 90: 1217-1226.

Kobak, J, T Kakareko, and M Poznanska. 2010. Changes in attachment strength and aggregation of zebra mussel, *Dreissena polymorpha* in the presence of potential fish predators of various species and size. *Hydrobiologia*, 644:195-206.

Koch, N, B Lynch, and R Rochette. 2007. Trade-off between mating and predation risk in the marine snail, Littorina plena. *Invertebrate Biology*, 126: 257-267.

Large, SI, and DL Smee. 2010. Type and nature of cues used by *Nucella lapillus* to evaluate predation risk. *Journal of Experimental Marine Biology and Ecology*, 396:10-17.

Legault, C, and JH Himmelman. 1993. Relation between escape behavior of benthic marine invertebrates and the risk of predation. *Journal of Experimental Marine Biology and Ecology*, 170:55-74.

Leonard, G, MD Bertness, and PO Yund. 1999. Crab predation, waterborne cues, and inducible defenses in the blue mussel *Mytilus edulis. Ecology*, 80:1-14.

Lewis, D, and LA Eby. 2002. Spatially heterogeneous refugia and predation risk in intertidal salt marshes. *OIKOS*, 96: 119-129.

Lima, SL, and PA Bednekoff. 1999. Temporal variation in danger drives antipredator behavior: the predation risk allocation hypothesis. *American Naturalist*, 153:649-659.

Lima, SL, and LM Dill. 1990. Behavioral decisions made under the risk of predation: a review and prospectus. *Canadian Journal of Zoology*, 68:619-640.

Matassa, CM, and GC Trussell. 2011. Landscape of fear influences the relative importance of consumptive and nonconsumptive predator effects. *Ecology*, 92: 2258-2266.

Menge, B, and J Lubchenco. 1981. Community organization in temperate and tropical rocky intertidal habitats: prey refuge in relation to consumer pressure gradients. *Ecological Monographs*, 51:429-450.

Menge, B, and AM Olson. 1990. Role of scale and environmental factors in regulation of community structure. *Trends in Ecology and Evolution*, 5: 52-57.

Moody, RM, and RB Aronson. 2012. Predator-induced defenses in a salt-marsh gastropod. *Journal of Experimental Marine Biology and Ecology*, 413: 78-86.

Mowles, SL, SD Rundle, and PA Cotton. 2011. Susceptibility to predation affects trait-mediated indirect interactions by reversing interspecific competition. *PLOS One*, 6:1-6.

Nakaoka, M. 2000. Nonlethal effects of predators on prey populations: predator-mediated change in bivalve growth. *Ecology*, 81:1031-1045.

Neo, ML, and PA Todd. 2011. Predator-induced changes in fluted giant clam (*Tridacna squamosa*) shell morphology. *Journal of Experimental Marine Biology and Ecology*, 397:21-26.

Newell, R. 2004. Ecosystem influences of natural and cultivated populations of suspension feeding bivalve molluscs: a review. *Journal of Shellfish Research*, 23: 51-61.

Ordzie, CJ, and GC Garafalo. 1980. Behavioral recognition of molluscan and echinoderm predators by the bay scallop, *Argopecten irradians* (Lamark) at two temperatures. *Journal of Experimental Marine Biology and Ecology*, 43:29-37.

Parker, J, DE Burkepile, and ME Hay. 2006. Opposing effects of native and exotic herbivores on plant invasions. *Science*, 311: 1459-1461.

Parrish, J, and L Edelstein-Keshet. 1999. Complexity, pattern, and evolutionary trade-offs in animal aggregation. *Science*, 284: 99-101.

Pawlik, J. 1992. Chemical ecology of the settlement of benthic marine invertebrates. *Oceanography and Marine Biology Annual Review*, 30: 273-335.

Peacor, SD, and EE Werner. 2001. The contribution of trait-mediated indirect effects to the net effects of a predator. *PNAS*, 98:3904-3908.

Peterson, C. 1982. Clam predation by whelks (Busycon spp.): Experimental tests of the importance of prey size, prey density and seagrass cover. *Marine Biology*, 66: 159-170.

Peterson, CH, WG Ambrose Jr., and JH Hunt. 1982. A field test of the swimming response of the bay scallop (*Argopecten irradians*) to changing biological factors. *Bulletin of Marine Science*, 32:939-944.

Powers, SP, and JN Kittinger. 2002. Hydrodynamic mediation of predator-prey interactions: differential patterns of prey susceptibility and predator success explained by variation in water flow. *Journal of Experimental Marine Biology and Ecology*, 273:171-187.

Preisser, EL, DI Bolnick, and MF Benard. 2005. Scared to death? The effects of intimidation and consumption in predator-prey interactions. *Ecology*, 86: 501-509.

Rahman, S, and DR Webster. 2005. The effect of bed roughness on scalar fluctuations in turbulent boundary layers. *Experiments in Fluids*, 38: 372-384.

Rahman, Y, RB Forward Jr., D Rittschof. 2000. Responses of mud snails and periwinkles to environmental odors and disaccharide mimics of fish odor. *Journal of Chemical Ecology*, 26: 679-696.

Rawlings, TA. 1994. Effect of elevated predation risk on the metabolic rate and spawning intensity of a rocky shore marine gastropod. *Journal of Experimental Marine Biology and Ecology*, 181: 67-79.

Relyea, R. 2004. Fine-tuned phenotypes: tadpole plasticity under 16 combinations of predators and competitors. *Ecology*, 85:172-179.

Rittschof, D. 1990. Peptide-mediated behaviors in marine organisms. *Journal of Chemical Ecology*, 16:261-272.

Roberts, D, D Rittschof, DJ Gerhart, AR Schmidt, and LG Hill. 1989. Vertical migration of the clam *Mercenaria mercenaria* (L.) (Mollusca: Bivalvia): environmental correlates and ecological significance. *Journal of Experimental Marine Biology and Ecology*, 126: 271-280.

Rochette, R, DJ Arsenault, and JH Himmelman. 1998. Chemically-mediated predator recognition learning in a marine gastropod. *Ecoscience*, 5: 353-60.

Schmitt, RJ. 1987. Indirect interactions between prey: apparent competition, predator aggregation, and habitat segregation. *Ecology*, 68:1887-1897.

Schmitz, O, AP Beckerman, and KM O'Brien. 1997. Behaviorally mediated trophic cascades: effects of predation risk on food web interactions. *Ecology*, 78: 1388-1399.

Sih, A, DI Bolnick, B Luttbeg, JL Orrock, SD Peacor, LM Pintor, E Preisser, JS Rehage, and JR Vonesh. 2010. Predator-prey naivete, antipredator behavior, and the ecology of predator invasions. *OIKOS*, 119: 610-621.

Silliman, B, and MD Bertness. 2002. A trophic cascade regulates salt marsh primary production. *PNAS*, 99: 10500-10505.

Smee, DL, and MJ Weissburg. 2006a. Hard clams (*Mercenaria mercenaria*) evaluate predation risk using chemical signals from predators and injured conspecifics. *Journal of Chemical Ecology*, 32: 605-619.

Smee, DL, and MJ Weissburg. 2006b. Clamming up: Environmental forces diminish the perceptive ability of bivalve prey. *Ecology*, 87: 1587-1598.

Smee, DL, and MJ Weissburg. 2008. Heightened prey responses in risky habitats: predation pressure increases prey sensitivity to predation risk. *Marine Ecology Progress Series*, 363: 39-50.

Smee, DL, MC Ferner, and MJ Weissburg. 2008. Alteration of sensory abilities regulates the spatial scale of nonlethal predator effects. *Oecologia*, 156: 399-409.

Smee, DL, MC Ferner, and MJ Weissburg. 2010. Hydrodynamic sensory stressors produce nonlinear predation patterns. *Ecology*, 91: 1391-1400.

Smith, L, and JA Jennings. 2000. Induced defensive responses by the bivalve *Mytilus edulis* to predators with different attack modes. *Marine Biology*, 136: 461-469.

Stenzler, D, and J Atema. 1977. Alarm response of the mud snail, *Nassarius obsoletus*: specificity and behavioral priority. *Journal of Chemical Ecology*, 3:159-171.

Tolley, S, and AK Volety. 2005. The role of oysters in habitat use of oyster reefs by resident fishes and decapod crustaceans. *Journal of Shellfish Research*, 24: 1007-1012.

Trussell, GC. 1996. Phenotypic plasticity in an intertidal snail: the role of a common crab predator. *Evolution*, 50: 448-454.

Trussell, GC, and LD Smith. 2000. Induced defenses in response to an invading crab predator: an explanation of historical and geographical phenotypic change. *PNAS*, 97:2123-2127.

Trussell, GC, and MO Nicklin. 2002. Cue sensitivity, inducible defense, and trade-offs in a marine snail. *Ecology*, 83: 1635-1647.

Trussell, GC, PJ Ewanchuk, and MD Bertness. 2003. Trait-mediated effects in rocky intertidal food chains: predator risk cues alter prey feeding rates. *Ecology*, 84: 629-640.

Trussell, GC, PJ Ewanchuk, MD Bertness, and BR Silliman. 2004. Trophic cascades in rocky shore tide pools: distinguishing lethal and nonlethal effects. *Oecologia*, 139: 427-432.

Trussell, GC, PJ Ewanchuk, and CM Matassa. 2006. The fear of being eaten reduces energy transfer in a simple food chain. *Ecology*, 87:2979-2984.

Trussell, GC, PJ Ewanchuk, and CM Matassa. 2008. Resource identity modifies the influence of predation risk on ecosystem function. *Ecology*, 89: 2798-2807.

Turner, AM, 1997. Contrasting short-term and long-term effects of predation risk on consumer habitat use and resources. *Behavioral Ecology*, 8:120-125.

Turner, AM, 2008. Predator diet and prey behavior: freshwater snails discriminate among closely related prey in a predator's diet. *Animal Behaviour*, 76: 1211-1217.

Turner, A, RJ Bernot, and CM Boes. 2000. Chemical cues modify species interactions: the ecological consequences of predator avoidance by freshwater snails. *OIKOS*, 88: 148-158.

Turner, AM, and SL Montgomery. 2003. Spatial and temporal scales of predator avoidance: experiments with fish and snails. *Ecology*, 84: 616-622.

Vadas, RL Sr., MT Burrows, and RN Hughes. 1994. Foraging strategies of dogwhelks, *Nucella lapillus* (L): interacting effects of age, diet and chemical cues to the threat of predation. *Oecologia*, 100: 439-450.

Wahl, M, and ME Hay. 1995. Associational resistance and shared doom – effects of epibiosis on herbivory. *Oecologia*, 102: 329-340.

Webster, DR, and MJ Weissburg. 2001. Chemosensory guidance cues in a turbulent chemical odor plume. *Limnology and Oceanography*, 46: 1034-1047.

Weissburg, MJ, and RK Zimmer-Faust. 1993. Life and death in moving fluids: hydrodynamic effects on chemosensory-mediated predation. *Ecology*, 74: 1428-1443.

Wilson, ML. 2011. Sensory landscape impacts on odor-mediated predator-prey interactions at multiple spatial scales in salt marsh communities. PhD Thesis, Georgia Institute of Technology.

Wilson, ML, and MJ Weissburg. 2013. Biotic structural elements indirectly affect associated prey in a predator-specific manner via positive and negative modifications of the sensory environment. *Oecologia* 171:427-438 DOI 10.1007/s00442-012-2413-x

Vermeij, GJ, and AP Covich. 1978. Coevolution of freshwater gastropods and their predators. *American Naturalist*, 112:833-843.

In: Neuroecology and Neuroethology in Molluscs
Editors: Anna Di Cosmo and William Winlow
ISBN: 978-1-62948-970-4

Chapter 6

Cephalopods Meet Neuroecology: The Role of Chemoreception in *Octopus vulgaris* Reproductive Behaviour

Anna Di Cosmo[*] *and Gianluca Polese*
Dipartimento di Biologia, Università degli Studi di Napoli, Federico II, Napoli, Italia

Abstract

The aim of this chapter is to fill the gap regarding the potential of cephalopods to affect community structure and function in the sea, giving a neuroecological view of the reproductive behaviour of *Octopus vulgaris.*

Cephalopods are known for their ability to learn, memorize and show sophisticated behaviours in the wild when enacting strategies of predation, defence and reproduction. Furthermore they change colour pattern and texture quickly due to the presence of chromatophores controlled directly by the nervous system. Their boneless bodies, inspired the creation of a soft-bodied robot. They sleep, show a "personality" and the ability to use tools in the wild. Octopuses, in particular, show adaptive behaviours implying the existence of a "consciousness" and appear to perceive the pain; properties that strongly contribute to shaping of their behaviour in the sea. This results in the inclusion of all the species of living cephalopods, in the EU Directive covering the "protection of Animals used for scientific purposes" (1st January, 2013). Attention to the capabilities of these "advanced invertebrates" has been developed, for more than 100 years through behavioural and neurophysiological investigations. The majority of these studies have been conducted on *Loligo vulgaris*, *Sepia officinalis* and *Octopus vulgaris*, which are considered to be good animal models by neuroethologists. The cephalopods live in all oceans of the world with the exception of the Black Sea, spreading from the surface waters down into the deep sea and occupying a wide range of ecological niches. Even though they are the smallest class of the phylum Mollusca, because of their wide

[*] Email: dicosmo@unina.it.

distribution they play an important role in the community organization and function of the ecosystems that they inhabit. Much of the language that regulates life of the sea consists of chemical cues. These chemical signals determine feeding, habitat and mating choice. Some cephalopods show social behaviour exclusively during reproduction. The reproductive behaviour of *Octopus vulgaris* is under the control of a complex set of signal molecules such as sex steroids, neuropeptides and neurotransmitters that guide the behaviour from the level of individuals in evaluating mates, to stimulating or deterring copulation, to sperm-egg chemical signalling that promotes fertilization. These signals are intercepted by the olfactory organs and integrated in the olfactory lobes in the central nervous system. The olfactory organ plays a key role in the development of the switch in behaviour from growth to reproduction. In this context we propose a model in which the olfactory organ of *Octopus vulgaris* and the olfactory lobe could represent the on-off switch between food intake and reproduction.

Introduction

Cephalopods (squid, cuttlefish, octopods and nautilus) are the smallest, but the most complex class among Molluscs, and of invertebrates (Hanlon & Messenger 1998). They include exclusively marine animals that live in all the oceans of the world with the exception of the Black Sea, and are distributed from the surface waters down into the deep sea, occupying a wide range of ecological niches. They have been among the dominant large predators in the ocean during their long geological history. Cephalopods evolved during the Cambrian era and have become dominant predators and themselves prey to other top predators, thus playing an important role in the trophic chain of the marine ecosystems (Jereb & Roper 2010, Pierce et al. 2008).

What Do We Know about Cephalopods?

Scientists have been very attracted to their remarkable capabilities which allow them to do many things that one would not expect from a mollusc: for example their ability to learn, (Grasso & Basil 2009, Hochner et al. 2006) memorize and display sophisticated behaviours in the wild when enacting strategies of predation, defence and reproduction (De Lisa et al. 2012, De Lisa et al. 2013, Hanlon & Messenger 1998, Hvorecny et al. 2007). They can sleep (Frank et al. 2012), some show a "personality" (Pronk et al. 2010) and have the ability to use tools in the wild (Huffard 2013). Octopuses, in particular, are able to change their behaviour to suit different circumstances (Norman et al. 2001). This kind of adaptive behaviour suggests that these animals possess consciousness and the ability to perceive pain, a property that plays a fundamental part in shaping their behaviour in the sea (Crook & Walters 2011, Vitti et al. 2012). This has meant that since the 1st January 2013 the approximately 700 species of living cephalopods are included in the EU Directive covering the "protection of Animals used for scientific purposes" with the same legal protection previously adopted for vertebrates. (Directive 2010/63/EU, Polese et al. 2012b, Smith et al. 2013)

One of the most fascinating features of cephalopods is their ability to camouflage themselves by changing their skin colour and pattern quickly, faster than other animal, and body posture and texture despite themselves being blind to the colours. This ability is due to the chromatophores, iridophores, leucophores and reflector cells packed in a multilayer skin (Buresch et al. 2011, Grasso & Basil 2009, Messenger 2001, Packard 1972). Attention to these fascinating abilities has developed over more than 100 years of classical research dedicated to investigating both behavioural and neurophysiological aspects of these "advanced invertebrates" (Hochner et al. 2006). Neuroethology, combining as it does neuroscience and ethology, aims to give neural explanations to animal behaviour and is the most appropriate term to define the field of research developed around cephalopods (De Lisa et al. 2012, De Lisa et al. 2013). The majority of these studies have been conducted on *Loligo vulgaris*, *Sepia officinalis* and *Octopus vulgaris*, which are considered animal models.

Cephalopods have attracted attention not only from neuroethologists but also bioengineers, biotechnologists, marine biologists and fishery scientists. Their boneless bodies combined with the ability to change rapidly the skin colour pattern to mimic a range of species, shifting their shape and behaviour, as inspired by circumstance, led bioengineers to explore dynamic coloration creating an octopus-inspired soft-bodied robot that can also camouflage itself (Margheri et al. 2012). For the same reason the presence of a specific reflecting protein in the iridophore cells of squid skin in *Loligo pealeii* inspired material scientists to design and synthesize biomaterials for complex optical applications (Izumi et al. 2010). However, even in these developing fields of research, cephalopods are still thought as animal models, but if they are isolated from their environment, we will never understand how their sophisticated behaviour, "live fast, die young" life history strategies and trophic ecology, interact and shape the dynamics and structure of communities and ecosystems in which they live.

The increasing interest in cephalopods as food sources has recently induced the fishery scientists to make the first steps toward considering the ecological impact of cephalopods on communities and are developing a series of research and data collection studies on their fisheries (Jereb & Roper 2010). We can easily realize the high potential impact of cephalopods on natural communities knowing their well-developed sense organs, the structure of the motor apparatus, the anatomical-functional organization of their nervous system and some unusual effectors such as chromatophores, ink sac and arm appendages (Hanlon & Messenger 1998, Messenger & Tansey 1979).

The brain has a key role in decoding the signals coming from the sense organs and in selecting an appropriate motor program (Wells 1962, Young 1971). The brains of cephalopods are the largest of all invertebrate brains and across molluscs show the highest degree of centralization, although among cephalopods, there are different levels of centralization. Octopus brains are the most centralized and with the highest number of lobes, Nautilus brains are the least centralized and with the fewest lobes (Budelmann 1995, Grasso & Basil 2009). The brain comprises a central part, around the oesophagus and two large optic lobes laterally connected by an optic tract. The central part is divided in supra– and suboesophageal lobes linked by magnocellular lobes.

Interestingly most of the neurons are not in the brain; about two-thirds of an octopus's neuronal connections are distributed in the arms. The octopus flexible arms are considered muscular-hydrostats being composed almost entirely of muscles which confer the property of maintaining a constant volume (Gutfreund et al. 2006). Their movement is primary under control of the output of the nervous system which is hierarchically organised. In the cephalopods the most motoneurons are situated in the suboesophageal lobes and originating in the lower and intermediate motor centres innervate the effectors (Boycott 1961, Boycott &

Young 1956). The movement of the arms are under control of the nerve cords and ganglia that innervate them (Sumbre et al. 2001). These movements are fundamental in the reaching an object, defence and bipedal locomotion (Huffard 2007).

The lower motor centres are under control of the higher motor centres, the basal lobes and peduncle lobes. They receive and integrate gravitational inputs from the statocysts (Williamson & Chrachri 2004). All the signals and information collected from the environment via sense organs converge in the optic lobes and vertical-superior frontal system. They are the highest of brain centres, sites of visual memory and learning, where signals are integrated with other sensory inputs to drive the richest and flexible repertoire of movements, camouflage and behaviours. (De Lisa et al. 2012, Hanlon & Messenger 1998, Young 1939, Zullo & Hochner 2011). Within the cephalopods the Octopus brain is unique in possessing a chemo-tactile memory system which lies in the subfrontal and inferior frontal lobes and may have evolved in relation to its benthic life style (Hanlon & Messenger 1998). Octopods use the arms to explore and detect tactile and chemosensory information functioning as "natural biosensors" (Di Cosmo, personal observation). These proprieties are conferred on the arms by their appendages, the suckers that are considered in cephalopods as "other kinds of effectors" (Hanlon & Messenger 1998). These are complex organs that show a different organization, morphology and innervation between decapods and octopods (Hanlon & Messenger 1998). Usually they are organized in one or two rows, a small ganglion below each sucker coordinates motor and sensory information (Grasso & Basil 2009, Graziadei & Gagne 1973).

The ink sac, another unusual effector organ, which comprises the ink gland and a reservoir with sphincters, secretes melanin granules whose surfaces are covered by dopamine molecules, suggesting that melanin granules in the secreted ink serve as a sink/carrier for dopamine, preventing excessive dilution after ejection and ensuring efficient interactions with target organs in dopamine-mediated inter- and intra-species communication. The dopamine plays a key role, exerting strong impact on species and prey-predator interaction (Di Cosmo 2003, Fiore et al. 2004). This study was one of the first examples of the neuroecological approach in cephalopods (Derby et al. 2007).

Another example of the neuroecological approach arises from the knowledge that marine organisms possess bioactive factors as a common means of self-defence or for the protection of eggs and embryos (Nair et al. 2011). The only studies investigating cephalopods to identify bioactive factors with antimicrobial activity were those conducted on squids and cuttlefishes (Zatylny et al. 2000). These animals leave their eggs to fend for themselves and interestingly they do not get any infections. Their eggs release polypeptides, sexual pheromones and antimicrobial peptides trapped in the egg capsule, mainly expressed by female accessory sex glands, which confer on them an efficient protection against microorganisms (Laurencin et al. 2012). Recently our group characterized the immune system of *O. vulgaris* and identified bioactive compounds as antimicrobial peptides in order to assess responses to environmental signals and to identify how microorganisms affect the life of *O. vulgaris* (Troncone et al. 2013).

In aquatic systems, chemical cues determine feeding, habitat and mating choices affecting not only individual behaviour and population level-processes, but also community organization and ecosystem function. To detect chemical signals the animals are equipped with sophisticated olfactory organs able to sense the odour cues. Cephalopods are no different, they posses well developed olfactory organs but with the exception of molecular studies on the squid (Gilly & Lucero 1992, Lucero & Gilly 1995, Lucero et al. 1992, Lucero et al. 2000, Mobley et al. 2007, Piper & Lucero 1999), the role played by this organ in the control of feeding, growth, mating and behaviour and how its influence survival, fecundity, population sizes and predation rates is poorly investigated. In the next section we propose a model in which the olfactory organs of *Octopus vulgaris* that scan for environmental chemical cues, and the olfactory lobe (OL) that receives and integrates this information, could be considered the on-off switch between food intake and reproduction since the OL is the site on which converge other environmental signals (light, temperature) and internal stimuli (neuropeptides, hormones).

The Olfactory System of *Octopus vulgaris*

It is almost universally accepted that chemoreception is the most ancient sense in the animal kingdom, actually present even before that the first multicellular organisms appeared on the Earth. As happened in most animals, cephalopod chemoreception plays the main role in several ecological aspects of their life: searching for food, partner choice, parental care (Hanlon & Shashar 2003, Hanlon & Messenger 1998). But how and where do cephalopods perceive the chemical stimuli?

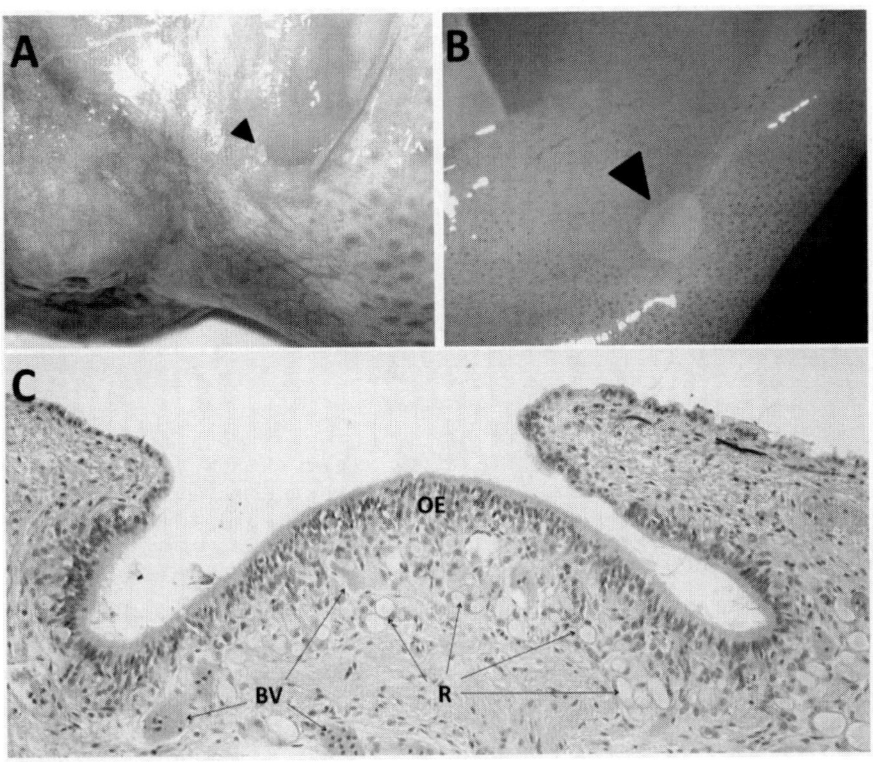

Figure 1. Olfactory organ of *Octopus vulgaris*: A, B ventral view of *O. vulgaris* mantle edge arrowhead indicates the olfactory pit; C, sagittal section of the olfactory organ. (OE: olfactory epithelium - R: ring shape cells - BV: blood vessels) (unpublished data).

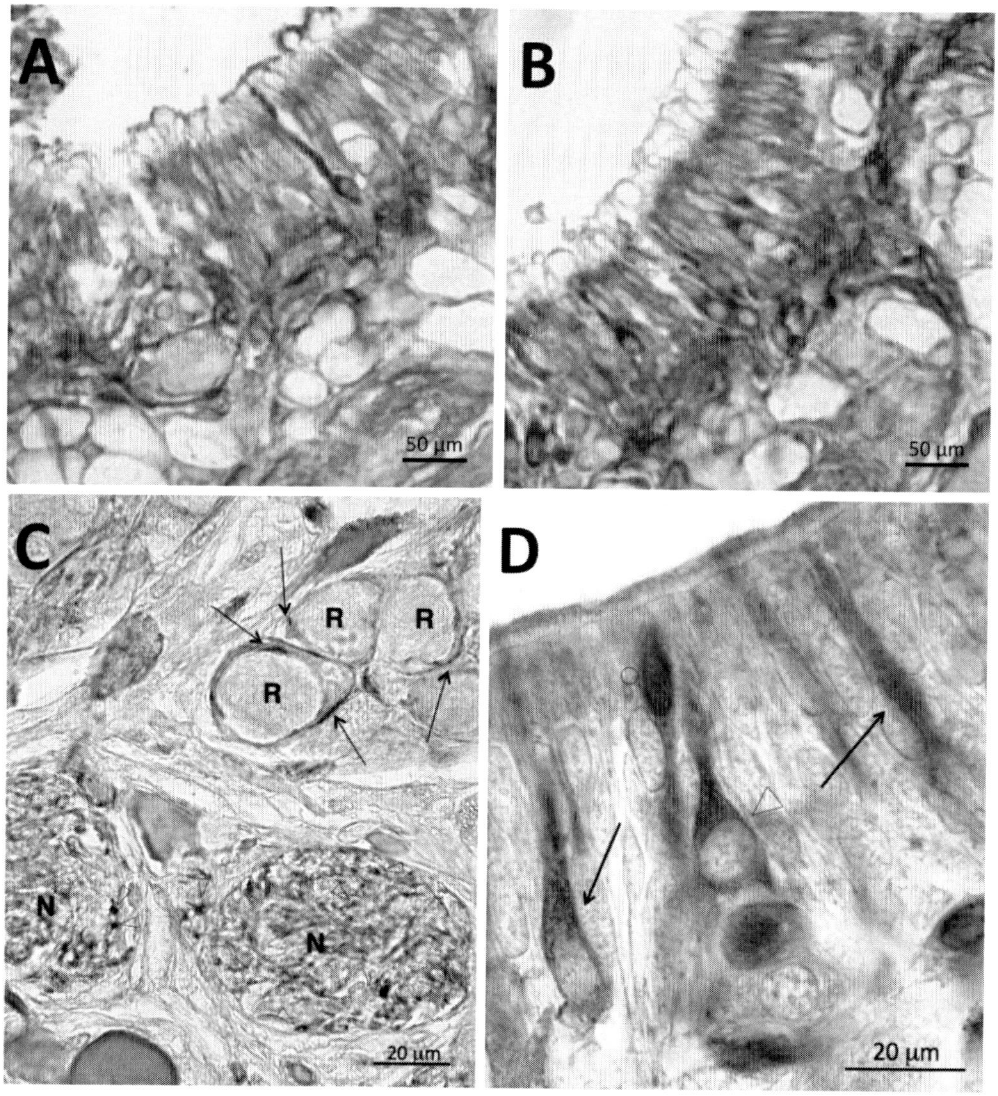

Figure 2. Immunoreactivity of peptides in the *Octopus vulgaris* olfactory epithelium: A – Olfactory sensory neuron FMRFamide immunoreactivity; B – APGWamide immunoreactive OSNs; C – GnRH immunoreactive fibres passing through the olfactory nerve (N) and reaches the ring shaped cells (R) (unpublished data).

In the mid 18th century Albert von Kölliker (von Kölliker 1844) was attracted by a pair of dimples found on both sides of the head of both octopus and squid. These openings in the skin were thought to be acoustic organs, but later it was suggested that they had a chemoreceptive function, perhaps analogous to the gastropod osphradium (see chapters 3, 8 and 9), although there was no experimental evidence in this regard. In 1974, Woodhams and Messenger using electron microscopy techniques and Cajal staining, described, though not exhaustively, the olfactory organ of *Octopus vulgaris* as a small pocket, or dimple, in the skin, presents on both sides of the head, ventral and posterior to the eye, and just anterior to the mantle cavity. In sagittal section, optical microscopy revealed a pseudo-stratified, columnar and ciliated epithelium. Below there were larger vacuolated cells, shaped as a ring.

Through the use of electron microscopy they were able to discriminate 3 cell types: one supporting cell type and two sensory cell types. More information about the olfactory epithelium has been gained in *Octopus joubini* in which Emery (1976) characterized six types of Olfactory Receptor Neurons (ORNs). There are few papers regarding the role of chemoreception in cephalopods to date (Walderon et al. 2011) and little is known about the mechanism and function of the olfactory system in octopods. What is known about the olfactory organ in cephalopods is mostly on decapods (Lucero & Gilly 1995, Lucero et al. 1992, Lucero et al. 2000, Mobley et al. 2007, Piper & Lucero 1999).

In our recent study of olfactory organ from adult *O. vulgaris* (Polese et al. 2012a) we found that it appears to be a tiny white invagination which is difficult to recognise (Figure 1). Using light microscopy the olfactory epithelium (OE) is shown to be organised as a bulging structure whose surface layer is made by ORNs and sustentacular cells, below which is a group of ring shaped cells that give turgor to the entire structure (Figure 1).

Using immunohistochemistry techniques we found immunoreactivity against the following peptides, GnRH, FMRFamide, APGWamide and NPY (Figure 2) (unpublished data) known to be involved in reproduction and energy balance (Di Cosmo & Polese 2013, Di Cristo et al. 2009a, Di Cristo et al. 2009b, Di Cristo et al. 2002a, Di Cristo et al. 2002b, Di Cristo et al. 2005, Minakata et al. 2009). In the olfactory epithelium, whilst FMRFamide, APGWamide and NPY immunoreactivities are found in both ORNs and fibres, GnRH immunoreactivity (-ir) is present, but only at the level of the fibres (Figure 2). Extending the observations to the olfactory lobe, it appears that the GnRH-ir fibres found in the mucosa come, indeed, from neurons located in the CNS.

A Neuroecological View of *Octopus vulgaris* Reproductive Behaviour

Coordination within and between organisms involves chemical messengers. The two systems responsible for this coordination are the nervous and the endocrine systems (see also Chapter 6), both coordinate various mechanisms and physiological processes finalized to optimize animal fitness. How do these two systems interact with each other to regulate reproduction? In the classical concept of input/output relations, different ecological stimuli as temperature, photoperiod, partners and food availability are perceived by sensory organs and delivery to specific areas of the CNS. These areas receive and integrate the information to generate internal responses that affect endocrine glands inducing gonad growth and maturation, egg laying and sexual behaviour. In *Octopus vulgaris*, females die just after their eggs hatch. They reproduce only once and put their available energy into generating and caring for their eggs. After finding a suitable den they do not eat and spend their whole time caring for the eggs, cleaning them and aerating them. For the last month of her life a mother octopus does not eat, she does not leave her den except to defend her eggs from predators. The last thing a mother octopus does for her young, when they are ready to hatch, is to blow them through her siphon and disperse them. It is not known which signals induce such anorexic behaviour or the neural pathways involved. *O. vulgaris* has third order neuroendocrine regulator system (La Font 2000) comparable to the vertebrates HPG axis (Hypothalamus, Pituitary, Gonad). What are the chemical messengers involved in the cross talk between nervous and endocrine systems? They can be divided into three main categories considering their chemical properties and mode of action: neurotransmitters, neuromodulators and hormones. To date, the chemical messengers known to play a role in the control of reproduction in *O. vulgaris* are summarized in Table 1.

Table 1. Chemical messengers involved in *O. vulgaris* reproduction

Chemical messengers	Source	Function	Reference
FMRF amide	Olfactory lobe, reproductive tract	Inhibits optic gland maturation	Di Cosmo et al., 1998; Di Cristo et al., 2005; Di Cosmo and Polese 2013
GnRH	Olfactory lobe, reproductive tract	Stimulates optic gland and oviduct muscle	Di Cosmo et al., 1998; Di Cristo et al., 2005; Di Cosmo and Polese, 2013
APGW amide	Olfactory lobe	Controls of male reproductive behaviour and ejaculation	Di Cosmo and Polese 2013
NPY	Olfactory lobe	Regulation of optic gland activity	Suzuki et al., 2002
CRF	Olfactory lobe	Regulation of optic gland activity	Suzuki et al., 2003
Galanin	Olfactory lobe	Regulation of optic gland activity	Suzuki et al., 2000
Serotonin	Olfactory lobe	Regulation of optic gland activity	Suzuki et al., 2000
NMDA	Nervous system	Increases GnRH mRNA level in the olfactory lobe	De Lisa et al., 2012
17ß-estradiol and its receptors	Nervous System, olfactory lobe, reproductive system	Increase GnRH and 17ß - estradiol receptor mRNA level in the olfactory lobe	De Lisa et al., 2012 Di Cosmo et al., 2002 Di Cosmo et al., 2001
Progesterone and its receptors	Nervous system, reproductive system	Stimulation of gametogenesis and vitellogenesis	Di Cosmo et al., 2001 Di Cristo et al., 2008; Di Cristo et al., 2010
Testosterone and its receptors	Nervous system, reproductive system	Stimulation of spermatogenesis	Di Cosmo et al., 2001 Di Cristo et al., 2010

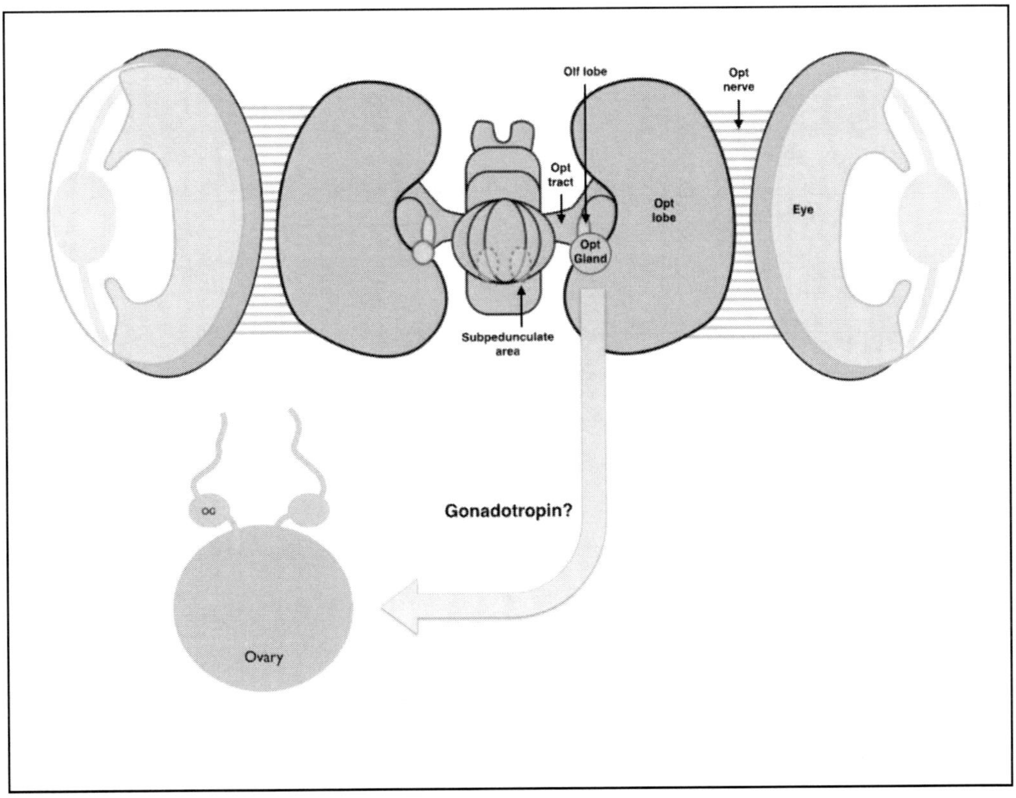

Figure 3. Darkness as a key in Wells and Wells (1959) model: pioneer work on *O. vulgaris* sexual maturation shows that the subpedunculate lobe negatively controls the activity of the optic gland, which is assumed to release an as yet unidentified gonadotropic hormone. Cutting the optic tract or making surgical lesions in the subpedunculate lobe resulted in an enlargement and subsequently hypertrophy of the gonads. Similar results were obtained by cutting the optic nerve or removing the optic lobes.

Until now, the control of reproduction in *O. vulgaris* was based on the Wells and Wells (1959) study that assigned a main role to darkness as the inducer of gonad maturation in female *O. vulgaris* as shown in Figure 3. After experimentally cutting the optic tract and optic nerves, they observed enlargement of the optic gland (a neuroendocrine gland) resulting in hypertrophy of the ovary and subsequently maturation. What is released from the optic gland that induce gonad maturation in *O. vulgaris* is still unknown, is it a gonadotropin? What are the other factors, if any, involved in the control of octopus reproduction? Le Gall et al. (1988) found FMRFamide in the optic gland nerve in *Sepia officinalis* suggesting inhibitory control of the optic gland activity. The findings of several neuropeptides, neurotransmitters and hormones in the nervous lobes controlling reproduction (Table 1) provide new insights to the Wells and Wells' model.

Furthermore, our group studies identify new lobes and sensory structures directly involved in the control of the activity of the optic gland: the olfactory lobe and the olfactory organ. The OL lies on the optic tract close to the optic gland; it is connected to subpedunculate area and, through the olfactory nerve, to the olfactory organ. Besides the finding of FMRFamide in the subpedunculate area and in the OL, we localized cGnRH-I isoform in neurons of the posterior olfactory lobule, close to the optic gland, that send fibres to the chief cells of the optic gland (Di Cristo et al. 2002a) (Table 1).

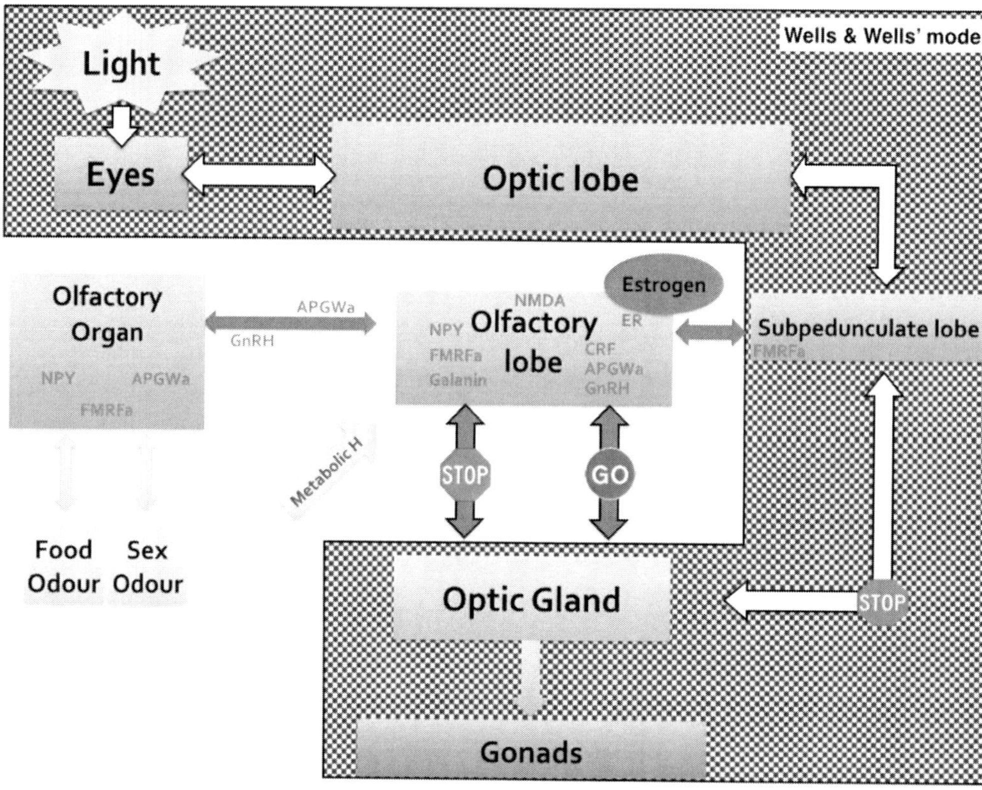

Figure 4. Control of reproduction based on chemical cues not just light. The Wells & Wells (1959) model is represented by the shaded area. The olfactory lobe acts as a switch between growth and reproduction. In addition the olfactory lobe modulates the olfactory organ to be more sensitive to food or sex odours via the molecular players identified (GnRH, FMRFamide, APGWaminde, NPY, CRF, Galanin, ER, and NMDA). The environmental chemical cues, perceived by the olfactory organ, affect the olfactory lobe inducing the behavioural choice between energy storage or reproduction.

These data allow us to hypothesize multipeptidergic control of the optic gland activity, not only in an inhibitory way (FMRFamide), but also excitatory (GnRH). Another peptide, APGWamide, known to work as a common neurotransmitter/neuromodulator peptide found in many species of molluscs and often related to sex organ growth or reproductive behaviour (Di Cosmo & Polese 2013), was localized in the same olfactory lobule in which we had previously localized the GnRH neurons (Table 1). Thus APGWamide is strictly linked to and associated with those systems and lobes that directly process both nearby and distant chemical signals (Table 1). Finally, other peptides, NPY, Galanin, Corticotrophin-Realising Factor (CRF), were found in the same nervous areas (Table 1). All these three peptides are known to play a key role in the balance between metabolism and reproduction (Crown et al. 2007, de Jong-Brink et al. 2001).

What happens at peripherally? In the fusiform ganglia, that directly innervate the male and female gonads, we discovered the presence of FMRFamide and GnRH neurons that, interestingly, send fibres to the circular muscle layers of the reproductive ducts, suggesting an involvement of these two neuropeptides in motor coordination, egg-laying, and spermatophore transport (Di Cosmo & Polese 2013, Di Cristo et al. 2002b). Extending our research at the peripheral level, we recently identified and characterized a chemoattractant peptide isolated from mature eggs of octopus females: Octo-Sap (octopus – sperm attracting peptide) (De Lisa et al. 2013). This bioactive peptide is able to increase sperm motility and induce chemotaxis by changing the octopus spermatozoa swimming behaviour in a dose-dependent manner. There are other chemical messengers that contribute to this scenario. Using the physiological range of 17ß-oestradiol concentrations, as determined by our previous studies on sex steroid hormone fluctuations during the annual reproductive cycle (Table 1), we observed a concentration-dependent increase of Oct-ER (octopus oestradiol receptor) mRNA levels in the olfactory lobes of *O. vulgaris* CNS, demonstrating that Oct-ER mRNA expression was up-regulated by 17ß-oestradiol. Moreover, the stimulation of the olfactory lobes with 17ß-oestradiol increased the Oct-GnRH mRNA confirming the existence of a strong coupling between 17ß-oestradiol and the transcriptional activity of Oct-GnRH neurones. The discovery of the role played by 17ß-oestradiol in the modulation of the transcriptional activity of both Oct-GnRH and Oct-ER genes in the olfactory lobes is in agreement with the existence of an axis 'area subpedunculate-olfactory lobe-optic gland' responsible for the reproductive behaviour of *O. vulgaris*, in which the activity of Oct-GnRH neurones is not only under the control of glutamatergic neurones (Di Cristo et al. 2009a), but also is regulated by 17ß-oestradiol (De Lisa et al. 2012).

In the Wells and Wells' model light plays a crucial role in gonad maturation. They demonstrated the inhibitory action of the light on the optic gland activity. Using surgical lesion of optic nerves or optic tract or subpedunculate area, they were able to induce gonad maturation in *O. vulgaris* females. Their model results in a simple loop where light switches reproduction on/off. In the wild, this kind of effect could be related to the habit of *O. vulgaris* females that leave shallow-water reaching deeper and darker horizons for spawning. All the data previously reported strongly suggest a different and integrative regulation of *O. vulgaris* reproductive behaviour involving another nervous structure, the olfactory lobe in which the multipeptidergic control exerted by GnRH, FMRFamide, APGWamide, NPY, Galanin, and CRF plays a crucial role in the regulation of the optic gland activity (Figure 4). Given that this lobe is also the site on which converges olfactory information, it could modulate the olfactory organ to be more sensitive to food odours, during the energy storage period, inhibiting the optic gland activity with a consequential switch off for reproduction. On the contrary, when the female is in the reproductive period the OL could modulate the olfactory organ to be more sensitive to sex odours resulting in a switch on for reproduction. A similar modulation of olfaction could be responsible for the food avoidance of the female during maternal care with a switch off for food intake (Polese et al. 2013) (Figure 4).

Conclusion

The behaviour of marine organisms is regulated by chemical cues. These chemically mediated interactions strongly affect population structure, community organizations and ecosystem function. The perception of the environment by *O. vulgaris* has been always attributed to its visual and chemo-tactile capabilities, but little information is available on its olfactory function. Chemical cues determine feeding choices, selection of mates and habitats, competitive interactions and transfer of energy within and among ecosystems. In *O. vulgaris* the organ dedicated to the perception of water borne stimuli coming from a distance is the olfactory organ. In this chapter we have assigned a crucial role to both the olfactory organ and the olfactory lobe (OL) in the chemical perception and integration of odour signals that allow the animal to balance their energy allocation between growth versus reproduction. Octopuses during the long non-reproductive adulthood, dedicate major attention in prey choice to improve the quality of sperm and eggs and exploration across the sea bottom. In contrast during the short reproductive period, *Octopus vulgaris* spends its time on multiple mating and maternal care behaviours.

In this chapter we have proposed a model for the control of the reproductive behaviour in *O. vulgaris* female, which puts it in a new neuroecological perspective. Although the model of the octopus reproductive behaviour is based on mediation by chemical cues, the language of life in the sea, many questions remain open:

- What kind of odours do the olfactory organs perceive?
- Is the OL the target organ for metabolic hormones such as leptin like and insulin like peptides?
- Has the olfactory organ any regulatory action on the OL?
- What is the nature of the factor released by the optic gland, is it a gonadotropin or a trophic factor that could modulate the OL?
- What else is in the OL, and what else does it do?
- Given that the molecules regulating reproduction are the same in both male and female, what exactly happens in male octopus?.
- How does the chemical communication system of octopus and other cephalopods affect the structure of cephalopod communities and marine populations?

References

Boycott BB. 1961. Functional organization of brain of cuttlefish *Sepia officinalis. Proceedings of the Royal Society Series B-Biological Sciences* 153: 503-+.

Boycott BB, Young JZ. 1956. The subpedunculate body and nerve and other organs associated with the optic tract of cephalopoda. *Hanstrom Festschrift Lund*: 76-105.

Budelmann BU. 1995. The cephalopod nervous system: what evolution has made of the molluscan design. *Experientia Supplementum (Basel)* 72: 115-38.

Buresch KC, Maethger LM, Allen JJ, Bennice C, Smith N, et al. 2011. The use of background matching vs. masquerade for camouflage in cuttlefish Sepia officinalis. *Vision Research* 51: 2362-68.

Crook RJ, Walters ET. 2011. Nociceptive Behavior and Physiology of Molluscs: Animal Welfare Implications. *Ilar Journal* 52: 185-95.

Crown A, Clifton DK, Steiner RA. 2007. Neuropeptide signaling in the integration of metabolism and reproduction. *Neuroendocrinology* 86: 175-82.

de Jong-Brink M, ter Maat A, Tensen CP. 2001. NPY in invertebrates: molecular answers to altered functions during evolution. *Peptides* 22: 309-15.

De Lisa E, Paolucci M, Di Cosmo A. 2012. Conservative Nature of Oestradiol Signalling Pathways in the Brain Lobes of Octopus vulgaris Involved in Reproduction, Learning and Motor Coordination. *Journal of Neuroendocrinology* 24: 275-84.

De Lisa E, Salzano AM, Moccia F, Scaloni A, Di Cosmo A. 2013. Sperm-attractant peptide influences the spermatozoa swimming behavior in internal fertilization in Octopus vulgaris. *Journal of Experimental Biology* 216: 2229-37.

Derby CD, Kicklighter CE, Johnson PM, Zhang X. 2007. Chemical composition of inks of diverse marine molluscs suggests convergent chemical defenses. *Journal of Chemical Ecology* 33: 1105-13.

Di Cosmo A. 2003. A new model to study the role and biochemical properties of nNOS: the nervous system of cephalopods. *Recent Research Developments in Biochemistry* 4: 405-21.

Di Cosmo A, Di Cristo C. 1998. Neuropeptidergic control of the optic gland of Octopus vulgaris: FMRF-amide and GnRH immunoreactivity. *Journal of Comparative Neurology* 398: 1-12.

Di Cosmo A, Di Cristo C, Paolucci M. 2001. Sex steroid hormone fluctuations and morphological changes of the reproductive system of the female of Octopus vulgaris throughout the annual cycle. *Journal of Experimental Zoology* 289: 33-47.

Di Cosmo A, Di Cristo C, Paolucci M. 2002. A estradiol-17beta receptor in the reproductive system of the female of *Octopus vulgaris*: characterization and immunolocalization. Mol Reprod Dev. 61:367-75.

Di Cosmo A, Polese G. 2013. Molluscan Bioactive Peptide In *Handbook of Biologically Active Peptides.*, ed. R Nachaman, pp. 276-86. Baton Rouge, LA: Abba Kastin.

Di Cristo C, De Lisa E, Di Cosmo A. 2009a. Control of GnRH expression in the olfactory lobe of Octopus vulgaris. *Peptides* 30: 538-44.

Di Cristo C, De Lisa E, Di Cosmo A. 2009b. GnRH in the brain and ovary of Sepia officinalis. *Peptides* 30: 531-37.

Di Cristo C, Di Donato P, Palumbo A, d'Ischia M, Paolucci M, Di Cosmo A. 2010. Steroidogenesis in the brain of Sepia officinalis and Octopus vulgaris. *Frontiers in bioscience* (Elite edition) 2: 673-83.

Di Cristo C, Paolucci M, Di Cosmo A. 2002a. GnRH and FMRFamide: two neuro-bioactive molecules involved in Octopus vulgaris reproduction. *Journal of Physiology-London* 543: 2P-3P.

Di Cristo C, Paolucci M, Iglesias J, Sanchez J, Di Cosmo A. 2002b. Presence of two neuropeptides in the fusiform ganglion and reproductive ducts of Octopus vulgaris: FMRFamide and gonadotropin-releasing hormone (GnRH). *Journal of Experimental Zoology* 292: 267-76.

Di Cristo C, Paolucci M, Di Cosmo A. 2008. Progesterone Affects Vitellogenesis in Octopus vulgaris. *The Open Zoology Journal* 1: 29-36.

Di Cristo C, Van Minnen J, Di Cosmo A. 2005. The presence of APGWamide in Octopus vulgaris: a possible role in the reproductive behavior. *Peptides* 26: 53-62.

Directive 2010/63/EU of the European Parliament and of the Council of 22 September 2010 on the Protection of Animals Used for Scientific Purposes., ed. Co Europe. Strasbourg.

Emery DG, 1976. Observations on the olfactory organ of adult and juvenile *Octopus joubini.* Tissue Cell, 8: 33 – 46.

Fiore G, Di Cosmo A, d'Ischia M, Palumbo A. 2004. Protein nitration in the cuttlefish Sepia officinalis. *Comparative Biochemistry and Physiology B-Biochemistry & Molecular Biology* 139: 145-45.

Frank MG, Waldrop RH, Dumoulin M, Aton S, Boal JG. 2012. A Preliminary Analysis of Sleep-Like States in the Cuttlefish Sepia officinalis. *Plos One* 7.

Gilly WF, Lucero MT. 1992. Behavioral-responses to chemical-stimulation of the olfactory organ in the squid *Loligo-opalescens*. *Journal of Experimental Biology* 162: 209-29.

Grasso FW, Basil JA. 2009. The Evolution of Flexible Behavioral Repertoires in Cephalopod Molluscs. *Brain Behavior and Evolution* 74: 231-45.

Graziadei PPC, Gagne HT. 1973. Neural components in octopus sucker. *Journal of Cell Biology* 59: A121-A21.

Gutfreund Y, Matzner H, Flash T, Hochner B. 2006. Patterns of motor activity in the isolated nerve cord of the octopus arm. *Biological Bulletin* 211: 212-22.

Hanlon R, Shashar N. 2003. Aspects of the Sensory Ecology of Cephalopods In *Sensory Processing in Aquatic Environments*, ed. S Collin, NJ Marshall, pp. 266-82: Springer New York.

Hanlon RT, Messenger JB. 1998. *Cephalopod Behaviour*. Cambridge University Press.

Hochner B, Shomrat T, Fiorito G. 2006. The octopus: A model for a comparative analysis of the evolution of learning and memory mechanisms. *Biological Bulletin* 210: 308-17.

Huffard CL. 2007. Ethogram of Abdopus aculeatus (d'Orbigny, 1834) (Cephalopods : Octopodidae): Can behavioural characters inform octopodid taxomony and systematics? *Journal of Molluscan Studies* 73: 185-93.

Huffard CL. 2013. Cephalopod neurobiology: an introduction for biologists working in other model systems. *Invertebrate Neuroscience* 13: 11-18.

Hvorecny LM, Grudowski JL, Blakeslee CJ, Simmons TL, Roy PR, et al. 2007. Octopuses (Octopus bimaculoides) and cuttlefishes (Sepia pharaonis, S-officinalis) can conditionally discriminate. *Animal Cognition* 10: 449-59.

Izumi M, Sweeney AM, DeMartini D, Weaver JC, Powers ML, et al. 2010. Changes in reflectin protein phosphorylation are associated with dynamic iridescence in squid. *Journal of the Royal Society Interface* 7: 549-60

Jereb P, Roper CFE. 2010. *Cephalopods of The World* Rome: Food and Agriculture Organization of the United Nations.

La Font R. 2000. The Endocrinology of Invertebrates. *Ecotoxicology* 9: 41-57.

Laurencin M, Legrand B, Duval E, Henry J, Baudy-Floc'h M, et al. 2012. From a Marine Neuropeptide to Antimicrobial Pseudopeptides Containing Aza-beta(3)-Amino Acids: Structure and Activity. *Journal of Medicinal Chemistry* 55: 2025-34.

Le Gall S, Feral C, Vanminnen J, Marchand CR. 1988. Evidence for peptidergic innervation of the endocrine optic gland in sepia by neurons showing fmrfamide-like immunoreactivity. *Brain Research* 462: 83-88.

Lucero MT, Gilly WF. 1995. *Physiology of squid olfaction*. 521-34 pp.

Lucero MT, Horrigan FT, Gilly WF. 1992. Electrical responses to chemical-stimulation of squid olfactory receptor-cells. *Journal of Experimental Biology* 162: 231-49.

Lucero MT, Huang W, Dang T. 2000. Immunohistochemical evidence for the Na+/Ca2+ exchanger in squid olfactory neurons. *Philosophical Transactions of the Royal Society of London Series B-Biological Sciences* 355: 1215-18.

Margheri L, Laschi C, Mazzolai B. 2012. Soft robotic arm inspired by the octopus: I. From biological functions to artificial requirements. *Bioinspiration & Biomimetics* 7.

Messenge JB. 2001. Cephalopodes chromatophores: neurobiology and natural history. *Biological Reviews of the Cambridge Philosophical Society* 76: 473-528.

Messenger JB, Tansey EM. 1979. Aminergic fluorescence in the cephalopod cerebellum. *Journal of Physiology-London* 287: P7-P8.

Minakata H, Shigeno S, Kano N, Haraguchi S, Osugi T, Tsutsui K. 2009. Octopus Gonadotrophin-Releasing Hormone: A Multifunctional Peptide in the Endocrine and Nervous Systems of the Cephalopod. *Journal of Neuroendocrinology* 21: 322-26.

Mobley AS, Mahendra G, Lucero MT. 2007. Evidence for multiple signaling pathways in single squid olfactory receptor neurons. *Journal of Comparative Neurology* 501: 231-42.

Nair JR, Pillai D, Joseph SM, Gomathi P, Senan PV, Sherief PM. 2011. Cephalopod research and bioactive substances. *Indian Journal of Geo-Marine Sciences* 40: 13-27.

Norman MD, Finn J, Tregenza T. 2001. Dynamic mimicry in an Indo-Malayan octopus. *Proceedings of the Royal Society B-Biological Sciences* 268: 1755-58.

Packard A. 1972. Cephalopods and fish - limits of convergence. *Biological Reviews of the Cambridge Philosophical Society* 47: 241-&.

Pierce GJ, Valavanis VD, Guerra A, Jereb P, Orsi-Relini L, et al. 2008. A review of cephalopod-environment interactions in European Seas. *Hydrobiologia* 612: 49-70.

Piper DR, Lucero MT. 1999. Calcium signalling in squid olfactory receptor neurons. *Biological Signals and Receptors* 8: 329-37.

Polese G, Ulloa Severino FP, Troncone L, Bertapelle C, Di Cosmo A. "Topology, morphology and function of olfactory organ in Octopus vulgaris", *Molluscan Neuroscience in the Genomic Era: From Gastropods to Cephalopods., Scripps, Florida, May,* 15 – 19, 2012a.

Polese G, Ulloa Severino FP, Winlow W, Di Cosmo A. "Anesthesia of Octopus vulgaris with the volatile general anesthetic isoflurane", *Cephalopod International Advisory Council Symposium, Majestic Palace Hotel | Florianópolis, SC, Brazil, 2012b*: 10.

Polese G, Troncone L. Bertapelle C. Di Cosmo A. "*Octopus vulgaris* as model of central and peripheral control of reproduction in molluscs", *17th ICCE, Universitade de Barcelona, Barcelona, Spain, July, 15 – 19, 2013.*

Pronk R, Wilson DR, Harcourt R. 2010. Video playback demonstrates episodic personality in the gloomy octopus. *Journal of Experimental Biology* 213: 1035-41.

Smith JA, Andrews PLR, Hawkins P, Louhimies S, Ponte G, Dickel L. 2013. Cephalopod research and EU Directive 2010/63/EU: Requirements, impacts and ethical review. *Journal of Experimental Marine Biology and Ecology* 447: 31-45.

Sumbre G, Gutfreund Y, Fiorito G, Flash T, Hochner B. 2001. Control of octopus arm extension by a peripheral motor program. *Science* 293: 1845-48.

Suzuki H, Muraoka T, Yamamoto T. 2003. Localization of corticotropin-releasing factor-immunoreactive nervous tissue and colocalization with neuropeptide Y-like substance in the optic lobe and peduncle complex of the octopus (Octopus vulgaris). *Cell and Tissue Research* 313: 129-38.

Suzuki H, Yamamoto T, Inenaga M, Uemura H. 2000. Galanin-immunoreactive neuronal system and colocalization with serotonin in the optic lobe and peduncle complex of the octopus (Octopus vulgaris). *Brain Research* 865: 168-76.

Suzuki H, Yamamoto T, Nakagawa M, Uemura H. 2002. Neuropeptide Y-immunoreactive neuronal system and colocalization with FMRFamide in the optic lobe and peduncle complex of the octopus (Octopus vulgaris). *Cell and Tissue Research* 307: 255-64.

Troncone L, De Lisa E, Porcellini A, Laccetti P, Polese G, Di Cosmo A. 2013. Morphofunctional characterization and antibacterial activity of haemocytes from Octopus vulgaris. *Journal of Natural History* in press: 19.

Vitti JJ, Cho MK, Tishkoff SA, Sabeti PC. 2012. Human evolutionary genomics: ethical and interpretive issues. *Trends in Genetics* 28: 137-45.

von Kölliker RA. 1844. *Entwicklungsgeschichte der Tintenfische.* German.

Walderon MD, Nolt KJ, Haas RE, Prosser KN, Holm JB, et al. 2011. Distance chemoreceptio and the detectionof conspecifics in *Octopus bimaculoider*. *Journal of Molluscan Studies* 77: 309-11.

Wells MJ. 1962. *Brain and behaviour in cephalopods.* 171-pp. viii + 71 pp.

Wells MJ, Wells J. 1959. Hormonal control of sexual maturity in octopus. *Journal of Experimental Biology* 36: 1-32.

Williamson R, Chrachri A. 2004. Cephalopod neural networks. *Neurosignals* 13: 87-98.

Woodhams PL and Messenger JB. 1974. A note on the ultrastructure of the *Octopus olfactory* organ. Cell Tissue Res 152:253-8.

Young JZ. 1939. Partial degeneration of the nerve supply of the adrenal. A study in autonomic innervation. *Journal of Anatomy* 73: 540-U5.

Young JZ. 1971. The Anathomy of the Nervous System of *Octopus vulgaris.* Oxford University Press, London.

Zatylny C, Gagnon J, Boucaud-Camou E, Henry J. 2000. ILME: A waterborne pheromonal peptide released by the eggs of Sepia officinalis. *Biochemical and Biophysical Research Communications* 275: 217-22.

Zullo L, Hochner B. 2011. A new perspective on the organization of an invertebrate brain. *Communicative & integrative biology* 4: 26-9.

Part 3. Evolutionary Conservation and Plasticity

In: Neuroecology and Neuroethology in Molluscs
Editors: Anna Di Cosmo and William Winlow
ISBN: 978-1-62948-970-4

Chapter 7

The Neuroendocrine-Immune Orchestra in Molluscs

Enzo Ottaviani*
Department of Biology, University of
Modena and Reggio Emilia, Modena, Italy

Abstract

The present chapter focuses on information gained in molluscs demonstrating that immune and neuroendocrine systems share a common evolutionary origin. It appears that Nature followed an economic conservative strategy based on the systemic re-utilization of a common pool of signal molecules (proopiomelanocortin-derived peptides, corticotrophin releasing-hormone, cytokines) which have been highly conserved throughout evolution.

Keywords: Molluscs; immune-neuroendocrine functions

Introduction

Numerous experiments mainly performed in mammals have shown that the immune and neuroendocrine systems are deeply connected and interrelated (Blalock, 1989). At the same time evidence was also found in other vertebrate taxa (Ottaviani et al., 1992, 1995; see for review Ottaviani et al., 1997).

However, few data are available on the evolutionary origin of the cooperation between the immune and neuroendocrine systems. In this regard a significant contribution was

* Corresponding author: Prof. Enzo Ottaviani, Department of Biology, University of Modena and Reggio Emilia, Via Campi 231/D, 41125 Modena, Italy. Tel: +39.059.2055536, Fax: +39.059.2055548, E-mail: enzo.ottaviani @unimore.it.

provided by the reported findings in invertebrates (see for reviews Ottaviani and Franceschi, 1997; Ottaviani et al., 2007; Adamo, 2008).

Biologists have always been fascinated by the extraordinary variety of living forms. Even Lucretius wondered what lay behind this incredible variety of sizes and shapes ("Nam cur tam varie res possent esse requiro"). This question stimulated philosophers and scientists to search for unifying laws and principles, and in the case of biology the unifying principle was provided by the cell. All living organisms are made up of cells that work in basically the same way, at least for the main functions and properties, such as the production of protein and energy, cell proliferation, etc. How can the unity of the living organisms and the variety of forms, species and breeds be reconciled? The result has been obtained through evolution as a consequence of extremely parsimonious strategies. In the present chapter the immune and neuroendocrine responses in molluscs will be examined, showing the existence of a combinatorial strategy where the systematic use and re-use of a small group of signalling molecules is shared by the immune and neuroendocrine systems for different purposes and functions from the beginning of life.

Which Molecules Are Involved in the Immune and Neuroendocrine Responses?

By using different methodological approaches, the presence of proopiomelanocortin (POMC)-derived peptides such as adrenocorticotrophic hormone (ACTH), α-melanocyte-stimulating hormone (α-MSH),β-endorphin, MET-enkephalin and corticotrophin releasing-hormone (CRH) (Ottaviani et al., 1990, 1998; Stefano et al., 1999) were demonstrated in molluscs. Although there is no clear evidence for POMC genes in invertebrates, with a few exceptions (Salzet et al., 1997; Stefano et al., 1999), indirect studies confirmed their presence. It was demonstrated by *in situ* hybridization that the molluscs *Planorbarius corneus* and *Viviparus ater*, and the bivalve *Mytilus galloprovincialis* contain POMC mRNA (Ottaviani et al., 1995). Furthermore, another group of molecules, i.e., cytokine-like molecules, are present in various molluscan species. Their functions in invertebrates closely overlap those found in vertebrates. Indeed, they affect immune functions such as cell motility, chemotaxis, phagocytosis and cytotoxicity, intervene in stress response, regulate programmed cell death, mediate wound repair and show pleiotropicity, functional redundancy and receptor promiscuity. The presence of interleukin (IL)-1α and β, IL-2, IL-6, IL-8, tumour necrosis factor (TNF)-α, transforming growth factor (TGF)-β1, platelet-derived growth factor (PDGF)-AB and neurotrophic factor have been previously described in detail (see for review Ottaviani et al., 2004).

Immune Functions

Cell Shape Changes

According to Manske and Bade (1994) cell shape changes are the expression of cell motility, while chemotaxis (see below) is related to cell migration.

The incubation of the hemolymph of *Mytilus galloprovincialis* with ACTH (1-24) provokes cell shape changes of immunocytes via the adenylate cyclase/cAMP/protein kinase A pathway, as well as the activation of protein kinase C (Sassi et al., 1998), while CRH requires the synergistic action of the two second messengers, cAMP and IP_3 (Malagoli et al., 2000). The conformational cell modifications were detected using computer-assisted microscopy image analysis and evaluated as a shape factor that represents the degree of deviation of the cell from a rounded (inactive state) to an ameboid form (active state). This value was determined by measurements of the cellular area and the perimeter and mathematically expressed by using the shape factor formula of the American Innovision Analysis System (Schön et al., 1991).

Cell motility is also induced by PDGF-AB, TGF-β1 and IL-8. However, these molecules trigger the cells using different transduction signalling pathways. PDGF-AB and TGF-β1 transduce extracellular signals along the phosphoinositide signaling pathway. The action of PDGF-AB is extracellular Ca^{2+}-independent, while TGF-β1 is Ca^{2+}-dependent (Ottaviani et al., 1997; Kletsas et al., 1998).

IL-8 induces cell shape changes via protein kinase A and C pathways. Incubation with IL-8 induced immunocyte conformational changes, with a maximum modification observed after 5 min, which was then maintained for the rest of the experiment. The cell shape changes involved a redistribution of actin and phalloidin positivity was not uniformly distributed at the cell periphery, but concentrated in areas where the cells contacted the substrate (Ottaviani et al., 2000).

Chemotaxis

Chemotaxis, i.e., non random locomotion, is an important cell function which allows the cell to move towards a chemoattractant which has to be recognized. The transduction of the chemotactic signal is linked to complex molecular events involving the cytoskeleton and the cell moves in the direction of the concentration gradient of the chemotactic stimulus being able to distinguish the differences in concentration between two different points in space.

This function, being so important for cell survival is very well preserved and conserved throughout evolution, and integrated into more sophisticated types of responses which appeared later in evolution, such as acquired immunity mediated by antibodies.

The whole ACTH molecule (1-39) and its fragments (1-24), (1-4), (4-9), (1-13), (11-24) (Genedani et al., 1994b), CRH (Genedani et al., 1994) and the whole endorphin molecule (1-31) and its fragments (6-31), (8-31), (18-31), (1-17), (2-17) and acetyl-endorphin (1-31) (Genedani et al., 1994a) affect the molluscan immunocyte migration. A similar behaviour was also detected for cytokines, such as IL-1α, TNF-α, IL-8, PDGF and TGF-β1 (see for review Ottaviani et al., 2004; Steelman and Connors, 2009).

Phagocytosis

Phagocytosis is the first and vital defense mechanism of all living organisms. This phenomenon was observed for the first time in the larvae of starfish by Metchnikoff in 1882 (Tan and Dee, 2009).

The tests performed on molluscan immunocytes in the presence of POMC-derived peptides, CRH (Ottaviani et al., 1991, 1994), and different cytokines (Ottaviani et al., 1995, 1997) revealed a significant increase of phagocytic activity with respect to control immunocytes. However, not all the tested molecules influenced phagocytosis. For instance, while the whole endorphin molecule and its fragments exerted a chemotactic activity they did not affect the phagocytic activity (Ottaviani et al., 1994). Overall, the findings on chemotaxis and phagocytosis reveal that the chemotactic action as well as the phagocytic activity of the above reported molecules are not clear. Indeed, some effects are species-specific, while others are dose-correlated. However, the effect on phagocytosis is more uniform, which makes the general assumption that chemotaxis and phagocytosis are correlated, untrue for invertebrates. For instance, IL-2 is able to increase the phagocytic activity, but is ineffective on cell migration (Ottaviani et al., 1995). The discrepancies observed may be attributed to several factors, also including the possibility that the cell may respond to the same stimulus on the basis of their contingent status (Malagoli and Ottaviani, 2010).

Cytotoxicity

Spontaneous cytotoxicity was observed in different invertebrate species starting with sponges (Ratcliffe et al., 1985). This activity was tested against xenogenic tumor cells, xenogenic erythrocytes and allogenic cells (Cooper et al., 1996). A natural killer-like activity was demonstrated in the mollusc *P. corneus* (Franceschi et al., 1991). The *P. corneus* cells that exert this activity, called round hemocytes (RH), show a morphology similar to that of vertebrate lymphocytes and participate in allo- and xenografts (see for review Ottaviani, 1992). The RH are able to exert cytotoxic activity on a K562 cell line, in the short term (4h) ^{51}Cr release test. Wittke and Renwrantz (1984) found that intact and viable hemocytes from *Mytilus edulis* are able to produce cytotoxic substances which cause lysis of human erythrocytes. Recently, a new member of the tambjamine family has been isolated in the mollusc *Tambja ceutae* and it shows a remarkable and concentration-dependent cytotoxic activity against both tumor and non-tumor mammalian cells (Carbone et al., 2010).

Stress Response

POMC-derived peptides together with CRH are the central mediators of the stress response. In mammals the main circuit involves the hypothalamic-pituitary-adrenal gland axis. From an evolutionary point of view this complex type of response raises some problems. Indeed, the above-mentioned organs are not present in invertebrates, so that we surmize and confirm that invertebrate immunocytes may be functionally described as neuroendocrine cells, and as observed in vertebrates they are capable of performing stress and immune responses, simultaneously (see for review Ottaviani et al., 1991; Ottaviani, 2011). In this context, we found that CRH and ACTH, when added to the molluscan hemolymph provoked an increased concentration of biogenic amines (epinephrine, norepinephrine and dopamine) in the hemolymph and a concomitant decrease of biogenic amines in immunocytes (see for review Ottaviani and Franceschi, 1996), suggesting that a release of biogenic amines from immunocytes occurred. It should be stressed that the release of biogenic amines provoked by CRH is mainly mediated by the release of endogenous ACTH (Ottaviani et al., 1992). From these findings a simplified type of stress response appears to emerge early in evolution, and basic mechanisms are likely to be very well-conserved. Indeed, the key mediator molecules are the same, i.e., CRH and ACTH, and act the same order, i.e., CRH →ACTH→biogenic amines. All these events occur within the immunocyte.

On the whole, it emerges that the immunocyte is involved in immune and neuroendocrine functions, being able to recognize a variety of stimuli and set up responses in which immune and neuroendocrine elements are intermixed, representing a model to unravel the predicted common origin of these integrated systems (see for review Ottaviani et al., 2007). Indeed, the immunocyte contains different bioactive neurotransmitters and hormones (Ottaviani and Cossarizza, 1990; see for review Ottaviani et al., 2007). Blalock and Smith (1985) proposed that the mammalian lymphocyte, which is a neuroendocrine cell with receptors for neuropeptides and hormones, including POMC products (Wiegent and Blalock, 1987), should be considered to be a “mobile-immune cell”. This term should also be extended to the immunocyte (Ottaviani et al., 1991).

Cytokines are essential mediators of the immune system, but they are also involved in neuroendocrine responses, and the first studies in vertebrates reported that IL-1 provoked the release of CRH (Sapolsky et al., 1987) and the activation of the hypothalamic-pituitary-adrenal axis (Imura et al., 1991; Besedovsky and del Rey, 1996). From these data it is not unexpected to find that cytokines also play a role in stress response in invertebrates. Using molluscan immunocytes as model, we found that human IL-2 not only by itself induced the release of biogenic amines from molluscan cells, but also challenged CRH in this function, suggesting that human CRH and IL-2 may compete for a common receptor on the immunocyte membrane. Similar competition was also observed for IL1-α, IL-1β, TNF-α and TNF-β receptors. The release of biogenic amines in molluscs was also provoked by PDGF-AB and TGF-β1, the former inhibiting and the latter stimulating this effect (Ottaviani, et al., 1995; see for review Ottaviani et al., 2004).

Conclusion

Traditionally the study of evolution is presented as the study of the changes developed in taxa and in species during time. However, the studies presented here on the signaling molecules (i.e., neuropeptides, cytokines) as well as the large amount of molecular data now available on the major protein superfamilies with different activities, reveal that these primary signaling molecules are exceptionally well-conserved. This indicates that despite the progressive modification of organs and systems, the mechanisms which govern the exchange of information between cells have been essentially conserved. Consequently, in complex organisms such as vertebrates, information is exchanged between the various cell types by means of the same pool of signaling molecules and in the same way as in apparently much less complex animals, i.e., molluscs (Figure 1). Even in this case it seems to be true as stated by Lucretius: "what happens with the words and verbs is also true for the constituents of matter: it is the combinatorial changes of their movement, order and position which explains the variety of beings and species".

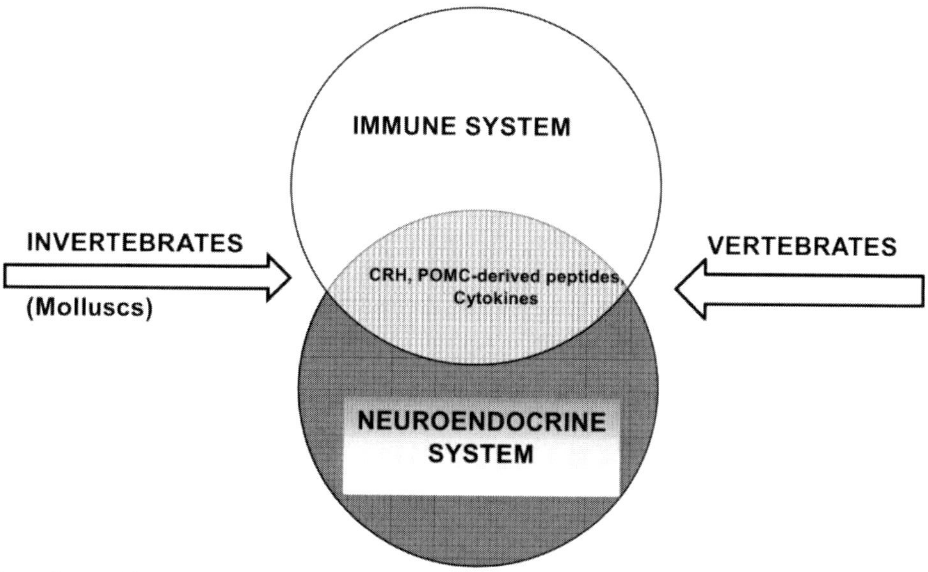

CRH, corticotrophin releasing-hormone; POMC, proopiomelanocortin.

Figure 1. A common pool of molecules is shared in immune-neuroendocrine responses from invertebrates to vertebrates.

Acknowledgments

This work was supported by MIUR (Italy) grants.

References

Adamo, S. A. Norepinephrine and octopamine: linking stress and immune function across phyla. *Inv. Surv. J.*, 2008, 5, 12-19.

Besedovsky, H. O., del Rey, A. Immune-neuro-endocrine interactions: facts and hypotheses. *Endocr. Rev.*, 1996, 17, 64-102.

Blalock, J. E. A molecular basis for bidirectional communication between the immune and neuroendocrine systems. *Physiol. Rev.*, 1989, 69, 1-32.

Blalock, J. E., Smith, E. M. The immune system: Our mobile brain? *Immunol. Today*, 1985, 6, 115-117.

Carbone, M., Irace, C., Costagliola, F., Castelluccio, F., Villani, G., Calado, G., Padula, V., Cimino, G., Lucas Cervera, J., Santamaria, R., Gavagnin, M. A new cytotoxic tambjamine alkaloid from the Azorean nudibranch *Tambja ceutae*. *Bioorg. Med. Chem. Lett.*, 2010, 20, 2668-2670.

Cooper, E. L., Suzuki, M. M., Cossarizza, A., Franceschi, C. Cytotoxic reactions in invertebrates: The earthworm model. In: *New directions in invertebrate immunobiology*, Söderhäll, K., Vasta, G., Iwanaga, S. (eds), Plenum, New York, 1996.

Franceschi, C., Cossarizza, A., Monti, D., Ottaviani, E. Cytotoxicity and immunocyte markers in cells from the freshwater snail *Planorbarius corneus* (L.) (*Gastropoda pulmonata*): implications for the evolution of natural killer cells. *Eur. J. Immunol.*, 1991, 21, 489-493.

Genedani, S., Bernardi, M., Ottaviani, E., Franceschi, C. Influence of endorphins on the migration of molluscan hemocytes. *Comp. Biochem. Physiol.*, 1994a, 15, 79-81.

Genedani, S., Bernardi, M., Ottaviani, E., Franceschi, C., Leung, M. K., Stefano, G. B. Differential modulation of invertebrate hemocyte motility by CRF, ACTH, and its fragments. *Peptides*, 1994b, 107C, 203-206.

Imura, H., Fukata, J., Mori, T. Cytokines and endocrine function: an interaction between the immune and neuroendocrine systems. *Clin. Endocrinol.* (Oxf.), 1991, 35, 107-115.

Kletsas, D., Sassi, D., Franchini, A., Ottaviani, E. PDGF and TGF-beta induce cell shape changes in invertebrate immunocytes via specific cell surface receptors. *Eur. J. Cell Biol.*, 1998, 75, 362-366.

Malagoli, D., Franchini, A., Ottaviani, E. Synergistic role of cAMP and IP_3 in corticotropin-releasing hormone-induced cell shape changes in invertebrate immunocytes. *Peptides*, 2000, 21, 175-182.

Malagoli, D., Ottaviani, E. Discrepant effects of mammalian factors on molluscan cell motility, chemotaxis and phagocytosis: divergent evolution or finely tuned contingency? *Cell Biol. Int.*, 2010, 34, 1091-1094.

Manske, M., Bade, E. G. Growth factor-induced cell migration: biology and methods of analysis. *Int. Rev. Cytol.*, 1994, 155, 49-96.

Ottaviani, E. Immunorecognition in the gastropod molluscs with particular reference to the freshwater snail *Planorbarius corneus* (L.) (Gastropoda, Pulmonata). *Boll. Zool.*, 1992, 59, 129-139.

Ottaviani, E. Immunocyte: the invertebrate counterpart of the vertebrate macrophage. *Inv. Surv. J.*, 2011, 8, 1-4.

Ottaviani, E., Capriglione, T., Franceschi, C. Invertebrate and vertebrate immune cells express pro-opiomelanocortin (POMC) mRNA. *Brain Behav. Immun.*, 1995c, 9, 1-8.

Ottaviani, E., Caselgrandi, E., Bondi, M., Cossarizza, A., Monti, D., Franceschi, C. The "immune-mobile brain": evolutionary evidence. *Adv. Neuroimmunol.*, 1991a, 1, 27-39.

Ottaviani, E., Caselgrandi, E., Franceschi, C. Cytokines and evolution: In vitro effects of IL-1α, IL-1β, TNF-α and TNF-β on an ancestral type of stress response. *Biochem. Biophys. Res. Commun.*, 1995a, 207, 288-292.

Ottaviani, E., Caselgrandi, E., Petraglia, F., Franceschi, C. Stress response in the freshwater snail *Planorbarius corneus* (L.) (Gastropoda, Pulmonata): Interaction between CRF, ACTH and biogenic amines. *Gen. Comp. Endocrinol.*, 1992a, 87, 354-360.

Ottaviani, E., Cossarizza, A. Immunocytochemical evidence of vertebrate bioactive peptide-like molecules in the immuno cell types of the freshwater snail *Planorbarius corneus* (L.) (Gastropoda, Pulmonata). *FEBS Lett.*,1990, 267, 250-252.

Ottaviani, E., Cossarizza, A., Ortolani, C., Monti, D., Franceschi, C. ACTH-like molecules in gastropod molluscs: a possible role in ancestral immune response and stress. *Proc. R. Soc. Lond. B*, 1991b, 245, 215-218.

Ottaviani, E., Franceschi, C. The neuroimmunology of the stress response from invertebrates to man. *Prog. Neurobiol.*, 1996, 48, 421-440.

Ottaviani, E., Franceschi, F. The invertebrate phagocytic immunocyte: clues to a common evolution of immune and neuroendocrine systems. *Immunol. Today*, 1997, 18, 169-174.

Ottaviani, E., Franchini, A., Cassanelli, S., Genedani, S. Cytokines and molluscan immune responses. *Biol. Cell*, 1995b, 85, 87-91.

Ottaviani, E., Franchini, A., Fontanili, P. The effect of corticotropin-releasing factor and pro-opiomelanocortin-derived peptides on the phagocytosis of molluscan hemocytes. *Experientia*, 1994, 50, 837-839.

Ottaviani, E., Franchini, A., Franceschi, C. Presence of immunoreactive molecules to CRH and cortisol in invertebrate haemocytes and lower and higher vertebrate thymus. *Histochem. J.,* 1998, 30, 61-67.

Ottaviani, E., Franchini, A., Kletsas, D., Bernardi, M., Genedani, S. Involvement of PDGF and TGF-β in cell migration and phagocytosis in invertebrate and human immunocytes. *Anim. Biol.*, 1997a, 6, 95-99.

Ottaviani, E., Franchini, A., Malagoli, D., Genedani, S. Immunomodulation by recombinant human interleukin-8 and its signal transduction pathways in invertebrate hemocytes. *Cell. Mol. Life Sci.*, 2000, 57, 506-513.

Ottaviani, E., Malagoli, D., Franceschi, C. Common evolutionary origin of the immune and neuroendocrine systems: from morphological and functional evidences to in silico approaches. *Trends Immunol.*, 2007; 28, 497-502.

Ottaviani, E., Malagoli, D., Franchini, A. Invertebrate humoral factors: cytokines as mediators of cell survival. *Prog. Mol. Subcell. Biol.*, 2004, 34, 1-25.

Ottaviani, E., Petraglia, F., Montagnani, G., Cossarizza, A., Monti, D., Franceschi, C. Presence of ACTH and beta-endorphin immunoreactive molecules in the freshwater snail *Planorbarius corneus* (L.) (Gastropoda, Pulmonata) and their possible role in phagocytosis. *Regul. Pept.*, 1990, 27, 1-9.

Ottaviani, E., Sassi, D., Kletsas, D. PDGF- and TGF-beta-induced changes in cell shape of invertebrate immunocytes: effect of calcium entry blockers. *Eur. J. Cell Biol.*, 1997b, 74, 336-341.

Ottaviani, E., Trevisan, P., Pederzoli, A. Immunocytochemical evidence for ACTH- and β-endorphin-like molecules in phagocytic blood cells of urodelan amphibians. *Peptides*, 1992b, 13, 227-231.

Ratcliffe, N. A., Rowley, A. F., Fitzgerald, S. W., Rhodes, C. P. Invertebrate immunity: Basic concepts and recent advances. *Int. Rev. Cytol.*, 1985, 97, 183-350.

Salzet, M., Salzet-Raveillon, B., Cocquerelle, C., Verger-Bocquet, M., Pryor, S. C., Rialas, C. M., Laurent, V., Stefano, G. B. Leech immunocytes contain proopiomelanocortin: nitric oxide mediates hemolymph proopiomelanocortin processing. *J. Immunol.*, 1997, 159, 5400-5411.

Sapolsky, R., Rivier, C., Yamamoto, G., Plotsky, P., Vale, W. Interleukin-1 stimulates the secretion of hypothalamic corticotropin-releasing factor. *Science*, 1987, 238, 522-524.

Sassi, D., Kletsas, D., Ottaviani, E. Interactions of signaling pathways in ACTH (1-24)-induced cell shape changes in invertebrate immunocytes. *Peptides*, 1998, 19, 1105-1110.

Schön, J. C., Torre-Bueno, J., Stefano, G. B. Microscopic computer-assisted analysis of conformational state: reference to neuroimmunology. *Adv. Neuroimmunol.,* 1991, 1, 252-259.

Steelman, B. N., Connors, V. A. Chemokinetic effect of interleukin-1 beta on cultured *Biomphalaria glabrata* embryonic cells. *J. Parasitol.*, 2009, 95, 772-774.

Stefano, G. B., Salzet-Raveillon, B., Salzet, M. *Mytilus edulis* hemolymph contains pro-opiomelanocortin: LPS and morphine stimulate differential processing. *Brain Res. Mol. Brain Res.*, 1999, 63, 340-350.

Tan, S. Y., Dee, M. K. Elie Metchnikoff (1845-1916): discoverer of phagocytosis. *Singapore Med. J.*, 2009, 50, 456-457.

Weigent, D. A., Blalock, J. E. Interactions between the neuroendocrine and immune systems: common hormones and receptors. *Immunol. Rev.*, 1987, 100, 79-108.

Wittke, M., Renwrantz, L. Quantification of cytotoxic hemocytes of Mytilus edulis using a cytotoxicity assay in agar. *J. Invertebr. Pathol.*, 1984, 43, 248-253.

In: Neuroecology and Neuroethology in Molluscs
Editors: Anna Di Cosmo and William Winlow
ISBN: 978-1-62948-970-4

Chapter 8

A Neuroplastic Network Underlying Behaviour and Seasonal Change in *Lymnaea stagnalis*: A Neuroecological Standpoint

William Winlow*[*1,2,3] and Gianluca Polese[1]
[1]Departimento di Biologia, Università degli Studi di Napoli, Federico II, Napoli, Italia
[2]School of Pharmacy and Biomedical Sciences, University of Central Lancashire, Preston, Lancashire, UK
[3]Honorary Research Fellow, Department of Veterinary Preclinical Sciences, University of Liverpool, UK

Abstract

The gastropod mollusc *Lymnaea stagnalis* (L.) is an advantageous model system for neureothological studies and is a suitable model for studying Neuroecology. Here we describe its well organised nervous system many of whose ganglia have been mapped, both electrophysiologically and morphologically, to characterise individual giant neurons and neuronal clusters. This has allowed us to speculate on the evolution and reorganisation of the brain in relation to torsion and detorsion and to show that many behavioural networks remain virtually intact during these processes, but reflect the evolutionary changes that the nervous system has undergone in adaptation of the animal to its environment.

We also describe the *Lymnaea* behavioural hierarchy in which the defensive whole body withdrawal reflex takes precedence over all other behaviours, all of which appear to be organised on a multiganglionic basis and underlain by a complex neural network in which several central pattern generators (CPGs) are embedded. The interactions of the CPGs with one another and with a number of wide-acting synaptic inputs are discussed with particular reference to the locomotor and respiratory systems of the animal.

* Address for Correspondence: 32 Hill Crescent, Newton, Preston, PR4 3TR, Lancashire, UK, e-mail: bill@winlow.myzen.co.uk.

Knowledge of these cellular interactions has allowed us to demonstrate seasonal changes within the respiratory CPG (RCPG) of *Lymnaea* with a loss of connectivity as winter approaches and a restoration of synaptic connectivity and the behavioural programme during the spring.

With its flexible and adaptable behavioural repertoire *Lymnaea* is well adapted to its freshwater environment. We suggest that it will prove to be a useful and sensitive model for studying the neuroecological effects of climate change with particular respect to invasive freshwater species (IFS) such as predators, freshwater plans or microbial species.

Introduction

"*A neural system is particularly likely to reflect the long and varied histories of its owner's ancestors and have all sorts of rather odd features*", Patrick Bateson (1987).

Integrated animal behaviour is the product of the whole nervous system. Thus the functions of systems of neurons and individual neurons may vary from time to time as circumstances change and different behaviour patterns emerge from complex neural networks. Morton and Chiel (1994) suggested that reorganisation of neural circuits is fundamental and widespread within nervous systems. Later, Chiel and Beer (1997) suggested that "*Changes in behavior over evolutionary time are associated with coordinated changes in both the periphery and the central nervous system*" so that co-evolution matches neural and peripheral properties. This has clearly been the case in pulmonate molluscs such as *Lymnaea stagnalis* (L.) and the effects of torsion on the nervous system are outlined below. More recently Briggman and Kristan (2008) reviewed multifunctional pattern generating circuits and demonstrated that some neurons can be active in multiple behaviours with major shifts in their activity patterns. What is more, many features of central pattern generator (CPG) output are probably generated by the properties of the individual neurons and by network interactions as is the case in the *Lymnaea* feeding CPG (FCPG) (Straub et al., 2002; Nikitin et al., 2013) supporting the view that motor networks are shifting coalitions of functional units whose ultimate configuration "depends on hormonal, modulatory and activity state, and the same neurons are configured in different ways by these forces for different functional roles" (Calabrese, 2007).

Lymnaea is very well-adapted to its aquatic environment and has been developed as a model system for neuroethological studies for three reasons. First, it has a clear hierarchy of adaptive behaviours (Winlow, 1989; Syed and Winlow, 1989; Winlow and Syed, 1991; Winlow, et al. 1992) in its wide behavioural repertoire, including feeding (Benjamin, 1983), whole-body withdrawal (Ferguson, 1984; Ferguson and Benjamin, 1991a,b; 1984; Benjamin et al. 1985), aversive behaviour (Kawai et al., 2004), locomotion (Winlow and Haydon 1986; Haydon and Winlow, 1986; Syed et al. 1988; Syed and Winlow, 1989b), respiration (van der Wilt et al. 1988; Syed et al. 1990) and learning (eg, Kemenes and Benjamin, 1989; Kemenes et al., 2002; Lukowiak et al., 1996; Nikitin et al., 2103). Second, it has a well developed central nervous system, or brain, comprised of a double ring of circumsoesophageal ganglia (Fig 1). Third, the central neurons and neuronal clusters are well defined and identifiable on the bases of colour, size and location (Mills & Winlow, 1979) and have been mapped in several studies, particularly those of the visceral and parietal ganglia (Fig 4) (Benjamin &

Winlow, 1981), the pedal ganglia (Fig 7) (Slade et al.,1981; Kyriakides et al., 1989) and the buccal ganglia (Benjamin, 1983; Benjamin, 2012) as well as a number of other neurones in specific locations (Fig 1) such as the paired cerebral giant cells (CGCs) (McCrohan & Benjamin, 1980) or the paired pleural B neurons (PlB) (Alania et al., 2004).

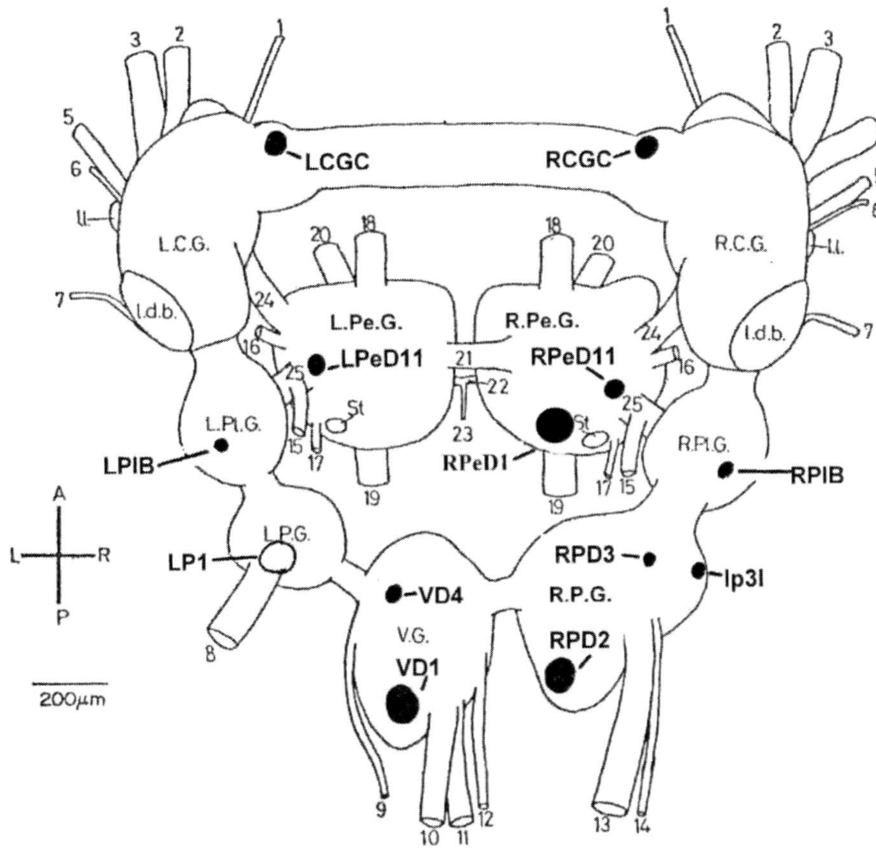

Figure 1. Dorsal view of the central nervous system of *Lymnaea stagnalis* (L.) with the exception of the paired buccal ganglia. The diagram shows locations of key identified neurons. Most are on the dorsal surface and are shown in black, while LP1 (left parietal 1) lies on the ventral surface of the left parietal ganglion and is only shown in outline. LCGC and RCGC, left and right cerebral giant cells; RPeD1, right pedal dorsal 1; LPeD11 and RPeD11 – left and right pedal dorsal 11; LPlB and RPlB, left and right pleural B cells; VD1 and VD4, visceral dorsal cells 1 and 4; RPD2 and RPD3, right parietal; dorsal cells 2 and 3; Ip3I, input 3 interneuron. A. anterior: P. posterior; L, left; R, right; L.C.G. and R.C.G. left and right cerebral ganglia; L.Pe.G. and R.Pe.G. left and right pedal ganglia; L.Pl.G. and R.Pl.G. left and right pleural ganglia; L.P.G. and R.P.G. left and right parietal ganglia: V.G. median visceral ganglion; 1.d.b. lateral dorsal body; 1.1. lateral lobe; St statocyst. (1) cerebro-buccal connective; (2) superior labial nerve; (3) median labial nerve; (4) penis nerve; (5) tentacle nerve; (6) optic nerve; (7) nuchal nerve; (8) left parietal nerve; (9) cutaneous pallial nerve; (10) intestinal nerve; (11) anal nerve; (12) genital nerve; (13) right internal parietal nerve; (14) right external parietal nerve; (15) inferior cervical nerve; (16) superior cervical nerve; (17) columellar nerve; (18) superior pedal nerve; (19) inferior pedal nerve; (20) medial pedal nerve; (21) dorsal pedal commissure; (22) ventral pedal commissure; (23) medial columellar nerve; (24) cerebro-pedal connective: (25) pedal-pleural connective.(Modified from Slade et al., 1981).

In this chapter we will first consider how the nervous system has developed into its asymmetrical form due to the process of torsion. We will then discuss some of the interactions within the complex multiganglionic network from which *Lymnaea* behaviour emerges. In *Lymnaea*, respiratory behaviour is driven by the respiratory central pattern generator (RCPG) which is embedded in multiganglionic network from which several behaviours have been shown to emerge, a form of distributed processing, and through which *Lymnaea* interacts both with its internal and external environment in an integrated manner. Finally, we will consider how the animal reacts to seasonal changes using as our example the respiratory system.

The Central Nervous System of *Lymnaea Stagnalis*

The brain of *Lymnaea* consists of a double ring of 9 ganglia (Fig. 1) and an anteriorly situated pair of buccal ganglia. The upper ring consists of the paired cerebral ganglia, paired pleural ganglia, paired parietal ganglia (though left and right ganglia are of unequal size) and the single median visceral ganglion. The paired pedal ganglia lie ventral to this upper ring and are connected to it via the cerebro-pedal connectives and pedal pleural connectives. The pedal ganglia are connected to one another by the dorsal pedal commissure and the ventral pedal commissure.

The overall colour of the brain is orange due to pigmentation of most of the cell bodies with a yellow carotenoprotein and a red haemoprotein (Benjamin and Walker, 1972) although some are white or cream, lacking this pigmentation. Karnaukhov (1971) suggested that carotenoid pigments, due to their many unsaturated double bonds, may help to provide an intracellular stock of oxygen, which would be advantageous to a freshwater pulmonate in times of environmental stress, such as poorly oxygenated water or the presence of surface predators. The cell bodies lie on the outer surfaces of the ganglia are mostly arranged in contralaterally paired clusters or as pairs of giant neurons with the exception of those in the medial visceral ganglion, but this and the left and right parietal ganglia may have been modified in evolution by the process of torsion (see below).

Evolution of the Brain of *Lymnaea*

Torsion in Gastropod molluscs. The major classes of molluscs appear to be derived from an archetypal bilaterally symmetrical "proto-mollusc" (Kandel, 1979). However, during evolution the gastropod body has become twisted due to the process of torsion during development (Chase 2002). The evolutionary pressures for torsion are open to conjecture, but it is commonly believed that torted conditions benefitted the larvae of pelagic ancestral forms, allowing the head to retract followed by the foot which closed the shell aperture, but of course not all untorted molluscs are pelagic (Chase, 2002). Another argument is that the gills were better aerated by being placed anteriorly (Morton, 1958, 1979). Furthermore, the osphradium which is adjacent to the pneumostome is also brought into an anterior position, where it has become part of the "anteroceptive sensory system" (Morton, 1958) sampling the environment into which the animal is progressing. This is important as the osphradium is now known to contain hypoxia sensitive peripheral chemoreceptor cells (PCRCs) which elicit activity in a neuron of the RCPG which is known to initiate ventilation (Bell, et al., 2007, 2008). In addition other osphradial cells have been shown to be sensitive to elevated PCO_2 and to three classes of odorants (Wedemeyer and Schild, 1995).

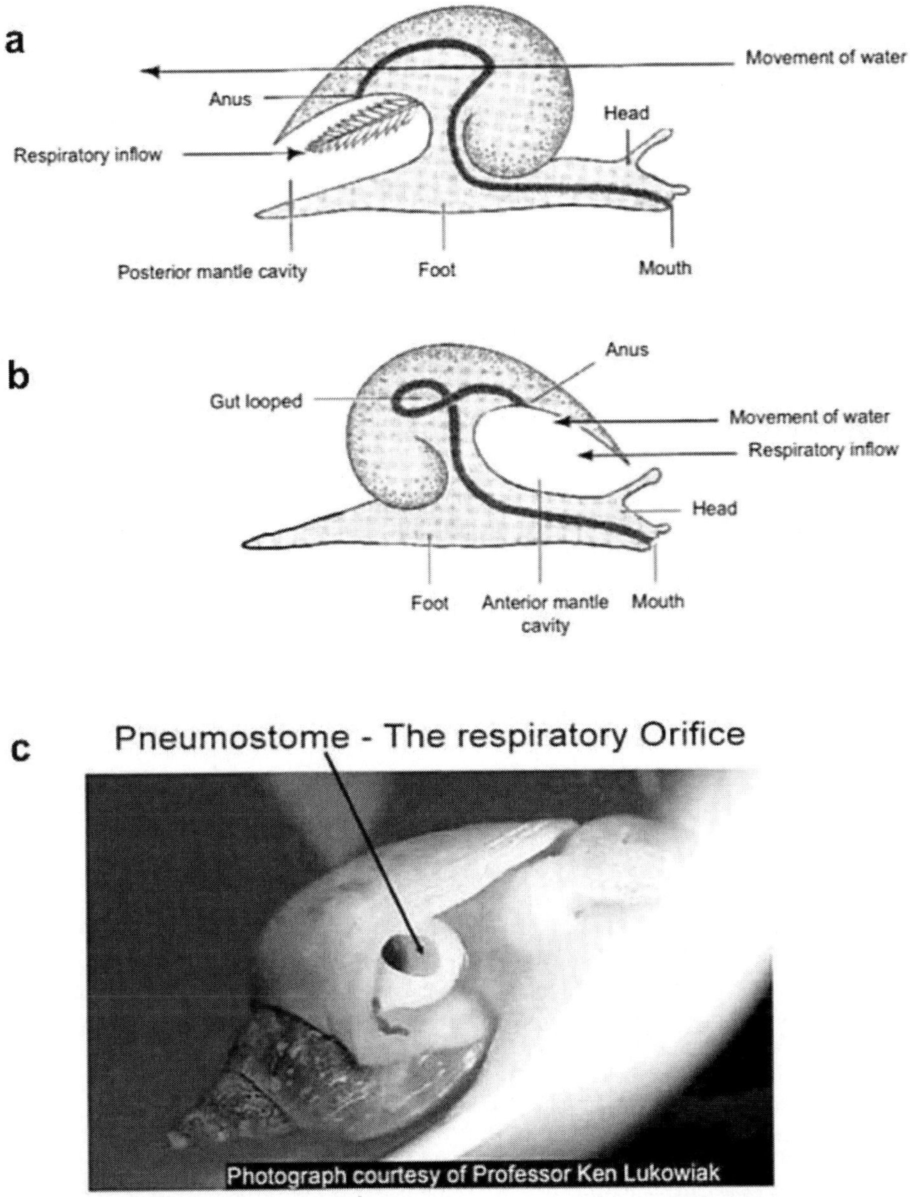

Figure 2. Evolution of pulmonates. a) Lateral view of a hypothetical untorted proto- gastropod with a posterior mantle cavity and gill. b) A similar view of a torted proto-pulmonate with the mantle cavity facing forward and modified to form a lung rather than a gill. The current due to the animal's locomotion augments the respiratory current. c) Lateral view of *Lymnaea* at the water surface showing the open pneumostome during a period of lung ventilation. (a & b after Borradaile and Potts, 1963; Kandel, 1979).

Torsion is unconnected with spiralling of the shell and according to Chase (2002) the fossil record indicates that early non-torted molluscs had coiled shells. At some point in the early Cambrian period (circa 550 million years ago), the first gastropods appeared in which the shell and mantle cavity were rotated 180° anti-clockwise towards the anterior end of the animal (Fig 2a,b) and the gills were replaced with a lung in pulmonates (Fig 2c). This had major effects on the relationships of the ganglia of the central nervous system (Fig 3a), which became twisted into a figure of eight (Fig 3b).

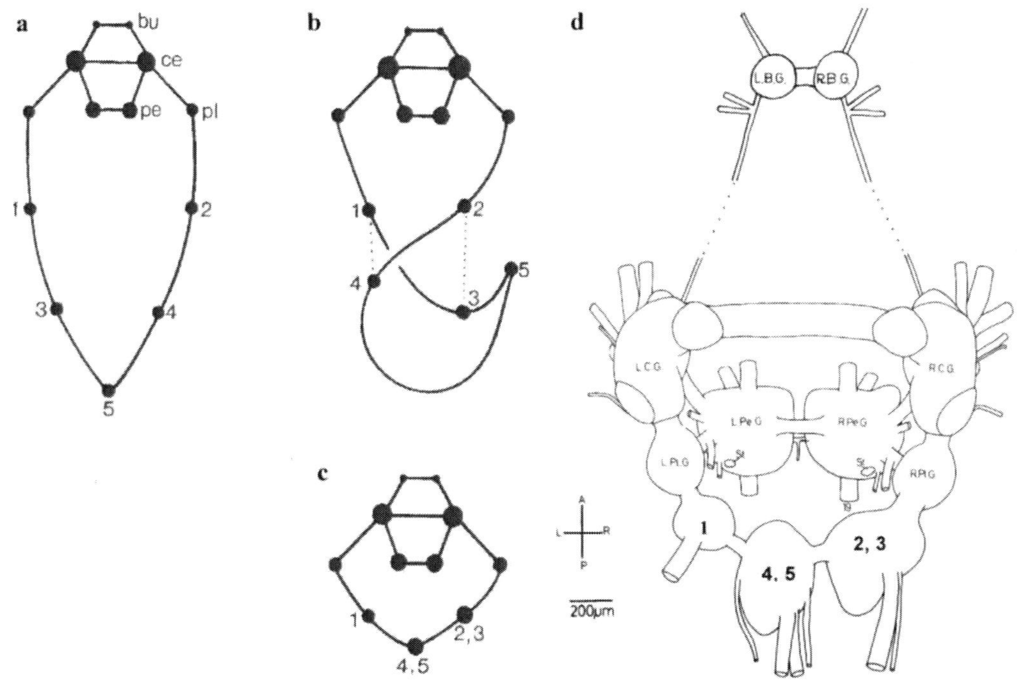

Figure 3. Evolution of the central nervous system of *Lymnaea.* a) Primitive untorted CNS; b) Post-torsion CNS of early gastropods; c) The torted nervous system of *Lymnaea* which appears to have been secondarily modified by the formation of zygoses (dotted lines, Fig 3b) between the left parietal ganglion (1, Fig 3a) and the supraintestinal ganglion (4) and between the right parietal ganglion (2) and the subintestinal ganglion (3). It is envisaged that this was followed by condensation of ganglia so that 2 and 3 become fused into the modern right parietal ganglion, which is paired with, but much larger than the left parietal ganglion (1 and see Fig 3d). The primitive supraintestinal (4) and visceral ganglia (5) then fused to form the modern visceral ganglion. Abbreviations: bu, buccal ganglia; ce, cerebral ganglia; pe, pedal ganglia; pl, pleural ganglia. d) Dorsal view of the central nervous system of *Lymnaea stagnalis*: A, anterior; P, posterior; L, left; R, right; BG, buccal ganglia; CG, cerebral ganglia; PeG, pedal ganglia; PlG, pleural ganglia; St, statocyst. (a to c after Kandel, 1979).

Evolution of the visceral and paired parietal ganglia of Lymnaea. Even in those species that remain torted (eg, basommatophoran pulmonates such as *Lymnaea*), the effects of torsion on the nervous system have become obscured over time due to condensation and fusion of ganglia (Fig 3d). McCrohan and Winlow (1985) have suggested that later in evolution the *Lymnaea* system was secondarily modified by the formation of zygoses (Fig 3b) between the left parietal ganglion (Fig 3a) and the supraintestinal ganglion and between the right parietal ganglion and the subintestinal ganglion and then by condensation of ganglia so that primitive right parietal ganglion fused with the primitive subintestinal ganglion and to become fused into the modern right parietal ganglion, which is paired with, but much larger than, the left parietal ganglion. The primitive supraintestinal and visceral ganglia then fused to form the modern visceral ganglion. The relative sizes of the modern parietal ganglia would support this theory, given that the left ganglion is much smaller than the right. Alternative explanations of parietal/visceral ganglionic fusion may be just as valid as has been suggested in the terrestrial pulmonate, *Achatina fulica* (Munoz, Pawson and Chase, 1983), but this may be accounted for by differences in the evolution of the two species, in accord with the ideas of Moroz (2009).

There is good electrophysiological evidence for the condensations of the ganglia that we have proposed above, based the receipt of common synaptic inputs by paired neurons and neural clusters and also on the appearance and properties of the neurons (Table 1). For example visceral B group neurons and parietal F group neurons appear to be paired clusters (Fig 4) with similar wide-acting synaptic inputs, action potential shapes and activity patterns (Benjamin and Winlow, 1981) while the giant neurons VD1 and RPD2 are strongly electrically coupled and receive common synaptic inputs. In addition the HIJK cells of the visceral ganglia share many common inputs with the A group of the right parietal ganglion and are known to be motor neurones in the ventilatory system (Syed et al., 1991 and see below for details). There are many other examples within the ganglia and the left and right parietal D group cells are clearly paired and are weakly electrically coupled to one another ipsilaterally and to neurons of other paired clusters as shown in Figure 5. Furthermore, LP1 (also known as LPaV1) and RPD3 both participate in local and whole body withdrawal (Moroz and Winlow, 1991; Inoue et al., 1996) and may be another functional pair of neurons, although they are very different in size, but this may reflect different evolutionary origins culminating in parallel functional evolution. Some cells and cell groups in the visceral ganglion are unpaired and are assumed to have originated from the primitive unpaired visceral ganglion. Thus, the present morphology (Fig 4) and electrophysiology of the visceral and parietal ganglia still partially reflects the paired origins of the parietal ganglia prior to torsion and subsequent ganglionic fusion (Table 1). In other words the original networks underlying behaviour remain largely intact, but reflect the evolutionary changes the nervous system has undergone in the adaptation of the animal to its environment.

Amphibious origins of basommatophoran pulmonates. During the course of evolution, it is believed that basommatophoran pulmonates such as *Lymnaea stagnalis* and *Chilina fluctuosa* invaded freshwater water habitats from estuarine niches (Brace, 1983) indicating that they may have been amphibious in origin rather than fully terrestrial. As compared with body plans shown in Fig 2a,b, the paired gills have been lost in *Lymnaea,* and replaced with a single lung closed by the erectile pneumostome (Fig 2c) suiting them to an amphibious lifestyle as they evolved. However, *Lymnaea* does not entirely depend on pulmonary ventilation, relying also on skin respiration when pO_2 levels are high in its environment thereby reducing the need for frequent trips to the water surface. Thus *Lymnaea*, whatever its evolutionary origin is well adapted to its freshwater environment.

Table 1. Derivation of the Modern Visceral and Parietal Ganglia in *Lymnaea*

Presumed Primitive ganglion	Identified Neurons	Modern ganglion	Paired With:	Behavioural actions where known	Reference(s)
Subintestinal	VD1	Visceral ganglion	RPD2	Part of pneumostome control system Responds to changes in PO_2 and to mechanical stimulation near the pneumostome and in the lung cavity	Van der Wilt et al., 1987, Moroz, 1991 Janse et al., 1985
	F group		B group	Inhibition by mechanical stimulation of lung, excitation by stimulation around pneumostome. Increased firing frequency on exposure to N_2	Janse et al., 1985
	HIJK cells		A group	Pneumostome and mantle Motor neurons	Syed et al., 1991 Moroz, 1991
	VV1 VV2		RPV1 RPV2 RPV3	Unknown	Benjamin and Winlow, 1981
Subintestinal	RPD2	Medial Right Parietal ganglion	VD1	Part of pneumostome control system Responds to changes in PO_2 and to mechanical stimulation near the pneumostome and in the lung cavity	Van der Wilt et al., 1987, Moroz, 1991 Janse et al., 1985
	B group		F group	Inhibition by mechanical stimulation of lung, excitation by stimulation around pneumostome. Increased firing frequency on exposure to N_2	Janse et al., 1985
	A group		HIJK cells	Pneumostome and mantle Motor neurons	Syed et al., 1991 Moroz, 1991
	RPV1 RPV2 RPV3		VV1 VV2	Unknown	Benjamin and Winlow, 1981
Right Parietal	RPD1 RPD3	Lateral Right Parietal ganglion	LP1	Withdrawal – mechanosensory neuron	Moroz and Winlow, 1991
	D group		D group	Withdrawal motor neurons	Ferguson and Benjamin, 1991a
Left Parietal	LP1	Left Parietal ganglion	RPD3	Withdrawal – mechanosensory neuron	Inoue et al., 1996
	D group		D group	Withdrawal motor neurons	Ferguson and Benjamin, 1991a
Visceral	G group	Visceral ganglion	Unpaired	Excitation by mechanical stimuli around the pneumostome, but inhibition by stimulation in lung cavity. Increased firing frequency on exposure to N_2	Janse et al., 1985
	M group		Unpaired	May contain pneumostome closer motor neurons – based on position	Benjamin and Winlow, 1981 Moroz, 1991
	VD2 VD3		Unpaired Unpaired	Respond to mechanical stimulation of mantle, pneumostome and lung wall and to hypoxia	Janse et al., 1985
	VD4		Unpaired	Part of respiratory CPG	Syed et al., 1990, Syed and Winlow, 1991a

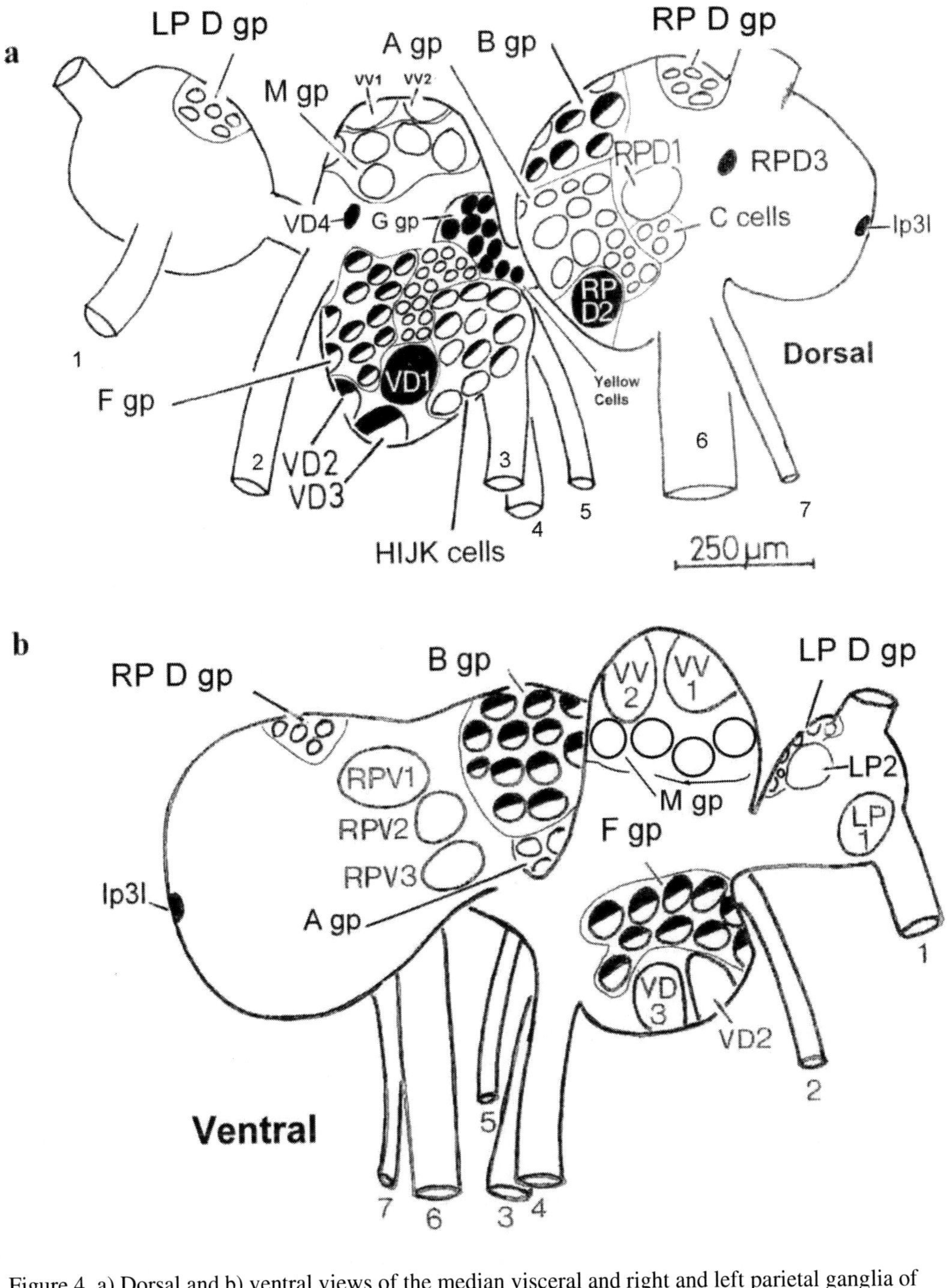

Figure 4. a) Dorsal and b) ventral views of the median visceral and right and left parietal ganglia of *Lymnaea* showing the positions of identified neurones and cell clusters. Shaded cells are white and unshaded cells are orange, with part shaded cells representing varying degrees of paleness of colour. Nerve trunks are as follows: 1 – left parietal nerve; 2 - cutaneous pallial nerve; 3 – intestinal nerve; 4 - anal nerve; 5- genital nerve; 6 - right internal parietal nerve; 7- right external parietal nerve.

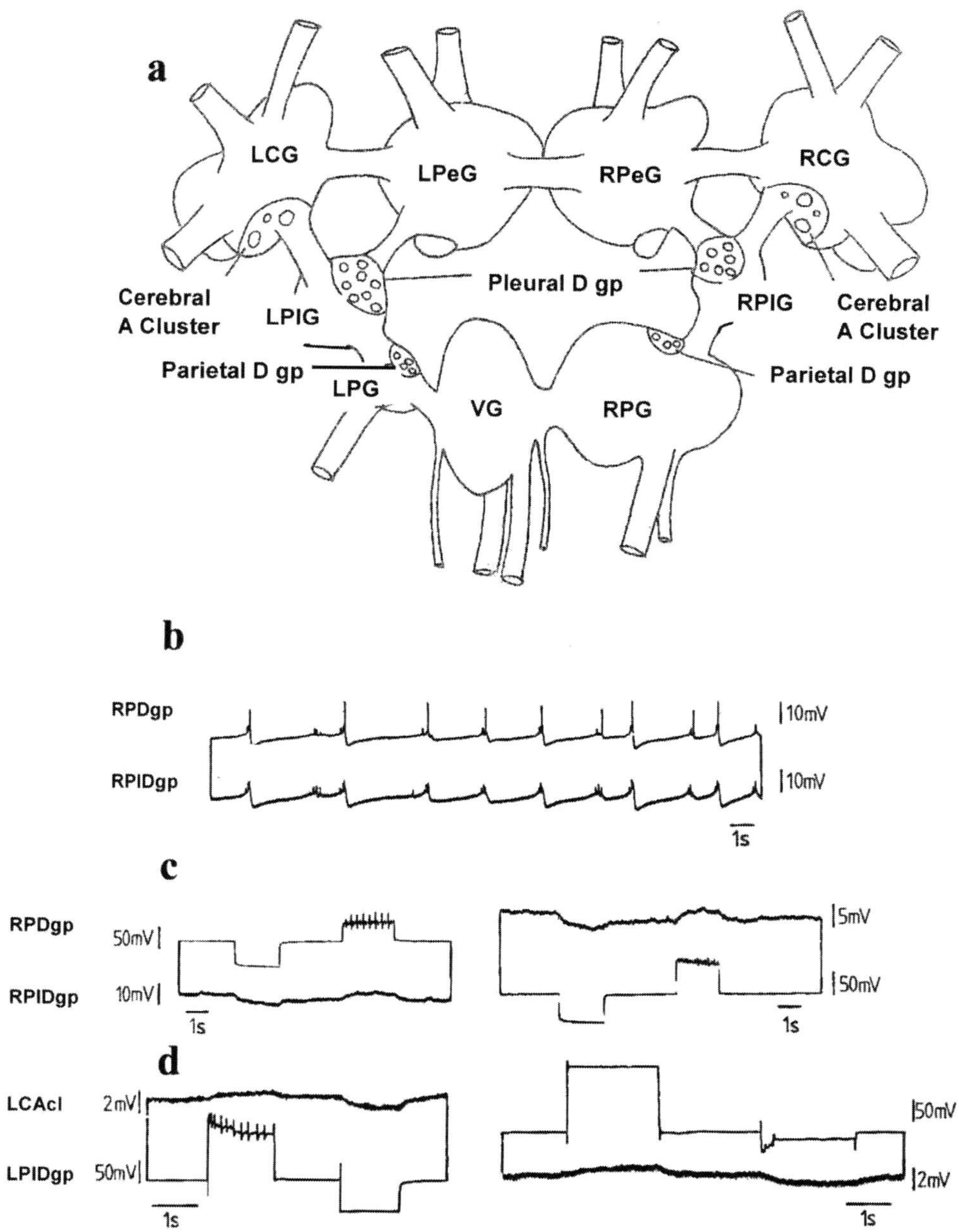

Figure 5. Symmetrically positioned groups of motor neurons that participate in the whole body withdrawal response. A) Location of the neuronal clusters. The cells within the clusters are ipsilaterally connected to one another. b) An electrically coupled right parietal D group (RPD) cell and a right pleural D group (RPlD). Note the similarity of spontaneous activity due in part to the weak electrical coupling demonstrated in c) by injection of depolarizing and hyperpolarizing current pulses first into the RPD group cell and then into the RPlD group cell. There are multiple amplitudes of action potentials in each of the cells due to multiple spike initiating sites in these multibranched neurons (Haydon, 1982; Haydon and Winlow, 1982). d) Weak electrical coupling between a left cerebral A cluster (LCAcl) cell and a left pleural D group (LPlD) cell, demonstrated by injection of depolarizing and hyperpolarizing current pulses first into the LPlD cluster cell and then into the LCA cluster cell. The coupling is observed to be weak since the cell bodies from which recordings were made are distant from the sites of the electrical connections. Abbreviations as in Figure 1.

Principles of Organisation of the *Lymnaea* Nervous System

Morphological and electrophysiological studies on the brain of *Lymnaea*, revealing the basic neural substrate from which behaviour emerges, have allowed the formulation of a set of general principles underlying motor control and coordination in the animal (McCrohan and Winlow, 1985). The brain is basically bilaterally symmetrical with bilaterally symmetrical pairs and clusters of neurones (e.g., Fig 5a) which are often synchronised by electrical coupling, receipt of common wide-acting synaptic inputs and by sharing of common patterns of endogenous activity (Fig 5b,c,d). Cyclical behaviours such as feeding or lung ventilation are driven by central pattern generators whose output may be modified by sensory inputs or neuromodulatory influences (Moroz, et al., 1993; Straub et al., 2007; Moccia et al., 2009). Motor neurons do not usually determine the periodicity of the motor discharge, but their endogenous properties can modify the basic rhythm (Straub et al., 2002) and may themselves be modified as part of learning and memory (Benjamin et al., 2008). The activities of CPGs may occur synchronously to produce complex behavioural sequences, as in feeding and lung ventilation or serial behaviours such as respiratory behaviour (see below).

Lymnaea Behaviour

To produce specific behaviours the nervous system interacts with the animal's environment and with its internal milieu via the sensory system. Behaviour is hierarchically organised, with the protective/defensive whole body withdrawal reflex (Ferguson and Benjamin, 1991a,b) at the top of the hierarchy (Fig 6), taking precedence over all other behaviours (Winlow et al. 1992): it is incompatible with arousal reactions and *vice versa*. The extent of withdrawal however depends on context and, as with vertebrate reflexes; it is not an all-or-none response and can be graded according to the type and intensity of the applied stimulus (Sunada, et al., 2010b). Behaviours such as respiration (Syed et al., 1990; Syed and Winlow, 1991a), feeding (Benjamin, 2012), control of neurosecretory cells (Benjamin, 1984), locomotion (Winlow and Haydon, 1986), withdrawal responses (Syed and Winlow, 1991b) and feeding (Kyriakides and McCrohan, 1988; Alania et al., 2004; 2008; Yeoman et al., 1994, 1996; Benjamin, 2012) are organised on a multiganglionic basis (Syed and Winlow, 1989a; Winlow and Syed, 1991) and the ganglia seem to be less autonomous than those of arthropods. Within this type of network individual neurons may be multifunctional (Haque et al., 2006; Benjamin, 2012). As a consequence many wide-acting synaptic inputs exist (Benjamin and Winlow, 1981), some of which have been localised to individual neurons and may participate in CPGs (Syed and Winlow, 1988, 1991a; Syed et al., 1990), although others are of unknown origin.

Lymnaea is capable learning (Andrew and Savage, 2000; Lukowiak et al., 1996) and both appetitive (Kemenes and Benjamin, 1989) and aversive conditioning (Kojima, et al., 1996; Kawai et al., 2004) have been demonstrated in the feeding system and aversive conditioning in the respiratory system (e.g., Orr et al., 2009). The animal has also been used to study the susceptibility of memory consolidation during lapses in recall (Marra et al., 2013) in the feeding motor neuron B4 in which a fictive feeding motor pattern was induced due to the

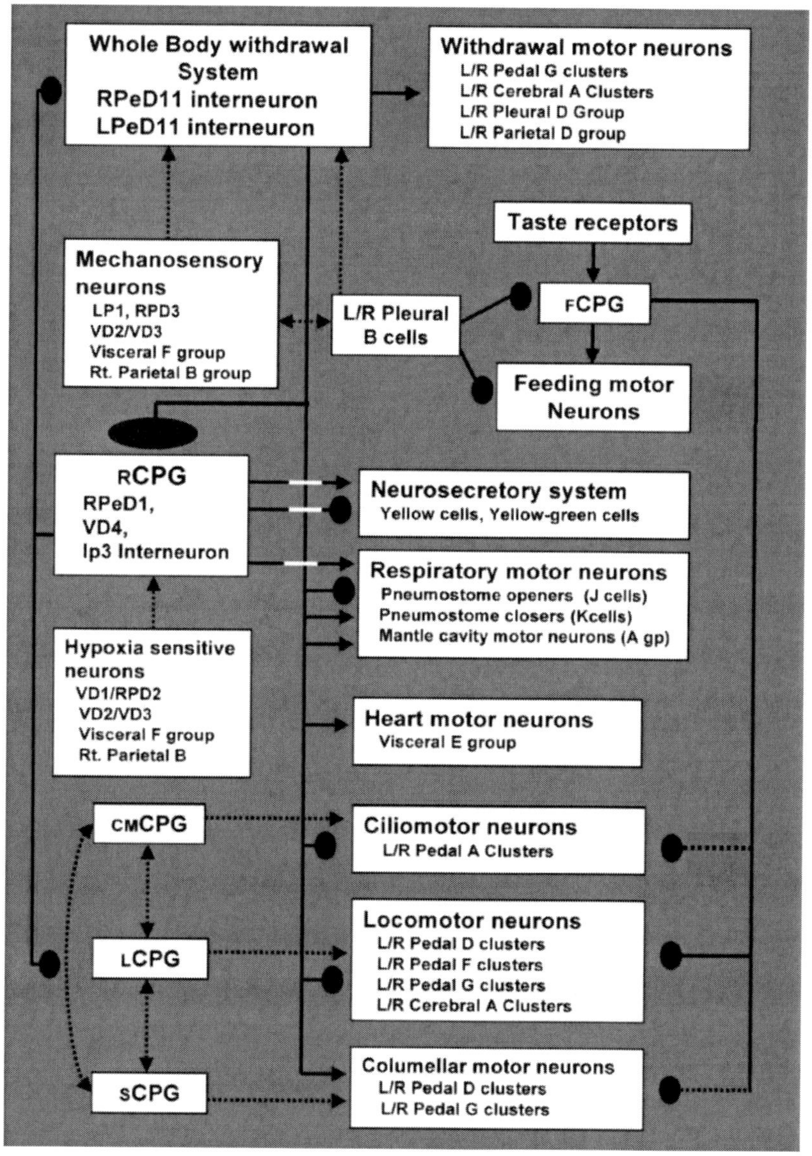

Figure 6. The whole body withdrawal system sits at head of behavioural hierarchy in *Lymnaea* and inhibits other behaviours. Interactions between the whole body withdrawal system and other behavioural systems are as follows: Excitatory connections: ⟶, Inhibitory connections: ⟶●, Presumed connections: ••••••••••••••, which require further investigation. The enlarged inhibitory connection between the whole body withdrawal system and the RCPG indicates that it completely overrides the RCPG. It also inhibits pneumostome opener motor neurons, ciliomotor neurons, and locomotor neurons, while exciting pneumostome closers, mantle cavity motor neurons, heart motor neurons and the columellar motor neurons which drive whole animal withdrawal. Feeding motor neurons and the FCPG are inhibited by the L/R PlB cells, which may be part of the whole body withdrawal system, but this remains to be demonstrated. The FCPG sends inhibitory inputs to locomotor motor neurons and the RCPG inhibits LCPG and may also inhibit the CMCPG and the SCPG as well as localised actions of the withdrawal system.

coordinated activity of the entire feeding network including the FCPG and its various modulatory interneurons. They propose that recall is weakened during periods of molecular transition from intermediate-term to long-term memory, allowing novel sensory cues to block the consolidation of long-term memory in a semi-intact preparation of the *Lymnaea* feeding system.

Thus in intact animals the functions of individual neurons or systems of neurons may vary allowing different behaviour patterns to emerge from the complex neuronal ensembles within the nervous system as circumstances change within the animal's ecological system. As the external environment alters, different behaviours are selected or the nervous system may have to coordinate synchronously occurring behaviours such as feeding and lung ventilation.

Distributed Processing in the Multiganglionic Neural Network

The complex multiganglionic neural network (Winlow and Syed, 1991; Winlow et al., 1992) underlying the behaviour in *Lymnaea* was revealed by combined electrophysiological and behavioural experiments in several laboratories (Fig 6). The network acts as the substrate from which the varying behaviours emerge (Winlow, et al., 1988; Syed et al., 1988; Syed and Winlow, 1989; Winlow, 1989a; Winlow and Syed, 1991; Winlow et al. 1992). Several CPGs are embedded within the network including the RCPG (Syed and Winlow, 1988, 1991), locomotor CPG (LCPG) (Winlow and Haydon, 1986; Syed and Winlow, 1988), the presumed ciliomotor CPG (CMCPG) (Syed and Winlow, 1989b), that for shell movements during locomotion (SCPG) (Haydon and Winlow, 1986) and the FCPG (Benjamin, 1983, 2012; Kyriakides and McCrohan, 1988). Experiments by these latter authors showed that cyclical synaptic inputs originating from neurons of the FCPG are also shared by neurons of the pedal, E,F and M clusters and also by nine pedal giant neurons (L/RPeD1, L/RPeD2, L/RPeD3, LPeD5 and L/RPeD10 – see Fig 7 for locations of neurons). Most of these connections were inhibitory and occurred predominantly during the radula retraction phase of the feeding cycle, perhaps to suppress locomotion at the precise time when the radula is in contact with the substrate. In addition, there is feedback to the FCPG during whole body withdrawal responses after strong tactile stimuli to the body. This inhibits firing of the serotonergic CGCs with a consequent cessation of feeding activity (Yeoman et al., 1994). Furthermore stimulation of the paired FMRFamidergic pleural interneurons (PlB) (Alania, et al., 2004, 2008) causes inhibition of all the neurons in the feeding network including the CGCs. They are thus part of the withdrawal system, perhaps connected with RPeD11 (see below).

Wide-acting Synaptic Inputs

In early electrophysiological studies (Winlow and Benjamin, 1976; Benjamin and Winlow, 1981) a series of wide-acting synaptic inputs were discovered to act within the network (Fig 6) and some have been localised to individual neurones. For example the giant dopamine-containing neuron (RPeD1 – right pedal dorsal 1) (Figs 1 and 7) (Winlow and Benjamin, 1977) VD4 (visceral dorsal 4, also known as VWI, the visceral white interneuron,

Benjamin 1984) and the input 3 interneuron (ip3I) (Fig 1 and 4) (Syed and Winlow, 1991a) generate wide-acting synaptic inputs and are themselves are part of the respiratory central pattern generator (RCPG) as described below (and see Fig 10).

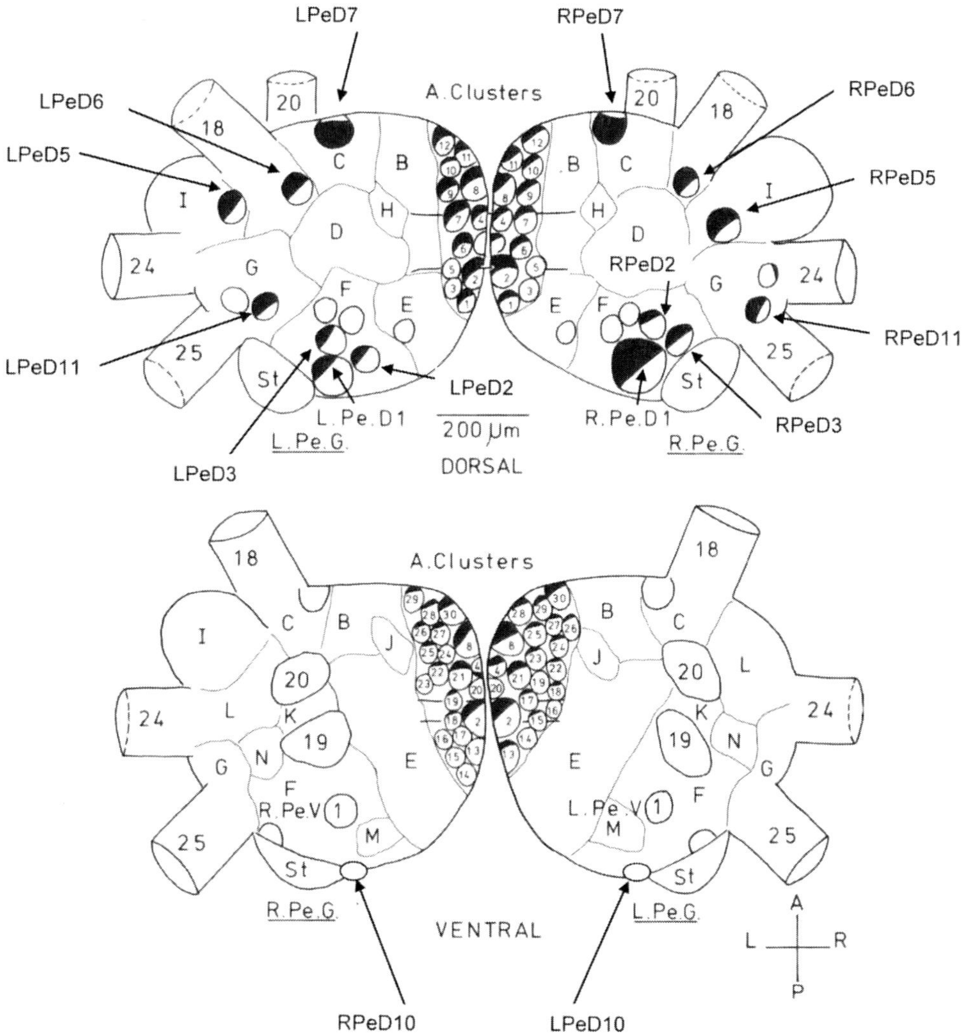

Figure 7. Dorsal and ventral maps of the left and right pedal ganglia of *Lymnaea*. Cell clusters are indicated by single letters and identified neurons are named. The cells of the paired pedal A clusters are individually identifiable and are each composed of 30 bilaterally symmetrical pairs of neurons lying on the medial faces of the pedal ganglia which we have numbered. With respect to the foot, all the A-cluster neurons have unipolar axons which are partially somatotopically organized. Cell shading is as in Figure 4.

The LCPG and the SCPG

Although basommatophoran pulmonates may be quite large animals, most species, including *Lymnaea,* are both aquatic and air-breathing. As a consequence they are positively buoyant (Henderson, 1963) and their cilia are effective locomotor structures. Thus, *Lymnaea* is a ciliary glider and rhythmic beating of the pedal cilia provides the prime locomotor force (Kaiser, 1960) during surface locomotion (Winlow and Haydon, 1986) with little overt muscular activity by the foot except when changing direction. In substrate locomotion ciliary beating is accompanied by alternate contractions and relaxations of the pedal musculature (Winlow and Haydon, 1986) which are driven by the LCPG (Syed and Winlow, 1988). Both forms of locomotion are accompanied by rhythmic shell movements driven by the SCPG via contractions of the columellar muscles (Haydon and Winlow, 1986; Winlow and Haydon, 1986). The rhythmic shell movements accompanying locomotion may aid cardiac pumping because *Lymnaea* has an open vascular system and it is suggested that they act as an aid to cardiac pumping for oxygen distribution around the body. The musculature to the foot is driven by motor neurons in the paired pedal D and G clusters (Fig 7) which innervate the body wall and column. Motor neurons from the paired pedal F clusters (Fig 7) innervate the body wall and foot and all the motor neurons participate in locomotor behaviour in semi-intact preparations, while neurons of the cerebral A clusters (see Fig 5 for location) innervate the ipsilateral body wall (Winlow and Haydon, 1986). With the exception of the electrically connected neurons of the cerebral A clusters, the motor neurons are not synaptically connected to one another, nor do they usually exhibit feedback to the CPG.

Ciliomotoneurons comprise about 30 paired neurons making up the left and right pedal A (L/RPeA) clusters (Syed, Harrison and Winlow, 1988; Syed and Winlow, 1989b) which are electrically coupled to one another (Fig 8a-d) and are partially somatotopically organised: neurons of the anterior group of the cluster mainly project out of posterior pedal nerve trunks, whilst those of the posterior group have a higher proportion of branches down the anterior pedal nerve trunks. Furthermore the neurons are individually identifiable. They receive a unique rhythmic, excitatory synaptic input, the A-cluster rhythm (Fig 8e-g), which is presumed to emanate from a CMCPG (Syed and Winlow, 1989b). They are known to contain serotonin (Casey and Winlow, 1985; Caunce et al., 1988; McKenzie et al., 1998) and have been shown to innervate both the pedal cilia and the mucus cells of the foot (McKenzie et al., 1998). All the L/RPeA cluster neurons are excited by stimulation of VD4, part of the RCPG, presumably to restart locomotion following respiration (Syed and Winlow, 1989b). The A-cluster neurones also receive delayed inhibition from the input 3 interneuron (Ip3I) which is part of the RCPG and which is known to drive pneumostome opening (Syed and Winlow, 1991a). Thus during the initial phase of respiratory behaviour, the locomotor and local withdrawal systems (see below) are temporarily inhibited, especially the foot, body wall neurons and the ciliomotoneurons.

In addition, all the elements of the RCPG are also involved in regulation of the neurosecretory yellow cells and yellow green cells which appear to have a role in ion and water regulation (Wendelaar Bonga, 1972; Soffe et al., 1978). According to Benjamin (1984) VD4 inhibits the Yellow Cells and Yellow-green Cells, and may convert Yellow Cells from single spiking to doubletting or bursting. RPeDl, has weak inhibitory effects on both types of neurosecretory cells but its main effect is to inhibit VD4 thus allowing the Yellow Cells to be strongly influenced by spontaneous excitatory inputs from the input 3 interneuron (Ip3I).

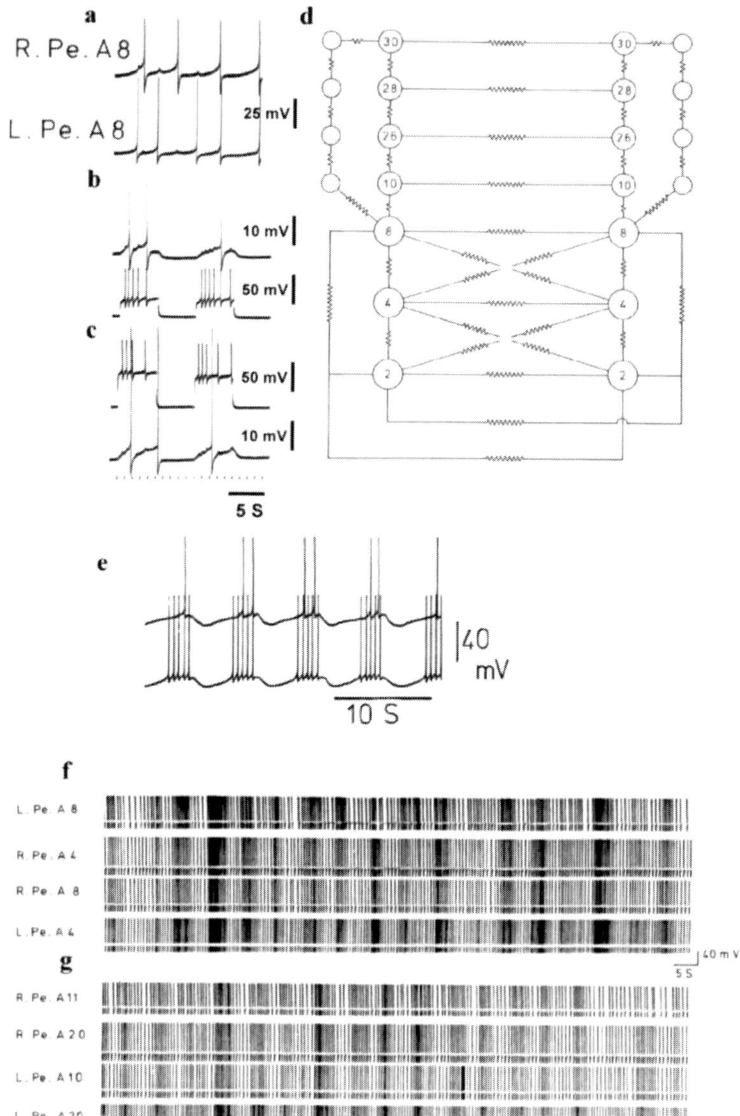

Figure 8. Cellular interactions of the Pedal A cluster neurons. All the cells of the left and right pedal A clusters are uniquely identifiable (see Fig 7) and are believed to be ciliomotor neurons. a) The largest cells within the cluster are electrically coupled to their contralateral homologues, for example LPeA8 and RPeA8 shown firing spontaneously. b,c) Square current pulses injected into one cell evoke electrical responses in the contralateral homologue. d) Summary of the electrical connections of the A cluster neurons. The largest neurons, L/RPeD2,4&8 are coupled to one another both ipsilaterally and contralaterally. The small A-cluster (40-50 μm soma diameter) neurons project ipsilaterally and are weakly electrically coupled to adjacent cells. The medium-sized (50-60 μm) and large (60-70 μm) cells project both ipsilaterally and contralaterally and make electrical connections with their contralateral homologues as well as their smaller near neighbours. The A-cluster neurons all have similar action potential shapes and receive several identifiable synaptic inputs whose sources are unknown. Some of these inputs have widespread synaptic actions in several ganglia whilst e) the "A-cluster rhythm" (ACR) is entirely restricted to the A-cluster neurons and all neurons in the cluster receive it whether large cells (f) or small and medium sized cells (g). Thus the activities of the ciliomotor pedal A cluster neurons are synchronised by the ACR and the electrical connections between the cells.

Respiratory Behaviour in *Lymnaea*

Respiratory behaviour is part of a complex behavioural sequence involving: the locomotor system; shell movements; erection of the pneumostome and lung ventilation (Fig 9). In their natural environment freshwater snails such as *Lymnaea* use both lung and skin respiration (Berg and Ockleman, I959; Jones, 1961). Lung ventilation is triggered when pO_2 is lowered either in the surrounding water or in the lung cavity as detected by oxygen-sensitive receptors mainly localised around the pneumostome and within the lung cavity itself (Janse, 1981, 1982; Janse et al., 1985; van der Wilt et al., 1988; Bell et al., 2007, 2008).

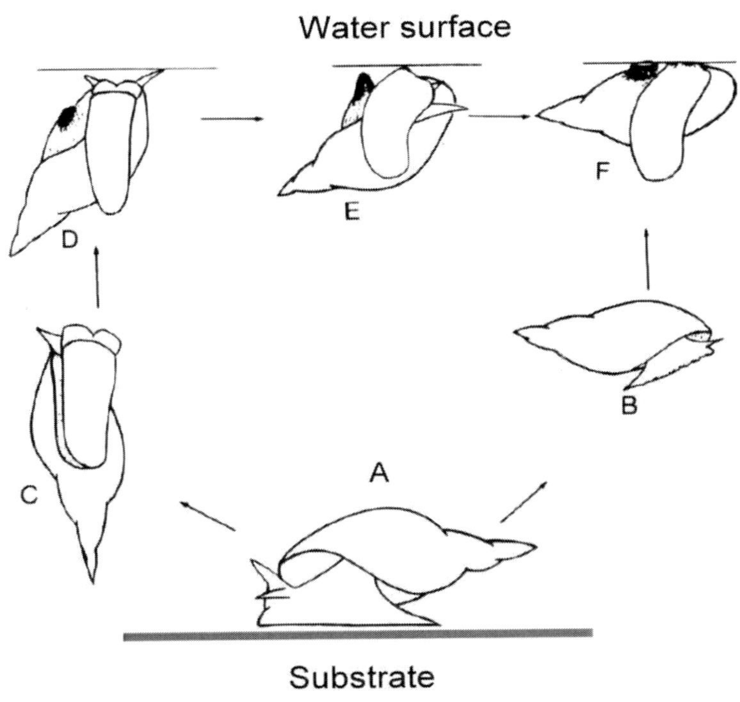

Figure 9. Stages of normal respiratory behaviour in *Lymnaea* (after Moroz, 1991). A, normal substrate locomotion; B – rapid surfacing behaviour when the animal ceases to adhere to the substrate and, because it is buoyant due to air in the lung , the animal floats to the water surface; C – the animal most usually comes to the surface by active locomotion (negative geotaxis) (Janse, 1981, 1982). D,E,F As the animal approaches the water surface pneumostome is erected and opened and locomotion is inhibited. Full details are given in Winlow, Moroz and Syed (1991). The duration of lung ventilation varies from several seconds to 5 minutes depending on environmental conditions.

Lung Ventilation in Lymnaea is via the open pneumostome at the water surface and while the animal is there several cycles of inspiration and expiration take place. These movements are driven by the RCPG which directly controls motor neurons to the muscles of the pneumostome and mantle cavity (Syed et al., 1991). The motor neurons to the pneumostome have been identified as visceral J cells (pneumostome openers) visceral K cells (pneumostome closers) and those causing contraction of the mantle cavity are the neurons of the right parietal A group (Fig 4a). During ventilation locomotion in the surface film usually ceases, but feeding at the surface may continue.

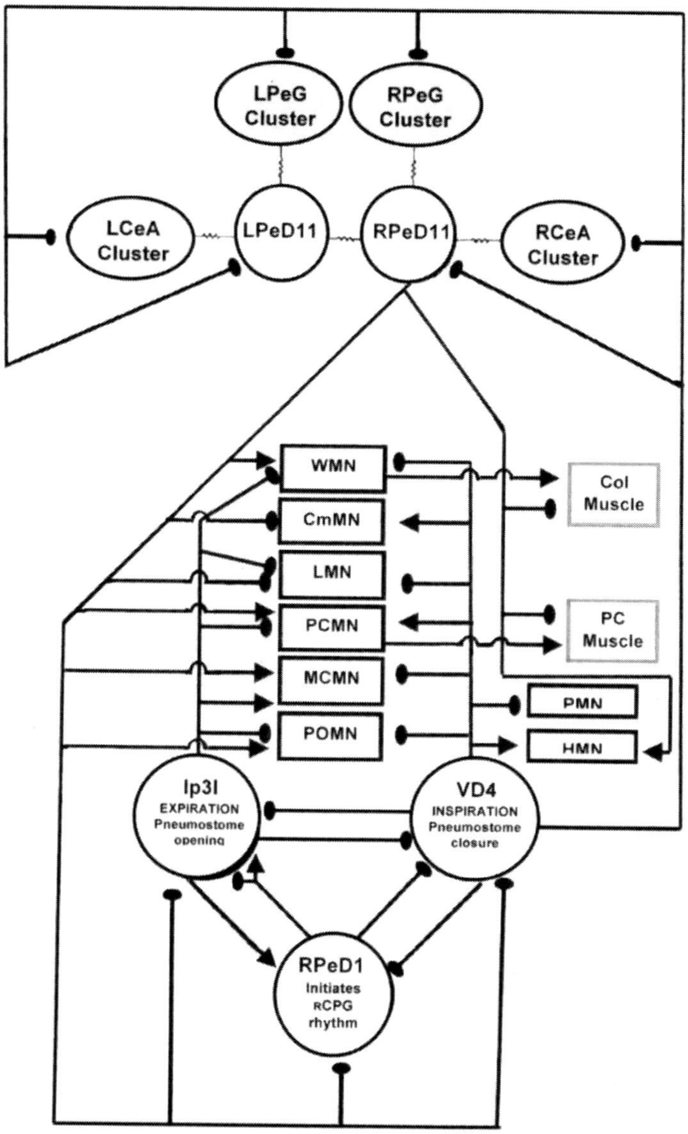

Figure 10. Interconnections of the RCPG and the whole body withdrawal interneuron RPeD11. The withdrawal system and the RCPG are mutually inhibitory, but this depends on context. Under normal circumstances the RCPG prevents localised withdrawal while the animal is breathing at the water surface, but when whole animal withdrawal takes place this overrides all other behavioural systems, strongly inhibiting the RCPG. Excitatory connections: ——▶, Inhibitory connections: ——●, Electrical connections —vvvv—, WMN, whole body withdrawal motor neurones; CmMN, ciliomotor neurons; LMN, locomotor motor neurons; PCMN, pneumostome closer motor neurons; MCMN, mantle cavity motor neurons; POMN, pneumostome opener motor neurons; PMN, penis retractor motor neurons; HMN, heart motor neurons; PC muscle, pneumostome closer muscle; Col muscle, columellar muscle.

The Respiratory Central Pattern Generator (RCPG). Of the three neurons comprising the RCPG (Fig 10), RPeD1 initiates the respiratory rhythm by exciting IP3I (pneumostome opening, expiration), whereas its effects on VD4 are inhibitory. Once excited by RPeD1, the IP3I activates RPeD1 while inhibiting the VD4 neuron. Dual inhibition of VD4 by IP3I and

RPeD1 results in rebound excitation of VD4 which in turn results in pneumostome closure (Inspiration) and the cycle repeats. The synaptic connections between RPeD1 and VD4 are mutually inhibitory (Fig 10) (Syed et al., 1990; Syed and Winlow, 1991).

In addition to being a multiaction neuron (Winlow and Benjamin, 1977, Winlow et al., 1981), initiating the RCPG discharge, RPeD1 is also a multifunctional neuron (Haque et al., 2006) which acts as a motor neuron capable of inducing partial opening of the pneumostome by direct actions on the pneumostome opener muscles. It also receives chemosensory and mechanosensory components of respiratory behaviour when the pneumostome is exposed to air. Furthermore, Bell et al. (2007, 2008) have demonstrated that hypoxia sensitive peripheral chemoreceptor cells (PCRCs) make direct contact with RPeD1. One branch of the neuron terminates in the osphradium, denervation of which severely disrupts respiratory activity by significantly reducing surface breathing activity in the freely behaving animal. The osphradial ganglion cells have been shown to be genuine respiratory chemoreceptors and make peripheral monosynaptic connections with RPeD1. They thus provide a hypoxia-induced, physiological respiratory drive to RPeD1 and are able to elicit activity in the RCPG. For further detail please see chapter 9 by Janes and Syed.

Coordination of the Locomotor and Cardiorespiratory Systems by Identified Neurons within the Multifunctional Network

Many of the complex types of behaviour exhibited by animals are a result of coordination between two or more motor centres of the central nervous system. In some cases neurons that coordinate various neural elements of a given behaviour may also play an integral role in coordinating seemingly unrelated behaviours, eg synchronous feeding and lung ventilation, or are coordinating neurones specified for each behaviour, eg synaptic inputs from the feeding system to inhibit locomotor motor neurons when the radula contacts the substrate (Kyriakides and McCrohan, 1988).

The functional integration of the behavioural repertoire of an animal is the determining factor in survival, particularly in the face of environmental stress. For example, if *Lymnaea* is subjected to a noxious stimulus during normal activity, behaviour such as respiration, feeding or locomotion is terminated immediately and the animal withdraws its entire head-foot complex into the shell. (Ferguson, 1984; Ferguson and Benjamin, 1991a,b). To satisfy its respiratory needs, however, a pulmonate snail must expose its pneumostome (respiratory orifice) to the air. This can only be achieved if the body is extended out of the shell, i.e., a behaviour that is in opposition to withdrawal behaviour. During respiration, locomotor behaviour is also suppressed, although, as mentioned above, feeding behaviour may not be suppressed. Furthermore, analysis of locomotion in *Lymnaea* has revealed that well-coordinated movements of the left and right body wall and foot musculature occur in the absence of whole-body withdrawal or respiratory activity (Haydon, 1984; Winlow and Haydon, 1986; Syed, 1988). Thus, a significant amount of coordination must exist among these three seemingly distinct behaviours.

A pair of bilateral, electrically coupled interneurons has been shown to coordinate the activities of locomotor and whole-body withdrawal motoneurones in *Lymnaea* (Syed and

Winlow, 1991) and are identified as left and right pedal dorsal 11 (L/RPeD11), which lie within the paired pedal G clusters (Slade et al., 1981). The neurones have similar morphologies to one another, encircle the lower ganglionic ring and also enter the cerebral ganglia where they are strongly connected to one another (Syed and Winlow, 1991b) and also to the cerebral A cluster neurons (Haydon, 1982; Winlow and Haydon, 1986; Ferguson, 1984) which are body wall neurons activated during withdrawal behaviour and have only ipsilateral projections to the periphery. However, they are indirectly electrically coupled to their contralateral homologues *via* their connections with L/RPeD11. Similarly, the activities of left and right pedal G cluster neurones, which surround L/RPeD11 and which are known to be involved in locomotion (Winlow and Haydon, 1986), are also coordinated by L/RPeD11 (Syed and Winlow, 1991b). Although the electrical coupling between LCeA cluster and RCeA cluster neurones is lost after selective ablation of L/RPeD11, the L/RCeA cluster neurones remain coupled to their ipsilateral neighbouring cells. Not only are these neurons electrically coupled to each other, but also to a large number of foot and body wall and column motor neurons, forming a fast-acting neural network which coordinates the activities of foot, column and body wall muscles (Fig 10). Although the main axons of L/RPeD11 do not exit via the pedal nerves to the foot, body wall and column musculature, electrical stimulation of the neurons causes contraction of the foot, column and body wall.

Interneurons L/RPeD11 are multifunctional since, in addition to coordinating motor neuronal activity, they also make chemical connections with heart motor neurons (Benjamin et al. 1988) and with neurons that participate in respiratory behaviour (Syed and Winlow, 1988a,b; Syed et al. 1991). They directly hyperpolarize motor neurons involved in pneumostome opening (expiration) and depolarize those involved in pneumostome closure (inspiration). In addition both L/RPeD11 have strong excitatory effects on another pair of electrically coupled neurones, visceral dorsal 1 and right parietal dorsal 2, which have previously been shown to be sensitive to changes in the partial pressure of environmental oxygen (PO_2) (Janse et al., 1985). Conversely, the interneuron visceral dorsal 4 (VD4), which controls the inspiratory phase of the respiratory behaviour (Syed *et al.* 1990) and is involved in pneumostome closure, has inhibitory effects on foot and body wall motoneurones as well as on L/RPeD11 and all their electrically coupled follower cells (Syed and Winlow, 1991b). This allows respiration to take place in the absence of locomotor movements. Thus L/RPeD11 coordinate the activities of locomotor motoneurones which innervate the left and right sides of the body, and they are also multifunctional interneurons, acting on neuronal networks that control the cardiorespiratory system of *Lymnaea* and are themselves inhibited by the RCPG via interneuron VD4. Interconnections of the RCPG and L/RPeD11 are summarised in Fig 10. *Whole animal withdrawal terminates motor commands for incompatible behaviours at all output levels.* Inoue et al. (1996) have convincingly demonstrated that that RPeD11, but not LPeD11, is capable of triggering whole body withdrawal and have shown that it terminates respiratory behaviour by directly inhibiting the neurons of the RCPG and also inducing contractions of the columellar muscles and the pneumostome closer muscles. Furthermore, Syed and Winlow (1991) demonstrated L/RPeD11 also directly inhibited the motor neurons. There are similar findings in the feeding system, as mentioned above, where the L/RPlB cells inhibit all the neurons of the feeding network including the CGCs (Alania et al., 2004, 2008) and may be part of the withdrawal system. Conversely, Kyriakides and McCrohan (1988) showed that locomotor neurons were inhibited by feeding motor neurons (see above). Thus, it seems that the neural circuit underlying whole animal withdrawal, which is at the top of the behavioural hierarchy, terminates "*incompatible motor commands simultaneously at all levels of motor output organization*" (Inoue et al., 1996).

Table 2. Activity patterns and membrane potentials of light green cells of the cerebral ganglia of *Lymnaea* in March of two successive years (previously unpublished)

LGC activity pattern	Membrane potential (mV)	Standard deviation	n	%
Silent inactivatable neurons	-63.7	±11.3	39	44.3
Silent neurons activated by depolarisation				26.1
Fast-adapting	-48.4	±8.9	15	
Slow adapting	-43.8	±6.3	8	
Spontaneously active neurons	-41.6	±7.7	21	23.9
Silent neurons activated by hyperpolarisation	-32.6	±12.6	5	5.7
Totals			88	

Table 3. Activity patterns of light green cells of the cerebral ganglia of *Lymnaea* in March and June (previously unpublished)

LGC activity pattern	March		June	
	n	%	n	%
Silent inactivatable neurons	39	44.3	0	0
Silent neurons activated by depolarisation	23	23.0	9	31.0
Spontaneously active neurons	21	23.9	18	62.1
Silent neurons activated by hyperpolarisation	5	5.7	2	6.9
Totals	88		29	

Evidence of Seasonal Changes in the *Lymnaea* CNS

Observations have been made on seasonal plastic changes in the *Lymnaea* CNS. For example, as the animal becomes increasingly active in the summer months there is a progressive increase in brain 5HT content accompanied by a progressive decline in dopamine content (Hetherington et al., 1994). A prolonged investigation over several annual cycles is required to determine the effects of photoperiod, water temperature, oxygenation and nutrient availability on biogenic amine levels and to determine their effects on the behaviour of the animal. There is also a clear relationship between action potential width and time of year in three identified neurons (RPeD1, RPD1 and RPD2) (Wood and Winlow, 1996). Studies on these neurones indicated that the half-width (Figure 11) was lessened and the slope of the regression line was reduced in winter snails as compared with those studied in July (Ahmed, Hopkins and Winlow, 1993). The major changes were in the calcium-dependent plateau phase of each of these action potentials. Similar changes to calcium conductances in nerve terminals would reduce the amount of transmitter released per action potential. Seasonal alterations in the activity of the light green cells (LGCs) of the cerebral ganglia have also been observed (Winlow, 1985, 1987). The activity patterns of the LGCs are very variable and may be classed as inactivatable, silent but activatable by depolarisation, spontaneously active, or silent but activatable following hyperpolarisation. These differences in LGC activity appear to be based on differences in membrane potential (Table 2) and in March 44.3% of neurons were inactivatable. However a further short study showed that in June no neurons were inactivatable and that the majority were spontaneously active (62.1%) and most of the remainder were activated by depolarisation (Table 3). Thus there is further evidence for seasonal change in the *Lymnaea* brain exhibited by the LGCs which are known to secrete growth hormone (Geraerts, 1976). Increased activity of the growth hormone secreting cells occurs at a time of increased food availability in the peak summer growth period.

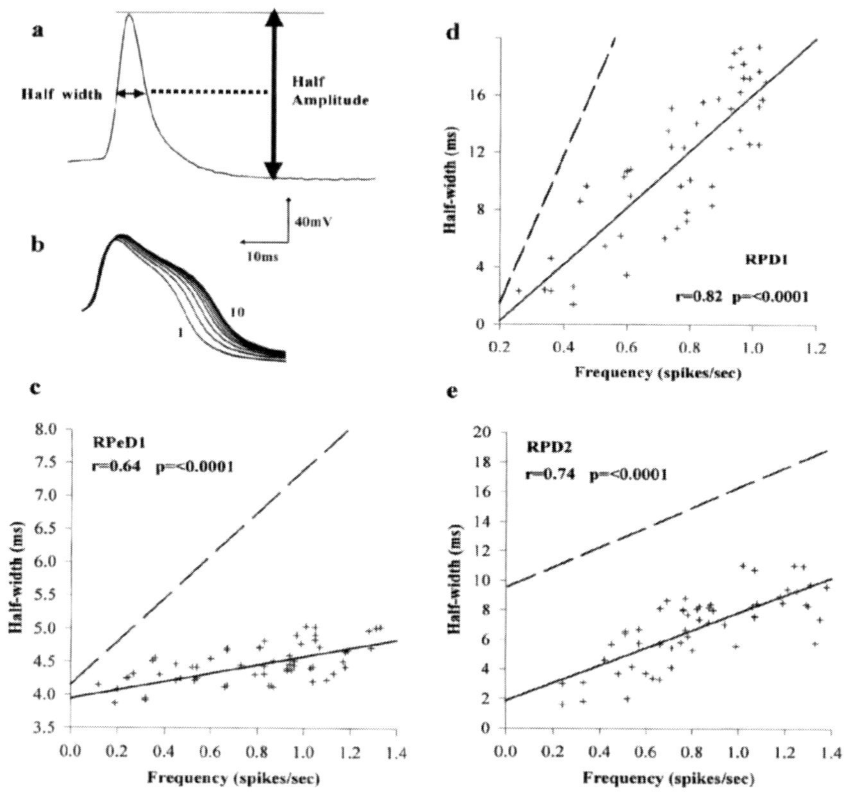

Figure 11. Seasonal changes in action potential shape in three identified *Lymnaea* neurons. Illustration of the measurement of action potential half-width, which is defined as the width of the action potential halfway between the peak of depolarisation and the peak of after-hyperpolarisation. Half-width is a reliable way to assess the duration of the calcium-dependent plateau phase of an action potential. b) Modification of a type 2 action potential (Benjamin and Winlow, 1981) by injection of positive current. Increasing spike frequency increases the duration of the plateau in the repolarisation phase. Thus the first spike (1) is narrower than the last spike (10). In 1993 Ahmed et al. reported on the frequency/half-width correlation coefficients (r ± standard deviations) of neurons recorded in July as follows: RPeD1, 0.63 ± 0.13; RPD1, 0.84 ± 0.04 and RPD2, 0.76 ± 0.11 (n=4 in all cases). These data were compared with data from February snails in the above diagrams. In each case the slope of the February regression line (solid line) is significantly reduced in winter as compared with July values (dashed line). Probability (p) values and correlation coefficients for the February snails are shown on the plots above as is the February half width frequency data for c) RPeD1; d) RPD1; e) RPD2. Thus the duration of the calcium dependent plateau of the action potentials in these three neurons is reduced in winter conditions.

Seasonal Changes in *Lymnaea* Behaviour

Behaviour in *Lymnaea* varies on a seasonal basis. Although many snails often live for only one season a significant number survive the winter, perhaps by hibernating in the mud in the bottom of their pond. During this period of low food supply, low oxygen content in the water, low temperature and short day length the animals neither feed nor attempt to breathe at the water surface. This suggests that the central pattern generators may be disabled or disconnected in winter. In the case of the RCPG this hypothesis is supported by unpublished data from Naweed Syed's laboratory at the University of Calgary. The Syed group demonstrated that animals can be successfully "winterized" by regulating temperature and day/night cycle. Specifically, they showed that at cooler temperatures, animals lacked behavioural responses (lack of activity, feeding, breeding, aerial respiration, etc.) that are reminiscent of hibernating/aestivating snails. Moreover, the isolated CNS from these animals (after 3 weeks of winter conditioning) did not exhibit the respiratory rhythm, nor were any synaptic connections present even though all recordings were performed at room temperature. Interestingly, Syed's group failed to reconstruct synapses between 'winter acclimatised' animals in a soma-soma configuration indicating that changes occurring in the conditioned animals are long-lasting. Syed's findings lead to the conclusion that the RCPG may be disabled or disconnected in winter and to support this conclusion we found that both the central (Park and Winlow, 1993) and peripheral synaptic connections of this circuit are plastic and exhibit seasonal modifications, concomitant with behavioural adaptability (Copping et al., 2000)

Seasonal changes in synaptic connections of the RCPG. Using electrophysiological recording techniques, seasonal plasticity of neuronal connectivity has been observed in the RCPG and the follower cells of RPeD1 (Copping et al., 2000). We originally saw that normal synaptic connections between the respiratory CPG neurons recorded throughout the summer months (Fig 12a) are either completely absent in animals captured during the winter months (Fig 12c) or exhibit a high degree of variability. It is reasonable to assume that winter animals that do not exhibit behavioural programmes, such as aerial respiration, are unlikely to maintain the neuronal connectivity that controls such a rhythmic behaviour. Our electrophysiological data showed that the restoration of synaptic connections from winter (late November to early March in animals from the UK) to summer months (May - September), is not an all-or-none phenomenon, but is a gradual process. For instance during the winter months, no synaptic connections are observed between the respiratory neurons, whereas during spring (mid February - mid May) biphasic synapses predominate, but during the summer months normal synaptic connections are fully restored with the appearance of the behavioural programme. Normally the follower cells of RPeD1 respond appropriately to its transmitter, dopamine, applied by a puffer pipette, but these responses are lost in winter snails (Fig 12c) suggesting that the dopamine receptors are down-regulated at this time (Copping et al., 2000) and are restored in summer snails (Fig 12d).

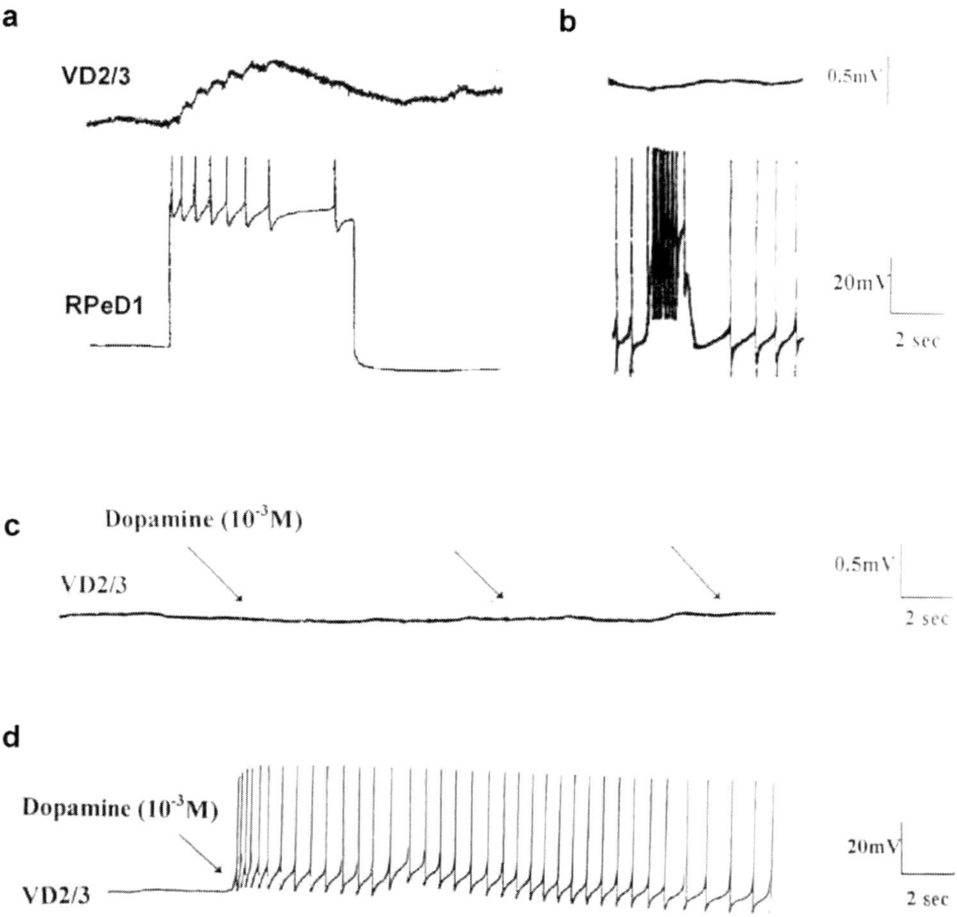

Figure 12. Seasonal variability in the chemical synaptic connections of RPeD1. a) Excitatory synaptic connections between the dopaminergic neuron RPeD1 and VD2/3 in a summer snail but no connections were found in winter snails (b). c) Application of 10^{-3} M dopamine to VD2/3 elicited no responses in a winter snail, but a very strong excitatory response in a summer snail (d), suggesting down-regulation of postsynaptic dopamine receptors in the winter months.

Lymnaea as a Model in Neuroecology

From these studies it is clear that seasonal changes modulate nervous activity and behaviour in *Lymnaea*. What is not clear is which environmental triggers induce and modulate these changes. Are the seasonal changes we have observed due to the interlinked factors of nutrient availability, day length and temperature? Only further experiments over several annual cycles using both laboratory bred animals and those taken from the wild will begin to solve this conundrum. Further questions will also arise as to the plastic changes in connectivity of the multiganglionic neural network and how they are modified by internal hormonal changes triggered by the environmental cues. This may be important in terms of understanding how climate change alters freshwater habitats, particularly as *Lymnaea stagnalis* has a widespread distribution in Europe, Asia (central, north and south and southeast), North America, North Africa and New Zealand.

Invasive freshwater species (IFS) may alter ecosystems and the behaviour of animals living in those ecosystems. Thus the local animal populations will need to be able to adapt to changes in their environment if they are to be fit to survive. For example aerial respiratory behaviour *Lymnaea* is able to be conditioned (Lukowiak, et, 2006) to extracts from sympatric predators and long term memory (LTM) may be formed at the level of RPeD1 (Orr et al., 2007) which reduces the frequency of aerial respiration. This is important since *Lymnaea* is at its most vulnerable to predation when breathing at the water surface. However different strains of *Lymnaea* from different locations have different learning abilities (Orr et al., 2008) and geographically isolated strains only respond to sympatric predators, not to allopatric predators (Orr, 2009a,b). However, juvenile snails could be conditioned to form LTM in the presence of extracts of sympatric predators, with a subsequent reduction in aerial respiratory behaviour. More recently, the excitability of RPeD11 has also been shown to be enhanced by predator induced stress using conditioning procedures (Sunada et al., 2011). These findings indicate that LTM may be distributed within the neurons of the multiganglionic neural network, which although hard-wired has innate flexibility of output, allowing the animal a wide range of responses as predators and circumstances change. They also show that in *Lymnaea* "*biologically relevant levels of stress associated with predator detection induce behavioral phenotypic alterations (i.e., LTM formation) in juveniles, which may increase their fitness*" (Orr et al., 2010) and thus their survivability.

More insidious IFS may be in the form of freshwater plant or microbial species whose metabolites may modify the feeding or reproductive behaviour of the animal and as a consequence its internal its neurohormonal milieu. For example the growth and functional maturation of the gonads in many invertebrates, as well as vertebrates, depends on the action of gonadotropic hormones (Engelmann 1994). These are generally synthesised and released by neuronal clusters in invertebrates and similar peptides are present in the cerebral ganglia of both *Lymnaea stagnalis* and the related species *Helisoma trivolvis* (Young et al., 1999). These cells innervate the male reproductive system and their cell bodies are close to neurons involved in the stimulation of egg laying. It is probable that the endogenous GnRH-like peptide plays a role in regulating reproduction in freshwater snails. Classically, changes in temperature, photoperiod, food intake and partner availability are integrated by the CNS to generate internal responses that affect endocrine glands and, subsequently, gonad growth and maturation, egg laying, sexual behaviour. Upsetting these feedback loops will modify the behaviour and survivability of species unless they are flexible enough to respond quickly to the changes induced by IFS. Past history suggests that *Lymnaea* has successfully evolved as circumstances have changed - its peripheral morphology, nervous system and behaviour have co-adjusted to environmental change to suit it well to its current environmental niche. Thus *Lymnaea* is likely prove a to be a useful and sensitive model for studying the neuroecological effects of climate change on animal behaviour and neural plasticity.

Conclusion

The nervous system of *Lymnaea* is believed to have coadapted to peripheral changes particularly due to torsion and rotation of the shell and mantle cavity by 180° and there is electrophysiological evidence that pairs and paired groups of neurons still exist within the apparently unpaired modern visceral and parietal ganglia.

Behaviour in *Lymnaea* is multiganglionic in origin and is underlain by a complex, distributed, neural network that underlies several behaviours and is capable of great flexibility of output so that individual behaviours fit into a behavioural hierarchy of compatible and incompatible actions. However, behaviour is the product of the whole nervous system in the intact animal, but certain key neurons fulfil important roles in coordinating or initiating behaviour – such neurons may be multifunctional and have multiple postsynaptic actions, eg RPeD1.

Several central pattern generators, such as the RCPG in *Lymnaea* are embedded in the neural network. The RCPG initiates and regulates lung ventilation via motor neurons to the pneumostome and mantle cavity. Respiratory behaviour involves several different motor systems, e.g., erection of the pneumostome, lung ventilation, inhibition of both locomotion and whole animal withdrawal, although compatible behaviours such as feeding may continue.

The connections of the giant dopamine containing neurone, RPeD1, a key neuron in the RCPG, exhibit seasonal plasticity concomitant with behavioural adaptability. It becomes disconnected from its follower cells in the winter months, and there is evidence for down-regulation of dopamine receptors during this period. Given the widespread interconnections of the RCPG, it is suggested that similar actions occur throughout the network and are reversed in those animals that survive the winter.

Lymnaea may well have originated in estuaries moved to freshwater during evolution, but whatever its origin, the animal is well adapted to freshwater ponds as is evidenced by its well studied, flexible, behavioural repertoire. It is likely prove to be a useful and sensitive model for studying the neuroecological effects of climate change on animal behaviour and neural plasticity.

References

Ahmed, IA, Hopkins, PM and Winlow, W (1993). Caffeine and ryanodine differentially modify a calcium-dependent component of soma action potentials in identified molluscan neurones in *situ. Comp. Biochem. Physiol., C*, 105C, 363-372.

Alania, M, Sakharov, DA and Elliott, CJH (2004). Multilevel inhibition of feeding by a peptidergic pleural intereneuron in the mollusc *Lymnaea stagnalis. J. comp. Physiol.,* 190, 379-390.

Alania, M. Vorontsov, DD and Sakharov, DA (2008). Higher-order contol of the feeding network in *Lymnaea. Acta Biologica Hungarica*, 59 (Suppl 1), 23-28.

Andrew RJ, Savage H (2000). Appetitive learning using visual conditioned stimuli in the pond snail, *Lymnaea. Neurobiol Learn Mem.* 73:258-73.

Bateson, P (1987). Epilogue: an ethological overview, In *Aims and Methods in Neuroethology*, 301-305 DM Guthrie (ed). Manchester University Press, Manchester

Bell, HJ, Inoue, T, Shum, K, Luk, C and Syed NI (2007). Peripheral oxygen-sensing cells directly modulate the output of an identified respiratory central pattern generating neuron. *European Journal of Neuroscience,* Vol. 25, pp. 3537–3550,

Bell, HJ, Inoue, T and Syed, NI (2008). A peripheral oxygen sensor provides direct activation of an identified respiratory CPG neuron in *Lymnaea*. In, M.J. Poulin and R.J.A. Wilson (eds.), *Integration in Respiratory Control: From Genes to Systems. Advances in Medicine and Biology,* 605, 25-29, Springer, New York

Benjamin, PR (1983). Gastropod feeding: behavioural and neural analysis of a complex multicomponent system. In *Neural Origins of Rhythmic Movements. Symposia of the Society for Experimental Biology.* Volume 37. Edited by: Roberts A, Roberts BL. Cambridge, UK: Cambridge University Press; 1983:159-193.

Benjamin, PR (1984). Interneuronal network acting on snail neurosecretory neurones (yellow cells and yellow-green cells of L*ymnaea*). *J. exp. Biol.* 113, 165-185.

Benjamin, PR, Elliott, CJH and Ferguson, GP (1985). Neural network analysis in the snail brain. In: *Model Neural Networks and Behavior*, 87-108, Ed AI Selverston, Plenum Press, New York.

Benjamin, PR (2012). Distributed network organization underlying feeding behavior in the mollusk *Lymnaea*. Neural Systems & Circuits 2012, 2:4. *http://www.neuralsystemsandcircuits.com/content/2/1/4*

Benjamin, PR, Kemenes, G and Kemenes, I (2008). Non-synaptic neuronal mechanisms of learning and memory in gastropod molluscs. *Front. Biosci.*, 13, 4051-4057.

Benjamin, PR and Walker TS (1972). Two pigments in the brain of a freshwater pulmonate snail. *Comp. Biochem. Physiol.*, 41B, 813-821.

Benjamin, PR and Winlow, W (1981). The distribution of three wide-acting synaptic inputs to identified neurones in the isolated brain of *Lymnaea stagnalis* (L). *Comp. Biochem. Physiol.* 70A, 293-307.

Berg, K and Ockleman, (1959). The respiration of freshwater snails. *J. Exp. Biol.*, 36, 690-708

Borradaile, LA and Potts, FA (1963). The Invertebrata: *A manual for the use of students,* 4th ed revised by GA Kerkut, Cambridge University Press.

Brace, RC (1983). Observations on the morphology and behaviour of *Chilina fluctuosa* Gray (Chilinidae), with a discussion on the early evolution of pulmonate gastropods. *Phil. Trans. R. Soc. Lond.* B 300, 463-491.

Briggman, KL and Kristan, WB (2008). Multifunctional pattern-generating circuits. *Annu. Rev. Neurosci.*, 31-271-294.

Calabrese, RL (2007). Motor networks: shifting coalitions. Current Biology, 17, R139-R141.

Casey, C and Winlow, W (1985). Cellular localisation of serotonin by glyoxylic acid condensation in Lymnaea. *J. Physiol.* 369, 169P.

Caunce, M, McKenzie, JD, Tripp, J and Winlow, W (1988). Serotonergic innervation of the pedal epidermis of *Lymnaea*. In: Neurobiology of Invertebrates: Transmitters, Modulators and Receptors, (eds.) J. Salanki and K. S.-Rosza. *Symp. Biol. Hung.* 36, 691-692.

Chase, RB (2002). *Behaviour and its neural control in gastropod molluscs.* Oxford University Press, New York, pp315.

Chiel, HJ and Beer, RD (1997). The brain has a body: adaptive behaviour emerges from interactions of nervous system, body and environment. *TINS*, 12, 553-557.

Copping, J, Syed, NI and Winlow, W (2000) Seasonal plasticity of synaptic connections between identified neurones in *Lymnaea*. In, *Neurobiology of Invertebrates. Membranes, Chemical Signalling and Systems Approach,* (eds) K. Elekes and J. Salanki, pp 205-210.

Engelmann F (1994) Invertebrates: hormone-regulated gonadal activity. In: *Comparative endocrinology*. National Research Council of Canada, pp 39–40.

Ferguson, GP (1984). *Neurophysiological analysis of whole body withdrawal in Lymnaea stagnalis* (L.). PhD Thesis, University of Sussex, UK.

Ferguson, GP and Benjamin, PR (1991a). The whole-body withdrawal reflex of *Lymnaea stagnalis*. I. Identification of central motoneurones and muscles. *J. exp. Biol.* 158, 63-95.

Ferguson, GP and Benjamin, PR (1991b). The whole-body withdrawal reflex of *Lymnaea stagnalis*. II. Activation of central motoneurones and muscles by sensory input. *J. exp. Biol.* 158, 97-116.

Geraerts, WPM (1976). The control of growth by the neurosecretory light green cell hormone in the freshwater snail *Lymnaea stagnalis*. *Gen. Comp. Endocrinol.*, 28, 61-71.

Haque, Z, Lee, TKM, Inoue, T, Luk, C, Hasan, SU, Lukowiak, K and Syed, NI (2006). An identified central pattern-generating neuron co-ordinates sensory-motor components of respiratory behavior in *Lymnaea*. *Eur. J. Neurosci.*, 23, 94–104.

Haydon, PG (1982). An electrophysiological study of the nervous control of locomotion in the pond snail, *Lymnaea stagnalis* (L.). PhD Thesis, University of Leeds

Haydon, PG (1984). An electrophysiological study of the nervous control of locomotion in the pond snail *Lymnaea stagnalis* (L). PhD Thesis, University of Leeds.

Haydon, PG and Winlow, W (1982). Multipolar neurones of Lymnaea stagnalis. I. Multiple spike initiation sites and propagation failure allow neuronal compartmentalisation. *J. Comp. Physiol.* 14A, 503-510.

Haydon, PG and Winlow, W (1986). Shell movements associated with locomotion of *Lymnaea stagnalis* are driven by a central pattern generator. *Comp. Biochem. Physiol.* 83A, 23-25.

Henderson A E (1963) On the underwater weights of freshwater snails. *Z. vergl. Physiol.* 46, 467-490.

Hetherington, MS, McKenzie, JD, Dean, HG and Winlow, W (1994). A quantitative analysis of the biogenic amines in the central ganglia of the pond snail *Lymnaea stagnalis* (L.). *Comp. Biochem. Physiol.* 107C, 83-93.

Inoue, T, Takasaki, M, Lukowiak, K and Syed, NI (1996). Identification of a putative mechanosensory neuron in *Lymnaea*: characterization of its synaptic and functional connections with the whole-body withdrawal interneuron. *J Neurophysiol* 76:3230-3238.

Janse, C (1981). The effect of oxygen on gravity orientation in the pulmonate snail *Lymnaea stagnalis*. *J. comp. Physiol.*, 142, 51-59.

Janse, C (1982). Sensory systems involved in gravity orientation in the pulmonate snail *Lymnaea stagnalis*. *J Comp Physiol A*, 145, 311-319.

Janse, C, van der Wilt GJ, van der Plas, J, and van der Roest, M (1985). Central and peripheral neurones involved in oxygen perception in the pulmonate snail *Lymnaea stagnalis* (Mollusca, Gastropoda). *Comp. Biochem. Physiol.* Vol. *82A, 45-69.*

Jones, JD (1961). Aspects of respiration in *Planorbis corneus* (L.) and *Lymnaea stagnalis* (L.) (*Gastropoda, Pulmonata*). *Comp. Biochem. Physiol.*, 4, 1-29.

Kaiser, P (1960). Die leistungen des flimmerepithels bei der fortbewegung der Basommatophoren. *Z. Wiss. Zool.*, 162, 368-39

Kemenes, G and Benjamin, P R (1989). Appetitive learning in snails shows characteristics of conditioning in vertebrates. *Brain Res*, 489, 163–166

Kemenes, I, Kemenes, G, Andrew, RJ, Benjamin, PR, and O'Shea, M (2002). Critical time-window for NO-cGMP-dependent long-term memory formation after one trial appetitive conditioning. *J. Neurosci.*, 22, 1414–1425.

Kandel, ER (1979). *Behavioral biology of Aplysia.* Freeman, San Francisco, pp463.

Karnaukhov, VN (1971). Carotenoids in oxidative metabolism of molluscoid neurons. *Exp Cell Res*, 64, 301-306.

Kawai, R, Sunada, H, Horikoshi, T and Sakakibara M (2004). Conditioned taste aversion with sucrose and tactile stimuli in the pond snail *Lymnaea stagnalis. Neurobiology of Learning and Memory*, 82, 164–168.

Kojima, S, Yamanaka, M, Fujito, Y and Ito, E (1996). Differential neuroethological effects of aversive and appetitive reinforcing stimuli on associative learning in *Lymnaea stagnalis. Zoological Science*, 13, 803–812.

Kyriakides, MA and McCrohan CR (1988) Central coordination of buccal and pedal neuronal activity in the pond snail *Lymnaea Stagnalis. J. Exp. Biol.* 136, 103-123.

Kyriakides, M, McCrohan, CR, Slade, CT, Syed, NI and Winlow, W (1989) The morphology and electrophysiology of the neurones of the paired pedal ganglia of *Lymnaea stagnalis* (L). *Comp. Biochem. Physiol.* 93A, 861-87.

Lukowiak, K, Ringseis, E, Spencer, G, Wildering, W, and Syed, NI (1996). Operant conditioning of aerial respiratory behaviour in *Lymnaea stagnalis. J. Exp. Biol.*, 199, 683–691.

Lukowiak, K, Martens, K, Orr, M, Parvez, K, Rosenegger and D, Sangha, S (2006). Modulation of aerial respiratory behaviour in a pond snail. *Resp. Physiol. & Neurobiol.*, 154, 61-72.

Marra, V, O'Shea, M, Benjamin, PR and Kemenes I (2013). Susceptibility of memory consolidation during lapses in recall. Nature Communications | 4:1578 | DOI: 10.1038/ncomms2591 |*www.nature.com/naturecommunications*

McCrohan CR and Benjamin PR (1980). Patterns of activity and axonal projections of the cerebral giant cells of the snail, *Lymnaea stagnalis*. J Exp Biol. 85:149-68

McCrohan, CR and Winlow, W (1985). Interganglionic co-ordination and bilateral symmetry in the nervous systems of gastropod molluscs. In: *Bilateral and Intersegmental Co-ordination*, (eds.) B.M.H. Bush and F. Clarac, 33-62. Cambridge University Press, Cambridge.

McKenzie, JD, Caunce, M Hetherington, MS and Winlow, W (1998). Serotonergic innervation of the foot of the pond snail *Lymnaea stagnalis*. J. Neurocytol., 27, 459-470.

Mills, J and Winlow, W (1979). The pond-snail brain - an answer to a neurophysiologist's prayer? *J. Physiol.* 290, 2-3.

Moccia F, Di Cristo C, Winlow W and Di Cosmo A. (2009). GABA(A)- and AMPA-like receptors modulate the Activity of an identified neuron within the central pattern generator of the pond snail Lymnaea stagnalis. *Invert Neurosci.* 9:29-41.

Moroz, LL (1991). Monoaminergic control of respiratory behaviour in the freshwater pulmonate snail, *Lymnaea stagnalis* (L.). In: *Signal Molecules and Behaviour*, eds. W.Winlow, O.Vinogradova and D.A. Sakharov, 101-123. Manchester University Press, Manchester.

Moroz, LL (2009). On the Independent Origins of ComplexBrains and Neurons. *Brain Behav Evol;*74:177–190.

Moroz, LL, Park J-H and Winlow W (1993). Nitric oxide activates buccal motor patterns in *Lymnaea stagnalis*. *NeuroReport,* 4, 643-646.

Moroz, LL and Winlow, W (1991). An identified, multifunctional, giant neurone of the left parietal ganglion is involved in control of defensive behaviour in *Lymnaea*. *J. Physiol.* 435, 117P.

Morton, DW and Chiel, HJ (1994). Neural architectures in adaptive behaviour. *TINS*, 17: 413-420.

Morton, JE, (1958). Torsion and the adult snail; a re-evaluation. *J. Mollus. Stud*,, 33, 2-10.

Morton, JE, 1979 *Molluscs*. 5th Ed, Hutchinson University Library, London, 264pp.

Munoz, DP, Pawson, PA and Chase, R (1983). Symmetrical giant neurones in asymmetrical ganglia: implications for evolution of the nervous system in pulmonate molluscs. J. *exp. Biol.* 107, 147-161.

Nikitin ES, Balaban PM and Kemenes G (2013). Nonsynaptic plasticity underlies a compartmentalized increase in synaptic efficacy after classical conditioning. *Curr Biol.* 23:614-619.

Orr MV, Hittel, K and Lukowiak, K (2010). Predator detection enables juvenile *Lymnaea* to form long-term memory. *J. exp. Biol.*, 213, 301-307

Orr MV, Hittel, K and Lukowiak, K (2008). Comparing memory-forming capabilities between laboratory-reared and wild *Lymnaea*: learning in the wild, a heritable component of snail memory. J. Exp. Biol., 211, 2807-2816.

Orr MV, Hittel, K and Lukowiak, K (2009a). "Different strokes for different folks": geographically isolated strains' of *Lymnaea stagnalis* only respond to sympatric predators and have different memory forming capabilities. *J. exp. Biol.* 212, 2237- 2247.

Orr, MV, Hittel, K., Lukowiak, KS, Han J and Lukowiak K.(2009b). Differences in LTM-forming capability between geographically different strains of Alberta *Lymnaea stagnalis* are maintained whether they are trained in the lab or in the wild. *J. Exp. Biol.,* 212, 3911-3918.

Orr, MV, El-Bekai, M, Lui, M, Wtason, K and Lukowiak, K (2007). Predator detection in *Lymnaea stagnalis. J exp Biol,* 210, 4150-4158.

Park, J-H and Winlow, W (1993). Variant synaptic connections in the respiratory central pattern generator in semi-intact preparations of *Lymnaea. J. Physiol.*, 473, 198P.

Slade, CT, Mills, J and Winlow, W (1981). The neuronal organisation of the paired pedal ganglia of *Lymnaea stagnalis* (L). *Comp. Biochem. Physiol.* 69A, 789-803.

Soffe, SR, Benjamin, PR and Slade, CT (1978). Effects of environmental osmolarity on the blood composition and light microscope appearance of neurosecretory neurones in the snail, *Lymnaea stagnalis.Comp. Biochem. Physiol.* 61A, 557-584.

Straub, VA, Grant J, O'Shea, M and Benjamin PR (2007). Modulation of Serotonergic Neurotransmission by Nitric Oxide. *J Neurophysiol* 97:1088-1099.

Straub, VA, Staras, K, Kemenes, G and Benjamin, PR (2002). Endogenous and network properties of *Lymnaea* feeding central pattern generator interneurons. *J. Neurophysiol.,* 88, 1569-1583.

Sunada, H, Horikoshi,T, Lukowiak, K and Sakakibara, M (2010a). Increase in excitability of RPeD11 results in memory enhancement of juvenile and adult Lymnaea stagnalis by predator-induced stress. *Neurobiology of Learning and Memory* 94, 269–277

Sunada, H, Sakaguchi, T, Horikoshi, T, Lukowiak, K and Sakakibara, M (2010b). The shadow-induced withdrawal response, dermal photoreceptors, and their input to the higher-order interneuron RPeD11 in the pond snail *Lymnaea stagnalis. J. Exp. Biol.* 213, 3409-3415.

Syed, NI, Bulloch, AGM, and Lukowiak, K (1990). In vitro reconstruction of the respiratory central pattern generator of the mollusc *Lymnaea. Science*, 250, 282-285.

Syed, NI, Harrison, D and Winlow, W (1988). Locomotion *Lymnaea* - role of serotonergic motoneurones controlling the pedal cilia. In: *Neurobiology of Invertebrates: Transmitters, Modulators and Receptors*, (eds.) J. Salanki and K. S.-Rosza. Symp. Biol. Hung., 36, 387-402.

Syed, NI, Harrison, D and Winlow, W (1991). Respiratory behaviour in the pond snail *Lymnaea stagnalis*. I. Behavioural analysis and identification of motor neurons. *Journal of Comparative Physiology,* A. 169, 541-555.

Syed, NI and Winlow, W (1988). A central pattern generator for locomotion in *Lymnaea. J.Physiol.* 406, 78P.

Syed, NI and Winlow, W (1989a). Evidence for a multiganglionic neuronal substrate underlying several behaviours in *Lymnaea stagnalis* (L.). *J. Physiol.* 412, 61P.

Syed, NI and Winlow, W (1989b) Morphology of and electrophysiology of neurons innervating the ciliated locomotor epithelium in *Lymnaea stagnalis* (L). *Comp. Biochem. Physiol.* 93A, 633-644.

Syed, NI and Winlow, W (1991a). Respiratory behaviour in the pond snail *Lymnaea stagnalis*. II. Neural elements of the central pattern generator. *Journal of Comparative Physiology,* A. 169, 557-568.

Syed, NI and Winlow, W (1991b). Coordination of locomotor and cardiorespiratory networks of *Lymnaea stagnalis* by a pair of identified interneurones. *Journal of Experimental Biology.*, 158, 37-62.

Van der Wilt, G, van der Roest, M and Janse, C (1988). The role of two peptidergic giant neurons in modulation of respiratory behaviour in the pond snail, *Lymnaea stagnalis. Symp. Biol. Hung.*, 36: 377-386.

Wedemeyer, H & Schild, D (1995). Chemosensitivity of the osphradium of the pond snail *Lymnaea stagnalis. J exp Biol,* 198: 1743-1754.

Wendelaar Bonga, S E (1972). Neuroendocrine involvement in osmoregulation in a freshwater mollusc, *Lymnaea stagnalis. Gen. comp. Endocrinol. (suppl.)* 3, 308—316.

Winlow, W (1985). Electrophysiological properties of neurosecretory cells in *Lymnaea stagnalis. Neuroscience Letters*, Suppl. 22, S365.

Winlow, W (1987). Action potential plasticity. In: *Growth and Plasticity of Neural Connections*, 127-144, (eds.) W. Winlow, and C.R. McCrohan. Manchester University Press, Manchester.

Winlow, W (1989). A multiganglionic neuronal substrate for behaviour in Lymnaea. *Neurosci. Letters, Suppl.* 36, S51.

Winlow, W and Benjamin, PR (1976). Neuronal mapping of the brain of the pond-snail, *Lymnaea stagnalis* (L). In: *Neurobiology of Invertebrates, Gastropoda Brain*, (ed. J. Salanki), 41-59. Akademiai Kiado, Budapest.

Winlow, W and Benjamin, PR (1977). Postsynaptic effects of a multi-action giant interneurone on identified snail neurones. *Nature*. 268, 263-265.

Winlow, W and Haydon, PG (1986). Locomotion of *Lymnaea stagnalis*: a behavioural and neuronal analysis. *Comp. Biochem.* Physiol. 83A, 13-21.

Winlow, W, Haydon, PG and Benjamin, PR (1981). Multiple postsynaptic actions of the giant dopamine-containing neurone R.Pe.D.1 of Lymnaea stagnalis (L). *J. exp. Biol.* 4, 137-148.

Winlow, W, Syed, NI and Harrison, D. (1988). An extensive neuronal substrate for molluscan behaviour. *Eur. J. Neurosci.* suppl. 1, 28.

Winlow, W and Syed, NI (1991). Modulation of behaviour by a multiganglionic neuronal ensemble. In: *Signal Molecules and Behaviour*, eds. W.Winlow, O.Vinogradova and D.A. Sakharov, 85-100. Manchester University Press, Manchester.

Winlow, W, Moroz, LL and Syed, NI (1992). Mechanisms of behavioural selection in *Lymnaea stagnalis*. In: *Neurobiology of Motor Programme Selection: New Approaches to Mechanisms of Behavioural Choic*e, eds. Kien, J., McCrohan, C.R. and Winlow, W., pp 52-87. Pergamon Press, Oxford.

Wood, G and Winlow, W (1996). Seasonal variation of spike width in identified neurones of *Lymnaea stagnalis. J. Physiol.*, 495, 38P.

Yeoman, MS, Pieneman, AW, Ferguson, GP, Ter Maat A and Benjamin PR (1994). Modulatory role for the serotonergic cerebral giant cells in the feeding system of the snail, *Lymnaea.* 1 Fine wire recording in the intact animal and pharmacology. *J. Neurophysiol., 72*:1357-1371.

Yeoman, MS, Brierly, MJ and Benjamin, PR (1996). Central pattern generator interneurons are targets for the modulatory serotonergic cerebral giant cells in the feeding system of *Lymnaea. J Neurophysiol*, 75, 11-25.

Young KG, Chang JP, Goldberg JI. (1999). Gonadotropin-releasing hormone neuronal system of the freshwater snails *Helisoma trivolvis* and *Lymnaea stagnalis*: possible involvement in reproduction. *J Comp Neurol.* 404, 427-437.

In: Neuroecology and Neuroethology in Molluscs
Editors: Anna Di Cosmo and William Winlow
ISBN: 978-1-62948-970-4

Chapter 9

Evolutionary Sophistication of Aerial Respiratory Behaviour in the Freshwater Mollusc *Lymnaea stagnalis*

Tara A. Janes and Naweed I. Syed*
Deptartment of Cell Biology and Anatomy and the Hotchkiss Brain Institute,
Faculty of Medicine, University of Calgary, Calgary, Alberta, Canada

Abstract

Respiration is a homeostatic behaviour that is essential for the survival of all Metazoans. However, maintaining oxygen levels within appropriate physiological limits is often not a straightforward task as the physical environment imposes specific and sometimes severe challenges for respiratory behaviour. This is especially true in soil and aquatic habitats where physical and biotic factors can result in local areas of low oxygen termed hypoxia. Therefore, animals must be able to make short-term alterations to their respiratory behaviour in order to overcome these hypoxic challenges. Over longer periods of time, environmental hypoxia has led to the evolution of various respiratory structures and behaviours that allow animals to be superbly adapted to their particular habitat. An excellent example of this adaptive process is the pond snail "*Lymnaea stagnalis*" which utilizes both cutaneous and aerial respiration to thrive in the stagnant and hypoxic environments that other species would find unsuitable. In addition, the neural networks underlying aerial respiration have been characterized allowing researchers to explore this mollusc's short- and long-term adaptation to hypoxia at the synaptic and molecular levels. The simple genetics and smaller nervous systems of many invertebrates have also proved invaluable for medical research purposes. For example, many forms of human disease affect metabolism or the respiratory, cardiovascular and nervous systems thus preventing the maintenance of appropriate oxygen levels through respiration. Understanding the cellular and system responses of invertebrates to hypoxia will therefore shed light on many human conditions and suggest novel treatments. Unfortunately, as humans have strived to improve their own lives, anthropomorphic activities have increased the incidence of environmental hypoxia and threatened the long-

* nisyed@ucalgary.ca.

term survival of *Lymnaea* and many other aquatic invertebrates. The loss of invertebrate biodiversity through local and global extinction threatens the stability of many ecosystems and the natural resources upon which humans depend. In summary, interdisciplinary research aimed at understanding the fundamental process of respiratory adaptation to hypoxia in invertebrates has direct benefits for conservation and medical research.

Introduction

Respiration is an Essential, Homeostatic Behaviour

The group "Metazoa" is comprised of all multicellular animals both living and extinct, and encompasses organisms with incredible diversity in both form and function. In spite of this variation, all metazoans are dependent on aerobic metabolism (also termed aerobic cellular respiration), which provides animals with the energy essential for life. This process utilizes oxygen (O_2) to transform glucose into cellular energy, with carbon dioxide (CO_2) and water as the resulting waste products. While aerobic metabolism is a highly efficient mechanism of energy production (hence its use by all metazoans), it is a 'fragile' pathway requiring appropriate levels of cellular O_2 and CO_2. Specifically, if cellular O_2 levels decline below their normal values, aerobic respiration ceases and cells switch to less efficient anaerobic mechanisms of energy production that can be sustained only in the short-term. However, O_2 is ironically also cytotoxic and excessive levels result in the production of reactive oxygen species which cause DNA and protein damage. In addition, high levels of waste CO_2 resulting from aerobic metabolism alter the cellular pH environment and interfere with proper protein and enzyme function. Therefore, a major challenge for animals is to keep tissue levels of O_2 and CO_2 within narrow, physiologically acceptable limits. This process of respiratory homeostasis is accomplished by organismal respiration (in conjunction with the cardio-vasculature), which evolved early in metazoan life to facilitate gas exchange and transport necessary to sustain aerobic metabolism.

For small bodied organisms such as nematodes and lower invertebrates (Cnidaria and Platyhelminthes), acquisition of O_2 and excretion of CO_2 can take place exclusively across the skin and cell membranes via simple diffusion (termed cutaneous respiration). As body size increases however, cutaneous respiration is no longer a practical means of gas exchange. Therefore, larger animals have evolved specialized respiratory structures (lungs or gills) and behaviours to ensure respiratory homeostasis is maintained.

Even in simple animals, respiratory behaviour is much more complicated than the simple exchange of respiratory gasses between the body and its environment. It can, however, be rapidly altered and fine-tuned to compensate for short- or long-term changes in environmental conditions, metabolic state or in response to disease. This intricate homeostatic control is made possible by the ability of the nervous system to produce respiratory behaviour that meets an animal's metabolic demands under a given set of environmental conditions. Specifically, a high degree of behavioural flexibility is accomplished by the coordinated efforts of neural circuits involved in the generation of the respiratory rhythm, sensory feedback and extensive modulation of respiratory behaviour. As discussed below, flexibility in respiratory behaviour is vital in allowing animals to adapt to an ever-changing environment.

In the following sections we will describe the intimate linkage between respiratory behaviour and the environment, and how an ecological approach to respiratory research is advancing this field in novel directions. Furthermore, we will describe common features of the metazoan respiratory system and discuss how these features can be adapted to meet particular environmental challenges using an invertebrate model. The model animal of choice for this discussion is the pond snail *Lymnaea stagnalis*, in which the central and peripheral neural circuits underlying respiratory behavior have been identified and well characterized. In this chapter, we have focused our discussion on hypoxia (low O_2), for which there is a vast body of literature describing the devastating effects from ecosystems to individual neurons. As a whole, this section will outline how invertebrates adapt their respiratory behavior to environmental hypoxia, and how this knowledge can assist with conservation efforts in addition to advancing our understanding of metazoan respiration and human disease.

A Multi-disciplinary Approach to the Study of Respiration

Importance of Invertebrate Models in Ecology

Invertebrates comprise a staggering 97% of all life on earth and therefore have a major influence on ecosystem health (Zhang and Shear, 2007). In the last 50 years, anthropomorphic threats to natural ecosystems have intensified resulting in the local or global extinction of many species that are subsequently replaced by animals that are more tolerant to the new conditions. Particularly worrying is the plight of aquatic invertebrates, which are abundant in ecosystems and comprise a large portion of the food chain. Many species are unable to rapidly adapt their respiration to hypoxic and polluted conditions imposed by fertilizers, sewage and water contamination in their habitat (Lotze et al., 2006). Upsetting the natural invertebrate community structure through species extinction and replacement will negatively impact biodiversity and deplete valuable natural resources. For example, a recent study examining marine ecosystems concluded that biodiversity was associated with increased ecosystem productivity and stability. Moreover, examination of long-term data collected over the past 1000 years from costal habitats found that since the start of industrialization, native species diversity has rapidly declined resulting in the loss of 33% of viable fisheries by the present day. Importantly, areas with higher regional species richness showed lower rates of collapse of commercially important fisheries. In addition, the authors estimate that up to 63% of oceanic filtering services (wetlands and submerged vegetation) have been lost, thus increasing the occurrence of local hypoxia and subsequent species extirpation in a positive feedback cycle (Worm et al., 2006). Therefore, understanding how invertebrates can adapt their respiratory behaviour to environmental change has important implications for ecology, conservation and the management of natural resources upon which the survival of the human species depends.

Remarkably, there are species of Cnidaria, Nematoda and Mollusca that have been able to adapt to polluted and hypoxic environments. The physiological and behavioural mechanisms underlying this tolerance are of great interest, not only for invertebrate conservation but also in terms of human evolution. Specifically, as our environment becomes more polluted with toxic metals, ultraviolet solar radiation, etc. what adaptive changes might we undergo to survive these new conditions? To this end, we can learn from simple systems that have already adapted to such environments.

Importance of Invertebrate Models in Medical Science

Invertebrates have a long history of use in medical science. For example, during the 1950's Hodgkin and Huxley performed their landmark studies to elucidate the ionic basis of neural action potentials utilizing the squid giant axon (Hodgkin and Huxley, 1952). Much of what we know about respiratory behaviour however comes from studies performed using mammalian models. This is due primarily to the implications of such research for the understanding and treatment of human disease. However, fundamental questions about respiratory behaviour remain unanswered due to the difficulty in studying the overwhelmingly complex mammalian nervous system. In contrast, invertebrates possess smaller nervous systems that can be thought of as 'scaled-down' versions of their vertebrate counterparts. Invertebrate neural circuits function in the same fundamental manner as in mammals but are easier to understand and manipulate experimentally both in vivo and in culture. Furthermore, invertebrates are useful model organisms because they are easy to maintain in the lab, exhibit short generation times and have relatively simple genetics. Since respiration is essential for all animals, it is not surprising that invertebrate respiratory behaviour appears to share fundamental similarities with that of vertebrates, making these 'simple' animals a good model for respiratory research. Despite these advantages, studies of respiration in invertebrates have been largely limited to behavioural observations including organismal responses to hypoxia, and the ability of different phyla to tolerate anoxia. Therefore, a clear understanding of the neural circuits and cellular/molecular mechanisms underlying invertebrate respiratory behaviour are yet to be elucidated. This research would provide valuable information for the understanding of fundamental mechanisms by which the nervous system controls respiratory behaviour in mammals and humans. A classic example illustrating this point is the contribution of invertebrate research to the understanding of specialized neural circuits termed central pattern generators (CPG's). A CPG is a group of neurons which together produce patterned output driving rhythmical motor activity underlying behaviours such as respiration. Much of what we know about CPG's in vertebrates comes from research investigating the anatomy and physiology of the crab stomatogastric ganglion (reviewed by Marder and Bucher, 2007). More current research utilizing the nematode worm Caenorhabditis elegans and fruit fly Drosphila melanogaster has provided a wealth of information regarding the Hypoxia Inducible Factor-1 (HIF-1) pathway, which has been highly conserved throughout metazoan evolution. In response to tissue hypoxia, HIF-1 facilitates the transcription of a myriad of genes involved in angiogenesis, erythropoiesis, cell survival, ventilation, etc. thus promoting survival and adaptation to hypoxia. A greater understanding of this pathway provides promise for the treatment of human disease and injury. Taken together, these studies illustrate the unique advantages of invertebrate models and how they can answer fundamental questions about animal physiology and, in particular, respiration.

Medical science and ecology are often considered widely divergent fields, however new research suggests that the two can have intimate linkages with each other. The human body can be thought of as a living environment for a multitude of different microbial species, some of which are important for health while others are pathogenic. Interestingly, these microbial populations have been found to grow and interact according to the fundamental processes of community ecology (Costello et al., 2012). Currently, community ecology theory is being applied towards the study of pathogenic bacteria inhabiting the lungs of cystic fibrosis patients to better understand the 'ecological' conditions that promote chronic and often fatal infections (Harrison, 2007).

Clearly, integrating concepts from different fields can provide a new perspective and insight into biological questions. In particular, we can learn a great deal about respiration by considering how the respiratory behaviour of different organisms is adapted to overcome the challenges posed by particular habitats. Species survival depends on the ability of animals to adapt their respiration to environmental changes in the short- and long-term. Respiratory behaviour is the most sophisticated measured index of an animal's ecological habitat and therefore we will now discuss how respiratory behaviour depends fundamentally on the environment.

The Link between Respiration and Ecology

Appropriate respiratory behaviour represents an intricate marriage between energy metabolism and the prevailing environmental conditions. Therefore, levels of environmental O_2 and CO_2, as well as the respiratory medium (water versus air) shape respiratory behaviour. Terrestrial animals living above the soil are surrounded by air which has a relatively high O_2 content (20.9% O_2) that does not fluctuate. Under these conditions, environmental hypoxia is encountered only at high altitude or as a result of disease. In contrast, elimination of CO_2 is a challenge in terrestrial habitats because this gas is much less soluble in air then it is in water. Therefore, respiration in these species is primarily driven by the need to eliminate tissue CO_2 on a minute to minute basis (under adequate O_2 conditions). Terrestrial animals also tend to be intolerant to hypoxia since these conditions are rarely encountered. However, diving mammals and those living at high altitudes have adaptations in place to survive under low O_2 conditions. In contrast, aquatic habitats present a very different set of challenges for animal respiration. Specifically, O_2 is much less soluble in water than it is in air, while CO_2 dissolves easily in water. Therefore, respiration in aquatic species is driven mainly by their need to acquire O_2. There is general consensus among researchers that aquatic hypoxia occurs when dissolved O_2 levels drop to 2 mg/L (48.2 torr at sea level) or less. At this level, invertebrates and fish begin to demonstrate behavioural symptoms of hypoxic stress including migration, abandoning sediment burrows, etc. Mass mortality of many species has been observed when dissolved O_2 levels drop below 0.5 mg/L (12.1 torr at sea level) (Rabalais et al., 2002; Diaz and Rosenberg, 2008). In addition, the O_2 content of water is negatively affected by increased temperature, salinity, turbidity, and degree of stratification. These variables can show wide fluctuations on a daily basis, seasonally or from year to year. This variation is enhanced further by biotic factors such as the presence of aquatic plants, seasonal algal blooms, and abundance of microorganisms involved in organic decomposition. Animals inhabiting the soil also encounter low and/or fluctuating O_2 conditions due to poor gas circulation as well as extensive and localized decomposition. Therefore soil and aquatic species face not only the challenge of hypoxic conditions, but also dramatic fluctuations in O_2 levels which may persist over the short- or long-term.

Because respiration is essential for survival, there is tremendous pressure for animals to adapt their respiratory behaviour and physiology to satisfy metabolic demands despite the environmental challenges that these animals face. While the nervous system of metazoans is capable of producing flexible respiratory behaviour to some degree, there are inherit limitations to how quickly natural selection can adapt respiration to dramatic changes in environmental conditions. This point is best illustrated by considering historic and current mass extinction events, which are hypothesized to be driven in part by hypoxia. For example, the Late Permian and Early Triassic periods are characterized by a drop in atmospheric O_2 to around 16%, which, in combination with global warming, may have contributed significantly to the extinction of 95% of the life on earth (Wignall and Twitchett, 1996; Huey and Ward, 2005). Today, anthropomorphic habitat degradation has produced similar conditions of global warming and aquatic hypoxia leading to species extirpation and extinction, albeit on a much smaller scale than that which occurred at the end Permian. Despite this grim outlook, there are some invertebrate groups that have an amazing ability to survive in very hypoxic or anoxic (zero O_2) habitats including cnidarians (Purcell et al., 2001), nematodes (Cook et al., 2000) and the molluscs, annelids and crustaceans inhabiting deep sea hydrothermal vents, the latter of which must represent one of the most challenging habitats on the planet (Hourdez and Lallier, 2007). This anoxia tolerance is not limited to invertebrates, but is also observed in freshwater turtles (Chrysemys) which survive for several months without O_2 during winter hibernation (Jackson, 2002).

Interestingly, humans can also adapt to moderate hypoxia encountered at altitude and populations of Tibetans, Andeans and Ethiopians have been found living up to 6000m where the amount of O_2 is half of that at sea level (Beall, 2006). While the physical and genetic adaptations that have taken place in these populations over generations have been studied extensively, it is not yet clear if humans can adapt to the sudden onset of hypoxic conditions that are frequently encountered as a result of disease. Specifically, identifying the pathological effects of hypoxia on the human body and whether respiratory behaviour can adapt to compensate for these changes remains a major challenge for medical science. Therefore, elucidating the mechanisms underlying adaptation to hypoxia and anoxia in metazoans will prove invaluable. For example, are there specific environmental conditions that promote or accelerate the adaptation of a species to hypoxia? It is possible there may exist cellular or molecular pathways for hypoxia tolerance that are likely conserved between invertebrates and mammals that could be exploited for medical treatment.

The Metazoan Respiratory System

In all animals, respiration supports aerobic metabolism by maintaining respiratory homeostasis. However, the anatomy of the respiratory system can be very different between phyla and even within more closely related groups. Variability can also be seen in respiratory behaviour and in the ways different species have become adapted to their particular environment. Still, there is a fundamental blueprint upon which the respiratory system of metazoans is based and includes the following key features: gas exchange surfaces, central pattern generators, chemoreceptors and respiratory plasticity.

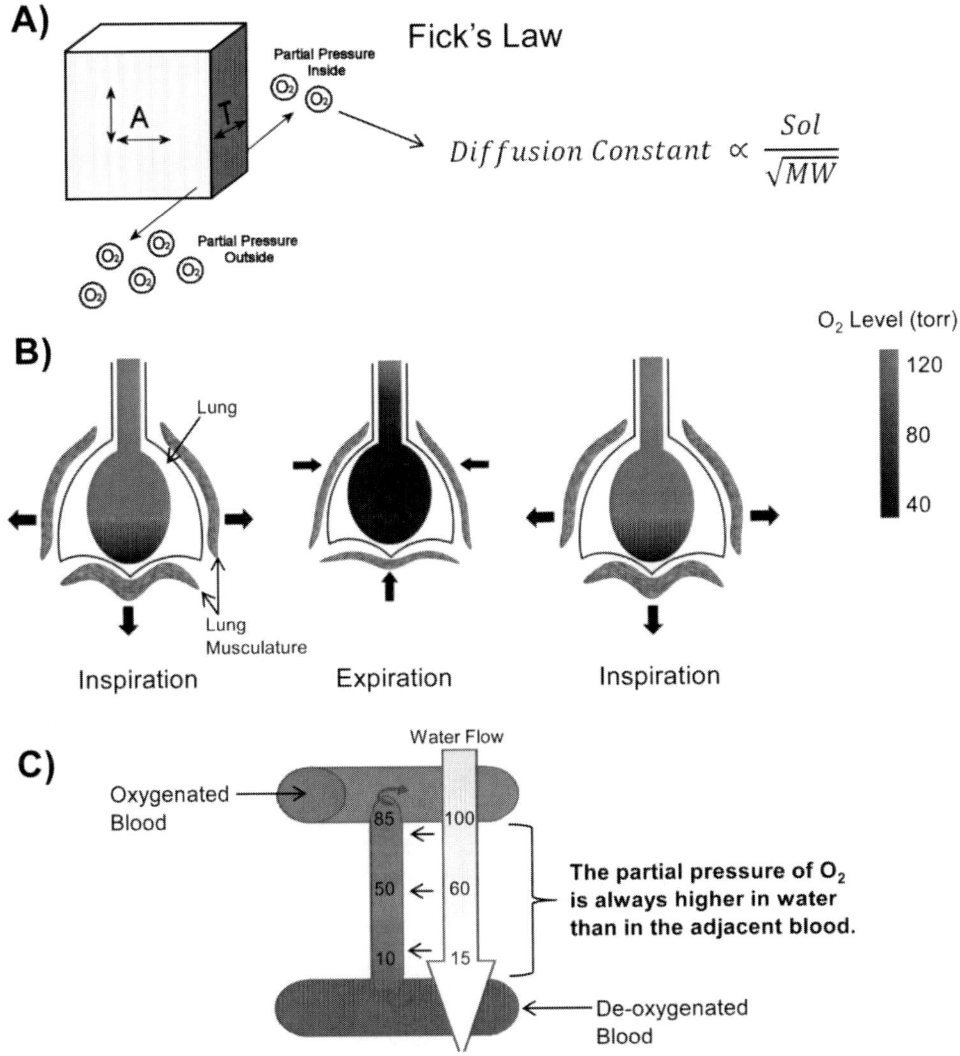

Figure 1. Differences in the efficiency of gas exchange in lungs versus gills. A) The diffusion rate of gasses across a surface depends on the area (A) and thickness (T) of the tissue, as well as the difference in gas partial pressure between the two sides. Gasses diffuse at different rates depending on their intrinsic solubility (Sol) and molecular weight (MW). B) During inspiration, muscles expand the lung cavity to draw in O_2-rich atmospheric air. O_2 diffuses across the lung tissue into the surrounding blood vessels and the remaining O_2-depleted air is exhaled. Note however that some O_2-depleted air remains in the lung following expiration. Therefore, mixing of atmospheric air with O_2-depleted air during the next inspiration cycle reduces the efficiency of lung ventilation. C) In counter-current exchange water is passed over the gills in a direction opposite to that of blood flow. This highly efficient mechanism maintains an O_2 partial pressure difference between the water and blood at all times that results in continuous diffusion of O_2 into the blood. (Figure modified from West, 1999 and Raven, 2001)

Gas Exchange Surfaces

Effective exchange of O_2 and CO_2 between the body and environment requires a suitable respiratory surface to maximize the diffusion rate of gasses. The physical parameters affecting gas diffusion are represented mathematically as Fick's Law:

$$\frac{dV}{dt} = \frac{A}{T} \text{ x } D\,(P_1 - P_2)$$

This equation states that the volume of gas diffusing across a tissue slice per unit time is a function of tissue area (A) and thickness (T), the diffusion constant (D) of the gas and the difference in gas partial pressure between the two sides ($P_1 - P_2$) (Figure 1a). From this relationship, we can see that diffusion rate will be maximized with a large tissue area, minimal thickness and high partial pressure gradient. Generally speaking, there are three major forms that respiratory surfaces can take: lungs, gills and skin.

In terrestrial habitats, lungs provide an ideal gas exchange surface because their internal location reduces evaporative water loss while providing the structural support necessary in an air-medium (one can imagine the fine filaments of gills collapsing when exposed to air). Lungs range from simple sac-like structures in fish, amphibians and terrestrial gastropods (less efficient) to those in mammals that contain millions of alveoli providing a huge surface area for efficient gas exchange (up to 100 m^2 in adult humans; West, 1999) (Figure 1b). The lung structure any given animal has evolved is determined by: 1) the rate at which oxygen is utilized (metabolic rate), 2) magnitude of gas exchange (a function of body size) and 3) whether secondary gas exchange surfaces are present. Flying insects, for example, have a very high demand for O_2 which is acquired using a unique tracheal system. The tracheal system originates in the body wall at valve-like openings (spiracles) and branches extensively throughout the body. It is highly efficient and delivers O_2 directly to the tissues via diffusion, but is only effective for insects because of their small body size (Bradley, 2007). However, during the Carboniferous and Permian, insect body sizes grew to 10x that of their current dimensions. Evidence suggests that this increase was made possible by higher atmospheric O_2 (27-34%), which would have resulted in more efficient gas exchange (Harrison et al., 2010).

Most aquatic species, including fish, amphibians and invertebrates utilize gills for respiration. In a few groups, including tadpoles, polychete worms, fish larvae and insect larvae, the gills are located externally. However, external gills are phylogenetically rare because of their fragile structure and relative inefficiency. Specifically, the large surface area of external gills results in high resistance to movement, but nevertheless must be constantly moved back and forth to circulate oxygenated water across the epithelium. In contrast, internal gills are a high efficiency gas exchange surface widely used by aquatic invertebrates and fish. Internal gills are stationary structures housed within a branchial chamber, through which freshwater is actively pumped. The location of the branchial chamber differs, depending on the group. For example, the stationary gills of molluscs are housed within the mantle cavity, while those of crustaceans are found between the body wall and the carapace at the base of the major appendages. In fish, the gills are located between the buccal and opercular cavities. Freshwater is actively pumped over the gills in a direction opposite to blood flow creating a highly efficient counter-current exchange which maximizes the O_2 partial pressure difference between the blood and water (Figure 1c).

Despite the advantages posed by specialized respiratory surfaces, there are a number of invertebrate groups which lack them altogether. This is especially true of certain invertebrate phyla (Cnidaria, Platyhelminthes, Nematoda) where gas exchange occurs cutaneously (Graham, 1988). One might imagine that cutaneous gas exchange imposes strict body size limitations for the animals that rely upon it. However, as with late Palaeozoic insects, diffusion limitations can be overcome to a greater extent by environmental factors. The Lions Mane jellyfish (Cyanea capillata) exemplifies this point by its ability to grow to an enormous size, reaching 8 feet in bell diameter with 100 foot tentacles. This large size is possible because these animals inhabit cold arctic waters with high levels of dissolved O_2 (Gröndahl, 1988).

Interestingly, it is not uncommon for animals to rely on multiple gas exchange structures to maintain respiratory homeostasis. For example, adult amphibians possess lungs and yet cutaneous respiration accounts for at least 20% of O_2 gas exchange in anuran species (Feder and Burggren, 1985). Another excellent example is the use of air-breathing organs by fish, a phenomenon which has evolved independently in many lineages in response to environmental hypoxia. Air-breathing in fish has been particularly well studied in the lungfish (Dipnoi) which utilize a structurally complex gas bladder connected to the esophagus (Burggren and Johansen, 1986). Finally, many invertebrates, both aquatic and terrestrial rely on cutaneous respiration in addition to lungs or gills (Graham, 1988).

The ventilation of respiratory gas exchange surfaces is made possible by the coordinated, rhythmic activity of the associated musculature. In turn, the precise timing and magnitude of respiratory muscle activity is controlled by the nervous system. Specifically, specialized central neural circuits, termed respiratory central pattern generators, are the source of the efferent motor activity driving respiration in Metazoans.

Central Pattern Generators

Respiration in the majority of metazoans does not occur passively; rather the animal actively moves the respiratory medium (water or air) over the gas exchange surface in order to enhance diffusion. This ventilation maximizes gas partial pressure differences between the body and environment thereby maximizing the diffusion rate across the surface (see Fick's equation). Ventilation of the gas exchange surface is accomplished by the activity of central pattern generators (CPG's). CPG's are central neural circuits consisting of relatively few neurons, which together generate the patterned activity underlying the rhythmic contraction of muscles.

Rhythm generation by CPG's can be accomplished via synaptic connections, pacemaker neurons or a combination thereof and these unique properties allow activated CPG's to continue to function in isolation (i.e., without input from other neural networks). The first CPG to be described was the cat locomotor CPG (Brown, 1911). Generally speaking, this CPG functions as a half-center model where two neurons with mutual inhibitory synapses alternate activity to drive antagonistic muscle groups (Figure 2a). However, many different patterns of excitatory and inhibitory synaptic connections can produce rhythmical output, a topic that has been well studied in the invertebrates (Selverston, 2010).

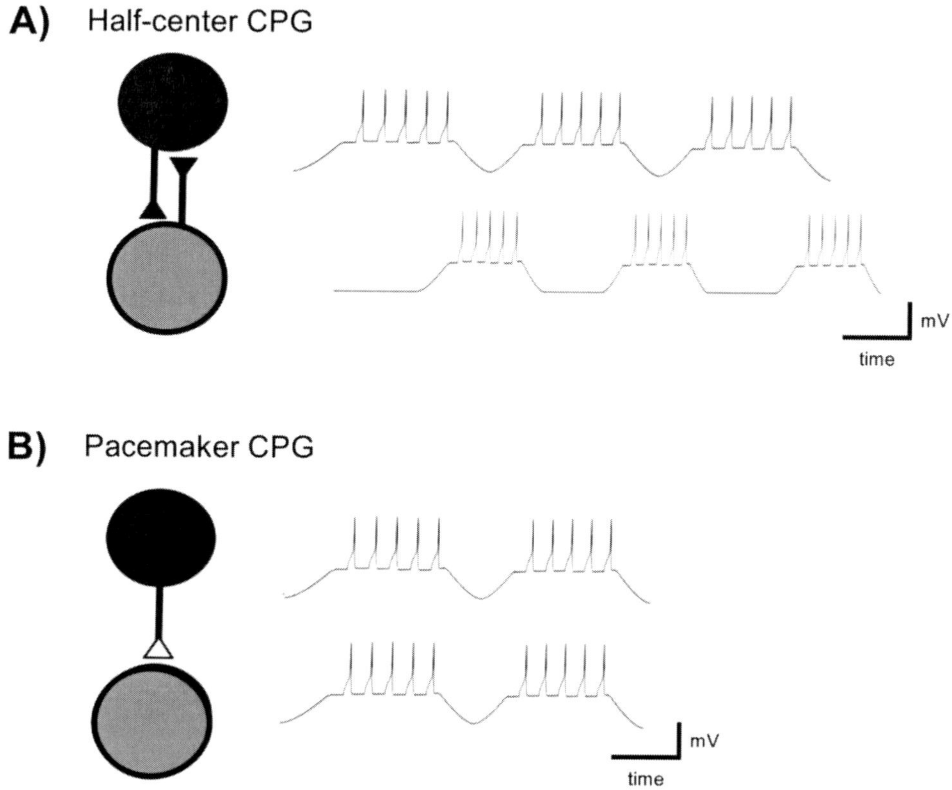

Figure 2. A model depicting the mechanisms underlying rhythm generation in central pattern generators (CPG's). A) In the half-center model, two neurons form inhibitory synapses (▲) with each other leading to alternating patterns of activity. Specifically, when the black neuron fires action potentials, activity in the grey neuron is inhibited. Once the black neuron stops firing action potentials, inhibition on the grey neuron is lifted. The grey neuron then starts firing action potentials and activity in the black neuron is inhibited. B) In pacemaker based CPG's, one neuron (black) cycles through periods of activity and inactivity. This pacemaker neuron then drives the activity of a follower neuron (grey) through an excitatory synapse (Δ) (Figure modified from Marder et al., 2001).

In many other CPG's, patterned output is generated by pacemaker neurons, which cycle through periods of activity and quiescence due to their intrinsic ionic currents (Figure 2b). This mechanism of rhythm generation was originally described for the mammalian respiratory CPG (rCPG) based on the intrinsic bursting properties of neurons within the Pre-Bötzinger Complex, located in the ventrolateral medulla (Smith et al., 1991). Current evidence however supports the interaction between pacemaker properties and excitatory synaptic connections underlying rhythm generation in the Pre-Bötzinger Complex (termed a "group pacemaker", Feldman and Del Negro, 2006). In addition, there is evidence that an additional population of neurons located within the Parafacial Respiratory Group is involved in rhythm generation, but the exact relationship between these neurons and those of the Pre-Bötzinger Complex has yet to be fully elucidated (Onimaru and Homma, 2006). This uncertainty vis-à-vis the exact location of the respiratory rhythm generator in mammals highlights the difficulty of unraveling the functions of complex neural networks.

In animals that lack a dedicated respiratory exchange surface, the CPG for locomotion can be re-configured to drive ventilation. Specifically, the peristaltic movements utilized during locomotion can also be used to create a stream of freshwater over the body surface, thus aiding in cutaneous respiration. An excellent example of this phenomenon is the lugworm Arenicola marina which inhabits the benthic zone of shallow coastal seas. A. marina resides in burrows in the often hypoxic bottom sediments. Ventilation in this species is accomplished by peristaltic movements of the body-wall musculature, which creates continuous currents of fresh sea-water over the skin (Wohlgemuth et al., 2000).

While CPG's can generate patterned activity, they do nevertheless require constant sensory feedback so that the appropriate rhythmic behaviour can be expressed in a given environment. Specifically, modulation of rCPG activity derives in large part from specialized sensory neurons termed chemoreceptors, which actively monitor O_2 levels.

Chemoreceptors

The environment is dynamic over both space and time. Therefore, animals (especially aquatic species) can be subjected to fluctuations in environmental O_2 levels. In addition, O_2 levels can be shifted outside of normal ranges by changes in metabolic state or by disease. Thus, adjusting rates of O_2 acquisition so that respiratory homeostasis is maintained represents a major challenge for the nervous system. Metazoans have evolved a common solution to this problem in the form of specialized sensory neurons, termed chemoreceptors, which continuously monitor O_2 levels and translate this information into appropriate modulation of respiratory behaviour (Figure 3a). Chemoreceptors for O_2 may be found in the central nervous system and in the peripheral nervous system innervating the skin, airways, gas exchange surfaces and arterial system (for vertebrate reviews see: Neubauer and Sunderram, 2004; Milsom and Burleson 2007).

In mammals, the main site of O_2 chemoreception is the carotid body, a small organ located at the bifurcation of the common carotid artery (De Castro, 1928; Heymans et al., 1933). During hypoxia, carotid body chemoreceptors become active leading to excitation of the rCPG and rapid increases in ventilation rate (Heymans and Bouckaert, 1930; Blain et al., 2009). However, the mechanisms underlying O_2 sensing in vertebrates remains poorly understood (reviewed by Lopez-Barneo et al., 2009). In contrast, evidence for O_2 chemoreception in invertebrates remains largely behavioural in all but a few species where chemoreceptors have been definitively identified. Specifically, there is a wealth of literature describing hypoxia avoidance and increased ventilation during hypoxic exposure in various invertebrates including cnidarians, annelids, nematodes, crustaceans, molluscs and insects (Taylor and Brand, 1975; Fredericks, 1976; Dusenbery, 1983; Massabuau and Burtin, 1984; Wohlgemuth et al., 2000; Wu, 2002; Bradley, 2007). However, the identity of O_2 chemoreceptors has only been elucidated in D. melanogaster (terminal sensory cones), C. elegans (BAG and URX neurons) and L. stagnalis (osphradial neurons) (Zimmer et al., 2009; Bell et al., 2007; Ribeiro et al., 2010; Vermehren-Schmaedick et al., 2010). The molecular mechanism underlying O_2 sensitivity in these species is an atypical soluble guanylyl cyclase (sGC), which does not bind nitric oxide (as in typical sGC's) but instead binds O_2. The dissociation of O_2 from the binding site of this enzyme (during hypoxia) causes a conformational change that activates the catalytic site leading to the production of cGMP, a powerful modulator of many molecular pathways (Morton and Vermehren, 2007; Figure 3b).

Since atypical sGC's function as O_2 sensors in different phyla, it is possible this represents a conserved molecular mechanism facilitating O_2 chemoreception among the invertebrates.

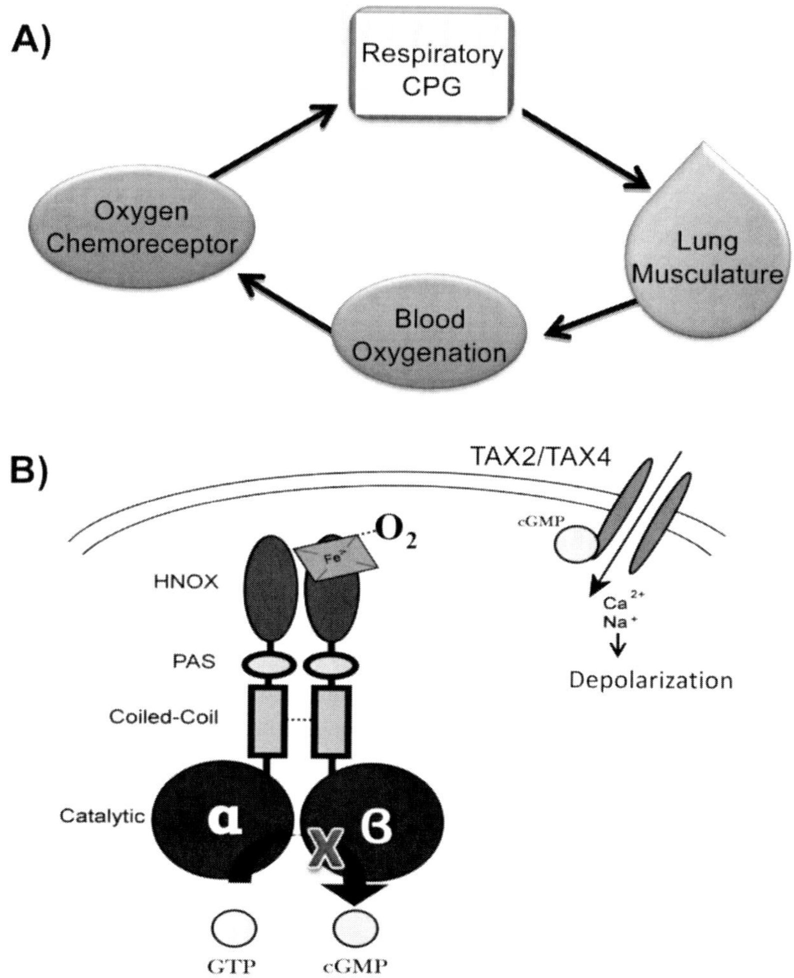

Figure 3. Oxygen chemoreception modulates respiration in metazoans but the underlying mechanism is known only for invertebrates. A) A simplified representation of the metazoan respiratory network demonstrates the role of O_2 chemoreceptors in translating blood oxygenation into respiratory activity. Specifically, as blood O_2 drops, chemoreceptors become activated. This neural activity speeds up the respiratory rhythm produced by rCPG neurons. The increased rhythm travels through motor nerves to the muscles surrounding the lungs, thus increasing the frequency of breathing and raising blood oxygenation. As blood O_2 increases, chemoreceptors de-activate, the respiratory rhythm slows, and so on. B) In invertebrates, O_2 chemoreception is mediated in sensory neurons by atypical soluble guanylyl cyclases (sGC's), a heterodimeric protein consisting of alpha and beta subunits. PAS and Coiled-coil domains are thought to mediate subunit interactions. The HNOX domain of the B-subunit contains a heme group (Fe^{2+}) that binds O_2 resulting in a conformation which inhibits the production of cGMP (an intracellular messenger) by the catalytic domain. During hypoxia, O_2 is not available to bind to the HNOX domain and the catalytic domain enters an active conformation. The cGMP produced may open cyclic nucleotide gated ion channels (TAX2/TAX4) resulting in the influx of positive ions, cellular depolarization and action potentials (Morton and Vermehren, 2007; Scott, 2011). (Figure modified from Luo et al., 2009).

While O_2 chemoreceptors certainly help shape the respiratory rhythm by linking metabolic demands with external O_2 levels, they represent only a fraction of the nervous systems ability for adaptation. Specifically, the nervous system of all animals functions as a dynamic system by constantly integrating information to change the cellular and synaptic properties of neurons resulting in new and different behaviours.

Respiratory Plasticity

Plasticity in the nervous system refers to the ability of neurons to enhance or depress the strength of their synaptic connections depending on previous or ongoing neural activity. As a result, neurons alter the way they receive and transmit information in both the short- and long-term, producing flexibility in neural function and behaviour that increases an animals scope for adaptation. Neural plasticity likely functions in most if not all neural networks and is a ubiquitous feature of the nervous system of all animals. Respiratory plasticity refers specifically to changes in synaptic strength between respiratory neurons, thereby changing the expression of respiratory behaviour.

In mammals, several different patterns of respiratory plasticity are induced by hypoxia depending on the severity and duration of the challenge. This plasticity affects many different neural networks, but the changes occurring in the sensory pathways between the carotid body O_2 chemoreceptors and their post-synaptic targets in the nucleus tractus solitarius (NTS) have been particularly well studied. The NTS, located in the medulla, is an important site involved in modulating respiratory behaviour. Specifically, the NTS integrates sensory information from the carotid body O_2 chemoreceptors (via the carotid sinus nerve), pulmonary stretch receptors and irritant receptors (i.e., smoke, dust) and relays this information to the rCPG (Alheid and McCrimmon, 2008). For example, exposure to acute hypoxia or artificial stimulation of the carotid sinus nerve produces a progressive increase in ventilation that is distinct from reflex activation of O_2 chemoreceptors (Eldridge and Millhorn, 1986; Engwall et al., 1991). However, when the animal is returned to normoxic conditions, ventilation returns to baseline levels within a few seconds. This transient increase in ventilation appears to be the result of short-term increases in the synaptic strength between carotid body chemoreceptors and post-synaptic neurons of the NTS and is termed short-term potentiation (Mifflin, 1997; Figure 4a). Specifically, short-term potentiation may stabilize respiratory activity during shifts between low and high respiratory drive (Menendez et al., 1999).

Over longer time periods, hypoxia produces distinct patterns of respiratory plasticity in mammals depending on whether the challenge occurs chronically or intermittently. Respiration becomes depressed during chronic hypoxia in a phenomenon termed hypoxic ventilatory decline (HVD; Figure 4b). In the central nervous system, increased inhibitory GABA-ergic neurotransmission appears to mediate HVD. Specifically, in freely behaving rats exposed to 45 minutes of hypoxia, GABA levels in the NTS were increased and GABA receptor antagonists reversed the expression of HVD (Tabata et al., 2001). This respiratory decline during sustained hypoxia is likely adaptive in that it allows central neurons to conserve energy and reduce their need to utilize anaerobic metabolism.

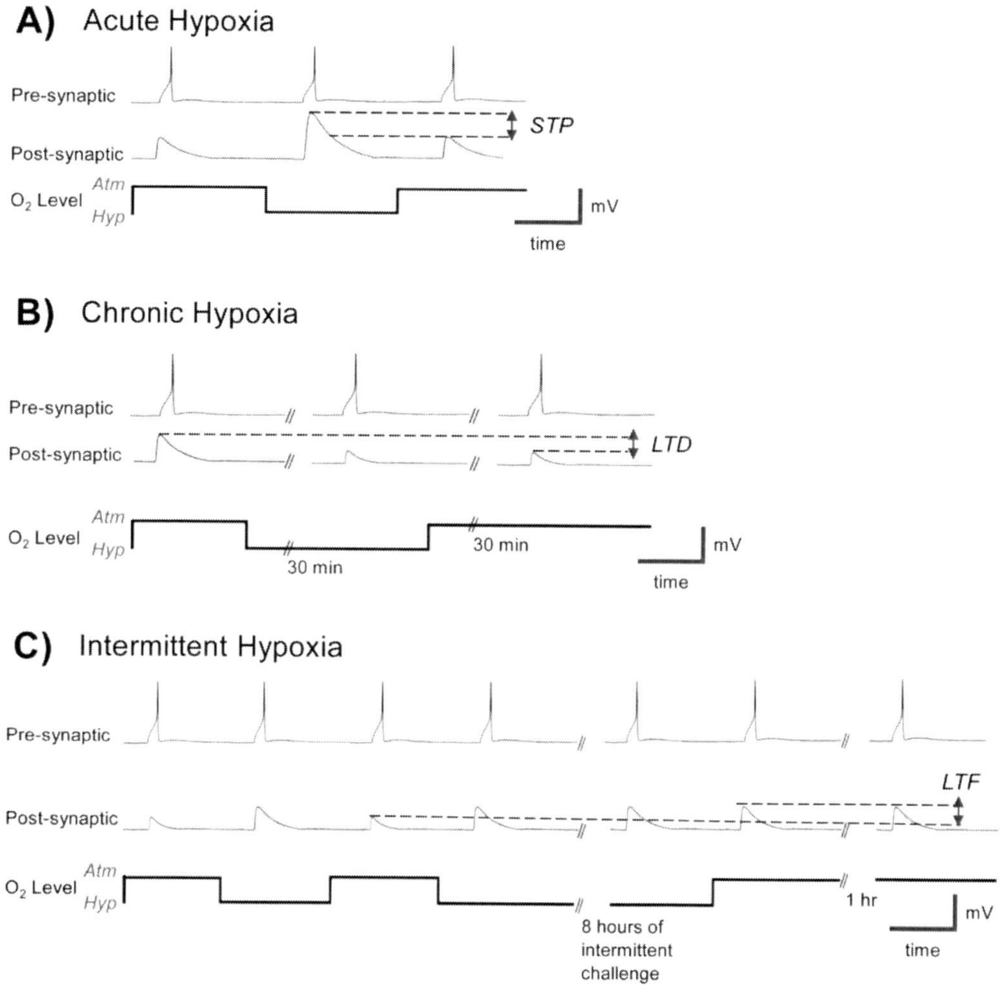

Figure 4. A model demonstrating respiratory plasticity at the synaptic level. A) During acute hypoxia (Hyp, 1 min) an action potential in the pre-synaptic neuron produces a post-synaptic potential (PSP) with a larger amplitude than that observed under atmospheric conditions (Atm). This transient increase in synaptic strength is termed short-term potentiation (STP). B) At longer time scales (30 min), hypoxia depresses the synaptic strength between neurons leading to a decrease in PSP amplitude elicited by pre-synaptic action potentials. The decrease in synaptic strength is still observable after exposure to 30 minutes of atmospheric O_2. This long-term depression (LTD) of synaptic strength underlies the expression of hypoxic ventilatory decline. C) In contrast, several hours of intermittent hypoxia produces a gradual increase in PSP amplitude that persists even after exposure to atmospheric conditions for an hour. This long-term increase in synaptic strength is termed long-term facilitation (LTF).

In contrast, intermittent hypoxia enhances respiration by increasing baseline respiratory drive (termed long-term facilitation, LTF; Prabhakar et al., 2007). Specifically, animals exposed to intermittent hypoxia continue to hyperventilate when returned to normoxic conditions (Figure 4c). Within the NTS, intermittent hypoxia may increase activation of excitatory AMPA receptors (de Paula et al., 2007), reduce hyperpolarizing K_{ATP} currents (Zhang et al., 2008), and increase spontaneous transmitter release (Kline et al., 2007) all of which lead to increased neural excitability underlying LTF. In recent years, there has been a surge in research activity aimed at understanding intermittent hypoxia in a clinical sense because of its link with obstructive sleep apnoea (OSA). It is thought that up to 30% of the population might suffer from OSA, a condition characterized by periodic cessation of breathing during sleep, effectively producing an intermittent hypoxia challenge. Unfortunately, intermittent hypoxia in humans is associated with significant pathology including disturbed sleep, high blood pressure, gray matter loss and cognitive impairment (Dempsey et al., 2010). Therefore, intermittent hypoxia is typically considered to be maladaptive (Prabhakar and Semenza, 2012).

While the effects of chronic and intermittent hypoxia on respiratory networks and behaviour have been studied extensively in mammals, little is known regarding how invertebrate respiratory neurons are altered by these challenges. Specifically, there is a wealth of information available on the ability of invertebrates to tolerate various degrees and durations of hypoxia, mostly relating to marine ecosystems (Vaquer-Sunyer and Duarte, 2008). There are also a few studies documenting the changes that occur in the respiratory behaviour of invertebrates in response to chronic hypoxia in terms of ventilation rate and metabolic demand for O_2 (Kanz and Quast, 1991; McMahon, 2001). In addition, several recent studies have investigated the effects of chronic and intermittent hypoxia on gene and protein expression in crustaceans (Brouwer et al., 2007; Zeis et al., 2009). However few, if any studies to date have explicitly examined changes in the invertebrate nervous system in response to chronic or intermittent hypoxia. This is an important consideration because while mammals may be exposed to chronic and intermittent hypoxia under some conditions (disease or high altitude), aquatic invertebrates regularly face these challenges as a consequence of the environment that they inhabit. Therefore, invertebrates may prove to be an invaluable tool in understanding hypoxia induced plasticity in neural networks. For example, does acute, chronic and intermittent hypoxia produce similar plastic changes in the sensory networks of invertebrates and are these changes adaptive under certain environmental conditions? In addition, the mechanisms underlying synaptic plasticity are likely similar among metazoans (reviewed by Glanzman, 2010), making invertebrates a good model for understanding patterns of respiratory plasticity.

The remainder of this chapter will be devoted to a discussion of respiration in the pond snail *Lymnaea stagnalis*. We will describe key features of the respiratory system (gas exchange surfaces, CPG, chemoreceptors and respiratory plasticity) and how they are adapted to meet specific environmental challenges in the short- and long-term. In addition, the following sections will outline how the behavioural adaptations of *Lymnaea* to hypoxia and the underlying neural mechanisms can advance our understanding of neuroethology and medicine.

Anatomy of the *Lymnaea* Respiratory System

Respiratory Exchange Surfaces: The Skin and Lung

The earliest form of respiration undoubtedly occurred as passive diffusion of gasses across the skin of small invertebrates. This cutaneous respiration is still exploited in a large number of extant invertebrate groups lacking respiratory structures, but also occurs commonly in higher phyla that possess gills or lungs (Graham, 1988). In *Lymnaea* stagnalis, cutaneous respiration is used in all stages of development for O_2 uptake. This is a practical form of gas exchange due to the relatively small size of these animals, with the largest specimens documented to reach a shell length of only 5 cm (Crabb, 1929).

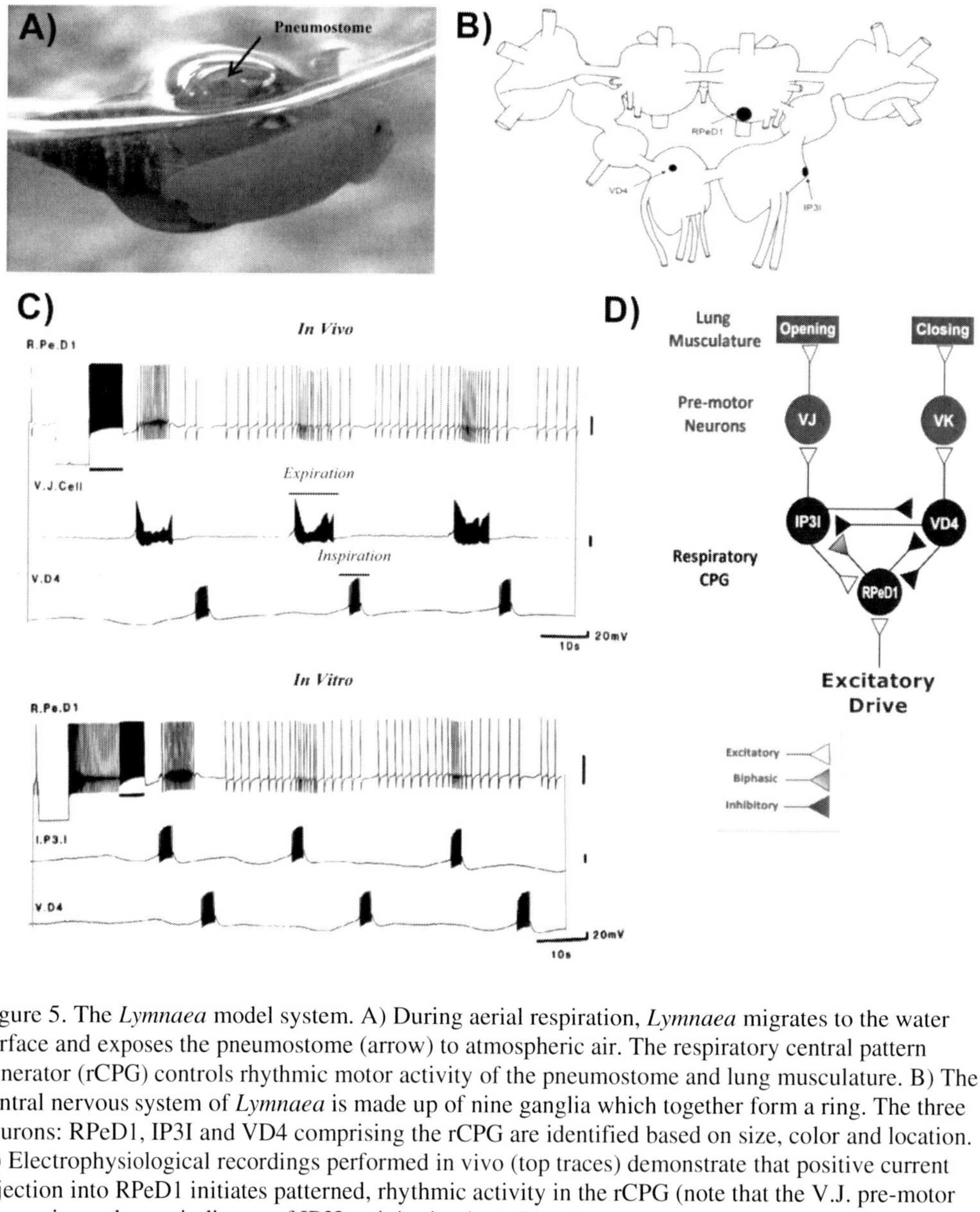

Figure 5. The *Lymnaea* model system. A) During aerial respiration, *Lymnaea* migrates to the water surface and exposes the pneumostome (arrow) to atmospheric air. The respiratory central pattern generator (rCPG) controls rhythmic motor activity of the pneumostome and lung musculature. B) The central nervous system of *Lymnaea* is made up of nine ganglia which together form a ring. The three neurons: RPeD1, IP3I and VD4 comprising the rCPG are identified based on size, color and location. C) Electrophysiological recordings performed in vivo (top traces) demonstrate that positive current injection into RPeD1 initiates patterned, rhythmic activity in the rCPG (note that the V.J. pre-motor neuron is used as an indicator of IP3I activity in vivo). The rhythm produced by these three neurons is maintained when they are cultured in vitro and allowed to reform their synaptic connections (bottom traces). D) Diagrammatic representation of the *Lymnaea* respiratory network. When RPeD1 receives excitatory drive, bursting activity is produced leading to rhythmical rCPG activity. IP3I and VD4 alternately excite pre-motor neurons leading to cycles of expiration and inspiration, respectively (Figure modified from Syed et al., 1990 and Janes and Syed 2012).

In addition to cutaneous respiration, *Lymnaea* can also breathe air. *Lymnaea* is a member of the Basommatophora, a suborder that includes freshwater pulmonate gastropods (Correa, 2010). The "pulmonate" classification refers to the presence of a highly vascularized lung cavity formed by a discrete portion of the mantle cavity. This cavity opens to the atmosphere under the right side of the shell via a modified section of the siphon termed the pneumostome. Thus, when *Lymnaea* is at the surface, it can breathe atmospheric air in a process known as "aerial respiration" (Noland, 1946). Aerial respiration is initiated when the snail comes to the water surface and exposes its pneumostome to the surrounding air (Figure 5a). As the pneumostome opens, muscles surrounding the lung cavity contract forcing "stale" air out (expiration). Subsequently, the lung musculature relaxes allowing passive re-inflation of the lung (inspiration) and the pneumostome closes trapping the atmospheric air inside (Syed et al., 1991a). Aerial respiration occurs episodically at intervals separated by 20 minutes or more when water oxygenation is close to atmospheric. In essence, this type of episodic surfacing and respiration is analogous to that performed by diving mammals. Jones (1961) performed a detailed analysis of bimodal respiratory behaviour in *Lymnaea* and concluded that aerial respiration of adult animals (shell length = 4 cm) accounts for 50-60% of O_2 uptake when the surrounding water is equilibrated with the atmosphere. It is important to note however that the contribution of aerial versus cutaneous respiration in these bimodal breathers varies with body size and water oxygenation.

Neural Networks Underlying Aerial Respiration

Lymnaea aerial respiration is a readily observable, rhythmic motor behaviour that provides an elegant model for the study of neural mechanisms underlying respiratory pattern generation. The central nervous system of *Lymnaea* is organized into a ring of nine ganglia that together contain approximately 20,000 neurons which are relatively large (40-90μm) and identifiable based on their location, size and color (Figure 5b). These characteristics facilitated the identification by Syed et al. (1990) of three central neurons involved in the generation of the motor activity producing aerial respiration (Figure 5b). Using in vivo preparations, electrophysiological recordings were performed on the two interneurons "Input 3 Interneuron" (IP3I) and "Visceral Dorsal 4" (VD4). Reciprocal inhibitory synapses between IP3I and VD4 produced an alternating pattern of excitation in these two neurons, each of which drove a distinct population of motor neurons leading to cycles of expiration and inspiration, respectively. A third neuron, the giant dopamine cell "Right Pedal Dorsal 1" (RPeD1), was hypothesized to be involved in the circuit function as it was found to cause excitation of IP3I through post-inhibitory rebound excitation while inhibiting VD4 (Figure 5c, top traces). In order to confirm that these three neurons are necessary and sufficient to produce the aerial respiratory rhythm of *Lymnaea*, Syed et al., (1990) reconstructed this network in cell culture and explored the resulting rhythmic activity. Remarkably, RPeD1, VD4 and IP3I reformed their appropriate synapses and produced activity that was not quantitatively different than that recorded in vivo (Figure 5c, bottom traces). Since positive current injection into RPeD1 (the "command neuron") was necessary to initiate rhythmic activity in the network, the authors concluded that this circuit relies on synaptic connections and not pacemaker neurons. Therefore, the rCPG neurons: RPeD1, VD4 and IP3I drive motor activity in the pneumostome and pulmonary musculature thus producing aerial respiratory behaviour (Syed and Winlow, 1991b; Janes and Syed, 2012; Figure 5d).

While the rCPG is essential for the production of the basic respiratory rhythm, this motor output must be flexible to allow *Lymnaea* to adapt its breathing to an ever-changing environment. For example, one can imagine that it would be advantageous for *Lymnaea* to increase the frequency of aerial respiration during periods of aquatic hypoxia in order to compensate for the decrease in cutaneous O_2 acquisition. This adaptive behaviour has indeed been observed in adult *Lymnaea* (Jones, 1961; Bell et al., 2007) and is mediated by the activity of O_2 chemoreceptors. Specifically, Inoue et al. (2001) performed experiments using *Lymnaea* semi-intact preparations (central nervous system connected to all peripheral organs) to determine where hypoxic excitation of the rCPG originates. When the central ring ganglia and peripheral regions (pneumostome, lung, etc.) were compartmentalized, perfusion of the periphery with hypoxia (67 torr O_2) initiated the respiratory rhythm by previously silent rCPG neurons. However, activity was not observed when the central ring ganglia were made hypoxic (Inoue et al., 2001; Figure 6a). These studies suggested that hypoxic drive to the rCPG originates at the periphery, but the putative O_2 chemoreceptors had yet to be identified. Interestingly, RPeD1 is the only rCPG neuron that projects to the peripheral tissues. Considering that positive current injection into RPeD1 was necessary to initiate rCPG activity in vitro, Inoue et al., (2001) hypothesized that O_2 chemoreceptors could either be projections of RPeD1 or peripheral neurons synapsing with RPeD1. To distinguish between these two possibilities, synaptic transmission was blocked by perfusion of the peripheral compartment with a zero Ca^{2+} saline solution. Concurrent perfusion of the periphery with hypoxia produced no change in rCPG activity, thus demonstrating that RPeD1 activity is modulated through synaptic connections with peripheral neurons (Figure 6a). Expanding on this early work, Bell et al. (2007) traced a number of RPeD1 projections to a small peripheral ganglion located adjacent to the pneumostome (Figure 6b). This structure, the osphradial ganglion, had been previously described in *Lymnaea* and proposed to contain neurons sensitive to environmental factors (Kamardin, 1976; Sokolov and Kamardin, 1977). To determine if osphradial ganglion neurons formed synapses with RPeD1, direct intracellular recordings were performed on the large ganglionic neurons of this structure in a semi-intact preparation. Direct stimulation of the osphradial neurons triggered bursting activity in RPeD1 that was blocked by high divalent cation saline ($6xCa^{2+}/6xMg^{2+}$) and zero Ca^{2+}/high Mg^{2+} saline indicating the presence of a monosynaptic, excitatory chemical synapse. The antagonist mecamylamine reversibly blocked synaptic transmission between osphradial neurons and RPeD1 demonstrating that this synapse is mediated by acetylcholine (Bell et al., 2007; Figure 6c). In a final experiment, the identified osphradial neurons were cultured in isolation or synapsed with RPeD1 and exposed to hypoxia ranging from 35 to 5 torr O_2. The osphradial neurons responded to hypoxia with bursting activity in both preparations and produced concurrent bursting in RPeD1 in the synapsed pairs (Figure 6d). Isolated RPeD1 showed no response to hypoxia (Bell et al., 2007). Together, these studies demonstrate that *Lymnaea* osphradial neurons represent true O_2 chemoreceptors which drive rCPG activity during hypoxia (Janes and Syed, 2012). Interestingly, a recent study reports that sGC's cloned from *Lymnaea* RNA are localized to the O_2-sensing ganglionic neurons of the osphradial ganglion (Ribeiro et al., 2010).

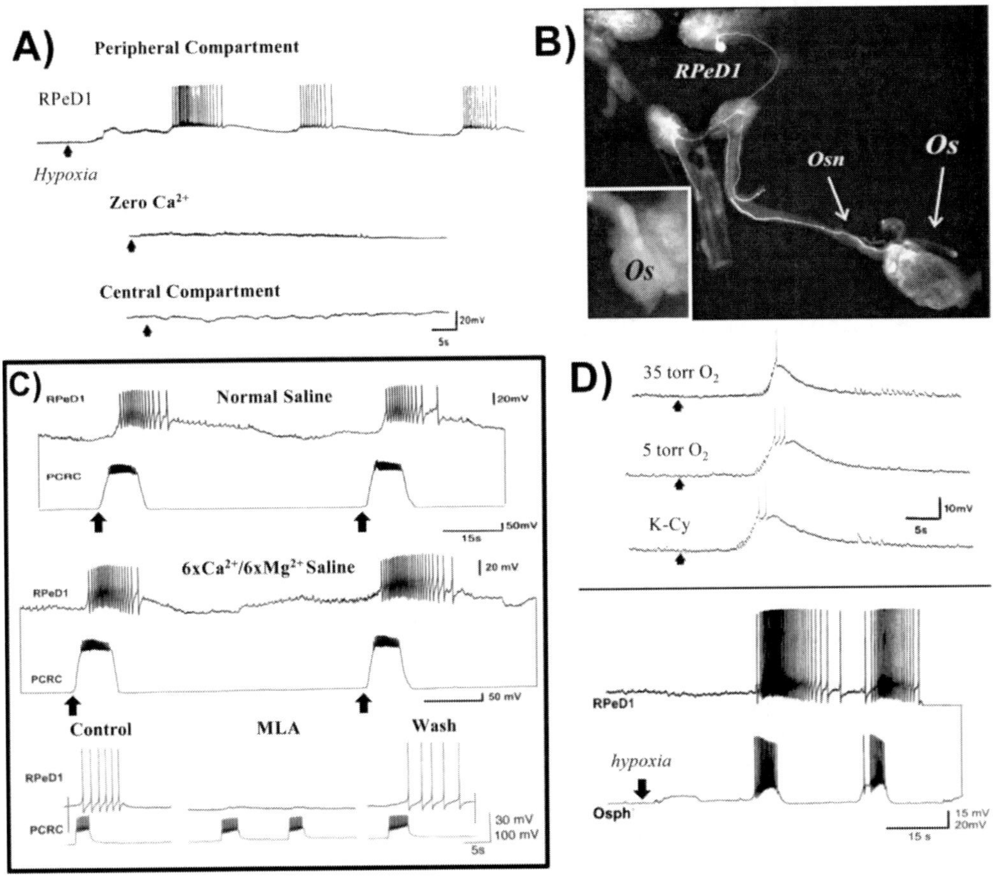

Figure 6. Identification of peripheral oxygen chemoreceptors in *Lymnaea*. A) Perfusion of the peripheral organs with hypoxic saline (67 torr O_2) produces rhythmic bursting activity recorded from the rCPG neuron, RPeD1. Zero Ca^{2+} saline, used to block synaptic function, abolishes this activity suggesting that hypoxic excitation of RPeD1 originates in the peripheral nervous system. There is no response from RPeD1 when the central nervous system is made hypoxic suggesting that O_2 chemoreceptors are not located centrally (Inoue et al., 2001). B) A semi-intact preparation showing that the central ring ganglia connects to the osphradial ganglion (Os) by a peripheral nerve. Injection of fluorescent Lucifer Yellow dye into RPeD1 labels a peripheral axonal projection that terminates at the osphradial ganglion. Insert: the osphradial ganglion shown at higher magnification. C) Stimulation of an osphradial neuron (PCRC) in a semi-intact preparation produces bursting activity in a previously quiescent RPeD1. This excitation persists in the presence of high Ca^{2+}/high Mg^{2+} saline indicating a monosynaptic connection, but is blocked by the cholinergic antagonist mecamylamine (MLA). D) Isolated osphradial neurons respond to acute hypoxia with dose-dependent increases in bursting activity. This response is also observed following the addition of potassium cyanide (K-Cy), which mimics hypoxia (top traces). Perfusion of isolated osphradial neurons with hypoxia also produces simultaneous excitation of RPeD1 (bottom traces) (Figure modified from Bell et al., 2007 and Janes and Syed 2012).

Adaptation of *Lymnaea* Respiration in Response to the Environment

Environmental Challenges

Species belonging to the family *Lymnaeidae* are distributed worldwide and inhabit freshwater systems ranging from rivers to stagnant swamps (Strong et al., 2008; Correa et al., 2010). *Lymnaea* stagnalis has a particularly wide range across the northern hemisphere, with occurrences reported from Asia, Europe, the Mediterranean and North Africa but is also found in New Zealand and throughout North America (Budha et al., 2010). Within this range, *Lymnaea* inhabits a variety of water bodies including rivers, streams, ponds and lakes where it can be found creeping around on the vegetation or along the benthic zone in shallow waters (Jones, 1961). Although these freshwater habitats might appear to be fairly stable and homogenous there is actually an incredible degree of spatial and temporal variation that occurs in these ecosystems (Figure 7a). Specifically, the dramatic short- and long-term changes in temperature and oxygen levels that occur in these habitats present significant challenges for invertebrate respiration.

Two of the most profound changes that can occur from year to year are changes in water level and flow rates, which primarily affect smaller ponds, rivers and streams. Hot, dry weather causes significant evaporation of these water bodies leading to stagnant, hypoxic conditions and high temperatures. In some cases, these conditions are further exacerbated by unsustainable draining of water bodies for irrigation, livestock and urban usage. Conversely, widespread flooding washes soil from the surrounding watershed into lakes and rivers creating turbid conditions that facilitate the establishment of hypoxia by limiting light penetration necessary for photosynthetic O_2 production. In addition, the high turbidity decreases the depth at which photosynthetic organisms can be found, thus altering the natural O_2 gradient and forcing benthic organisms into shallower water.

The seasons that occur in temperate climates can have profound effects on the water conditions found in lakes and ponds. Wetzel (1983) described in detail how the seasons affect temperature and O_2 gradients in lakes, as well as the confounding effects of lake productivity. During the winter for example, O_2 limitation may occur when ice prevents the diffusion of atmospheric O_2 into the surface waters (termed the epilimnion). In addition, snow-covered ice prevents light from reaching the water and abolishes photosynthesis. As organisms continue to respire under the ice, available O_2 is used up leading to anoxia and winter-kill of animals with higher metabolic rates (Figure 7b, Winter). In temperate climates, water temperatures typically range between 1 to 4°C during the winter months. Similarly, O_2 limitation can also be a serious problem for aquatic animals during the summer. As daytime temperatures increase, the epilimnion heats up while deeper waters near the lake bottom (hypolimnion) remain relatively cool. This thermal layering results in a density gradient, with the warm epilimnion layers "floating" on top of cooler waters below in a phenomenon termed stratification. The epilimnion can reach nearly 30°C during this period and supports abundant growth of algae and plants which saturate the water with O_2 during the day. In contrast, the hypolimnion and bottom sediments contain high densities of organisms performing cellular respiration as well as aerobic decomposition, both of which rapidly use up the available O_2. As a result of stratification, it is difficult for winds and currents to mix the warm, O_2-rich epilimnion with deeper water layers and therefore the hypolimnion may become hypoxic (Figure 7b, Summer). It is important to note as well that in addition to producing stratification, high water temperatures also reduce O_2 solubility. For example, freshwater at 5°C contains 12.8 mg/L O_2 but the same water at 30°C only contains 7.5 mg/L O_2. Clearly, this phenomenon has the potential to intensify hypoxic stress.

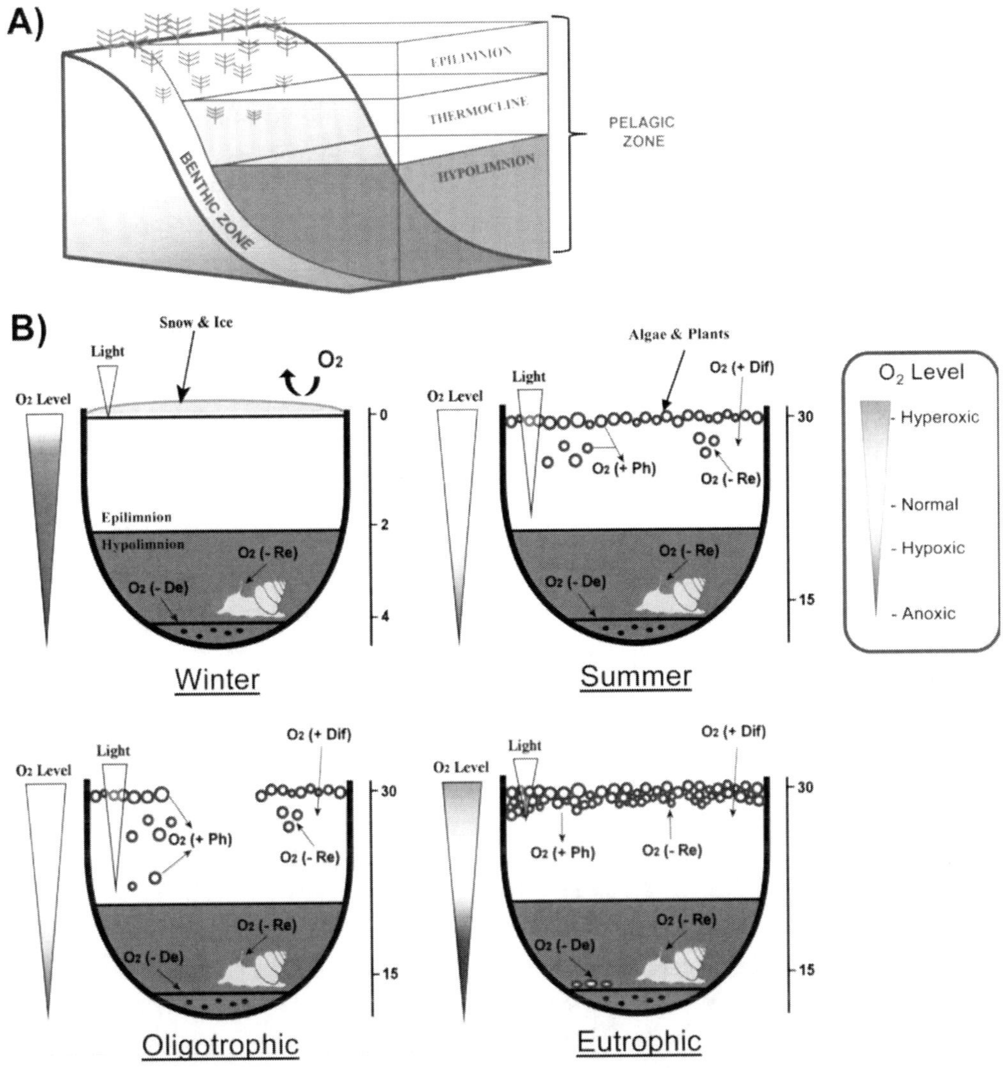

Figure 7. The effects of season and productivity on O_2 availability in lakes. A) Lake ecosystems can be broken down into the benthic zone, comprising the bottom sediments and the pelagic zone, comprising the water column. The pelagic zone can be further subdivided into distinct regions of epilimnion (surface waters), thermocline (where stratification occurs) and hypolimnion (bottom waters). (Figure modified from Pearson Education Inc. source) B) The availability of O_2 in lakes depends on the balance between diffusion of O_2 from the atmosphere (+ Dif) and photosynthesis (+ Ph) which add O_2 to the water and organismal respiration (- Re) and decomposition (- De) which deplete O_2. The seasons and enrichment level affects the balance of these processes and can lead to conditions of hypoxia or anoxia in the hypolimnion (see text for details and also Wetzel, 1983).

In low productivity lakes (oligotrophic), algae and plant growth at the surface is sparse allowing light to penetrate deeper into the water column. Photosynthesis occurring in deeper waters provides a local source of O_2 for benthic organisms (Figure 7b, Oligotrophic). In high productivity lakes (eutrophic) thick algae blooms produce hyperoxic conditions at the surface, but cut off light from deeper waters preventing additional photosynthesis. Furthermore, these algae blooms consume much of the available O_2 for their own respiration and intensify aerobic decomposition rates as they die off and sink down to the benthic zone. As a result, the excessive algae growth that occurs in eutrophic lakes facilitates the establishment of hypoxia or even anoxia within the hypolimnion (Figure 7b, Eutrophic). While lakes typically transition from oligotrophic to eutrophic over time via the build-up of nutrients and sediment, artificial eutrophication resulting from human activity is threatening natural ecosystems. Specifically, runoff from agricultural land and livestock areas as well as partially treated sewage contains high levels of phosphorus and nitrogen. As these two nutrients become concentrated in surrounding water bodies the resulting rapid eutrophication allows algal population growth to explode. The anoxic conditions produced by algal activity leads to summer die-offs as well as long-term changes in community composition and decreased biodiversity.

Thus far we have focused our discussion on the lake and pond systems that *Lymnaea* inhabits. However, it should be noted that the problem of eutrophication is not limited to freshwater ecosystems. Specifically, extreme eutrophication has led to the development of marine Dead Zones, which represent vast oceanic areas where the bottom waters are severely hypoxic or anoxic and animal life is limited to a few of the most tolerant species such as nematodes and priapulids (small bodied marine worms) (Diaz and Rosenberg, 2008; Vaquer-Sunyer and Duarte, 2008). One of the most dramatic examples of this phenomenon is currently occurring in the Gulf of Mexico. In the spring and summer, nutrient rich waters from the Mississippi River produce a marine Dead Zone along the continental shelf off the coast of Louisiana and Texas which grew to 20,000 km^2 in 2011 (Roman et al., 2012). Within this zone water O_2 levels near the bottom sediments can drop to 2 mg/L (~40 torr O_2) or less leading to the loss of many common taxa including crustaceans, bivalves, gastropods and echinoderms. Important fishery resources have also been affected by spreading dead zones within the Gulf of Mexico. For example, the catch rates of brown shrimp, one of the most economically important species in the Gulf, are negatively correlated with Dead Zone area. As the bottom waters become hypoxic, benthic fish and invertebrates either die or are forced to migrate away from their preferred habitat leading to increased susceptibility to predation and disruption of life cycles. Populations of pelagic species are also negatively affected following zooplankton community changes in response to the spreading hypoxia (reviewed by Rabalais et al., 2002).

The preceding examples illustrate the point that many aquatic invertebrates likely face conditions of hypoxia in their habitat over different timescales. This is more likely to occur in lakes, bays and other relatively 'still' bodies of water than in fast moving streams, rivers and open oceans where high turbulence and currents greatly improve the mixing of O_2 into the water column. When faced with hypoxic conditions, invertebrates can employ several different strategies including depressed metabolism to reduce O_2 demand, migration to areas with higher O_2 levels or changes in respiratory behaviour aimed at restoring homeostasis. However, when hypoxic conditions become more severe or occur more frequently these adaptations may not be sufficient for the animal to meet its O_2 demands. This hypoxic stress results in reduced growth and reproduction and in extreme cases, death.

Adaptation of *Lymnaea* to Hypoxia

The Acute Hypoxic Response

In aquatic ecosystems, environmental hypoxia poses one of the most severe challenges to invertebrate respiration. Jones (1961) collected environmental data from shallow drainage ditches north of Leiden (Netherlands) that *Lymnaea* were found inhibiting. During June, the bottom water (which was only 15 cm below the surface) O_2 levels ranged between approximately 1.32 mg/L (20 torr) before sunrise to 5.71 mg/L (100 torr) at midday. Conversely, the surface water O_2 ranged from approximately 0.66 mg/L (10 torr) before sunrise to 23.23 mg/L (490 torr) at midday. The abundant plant life found in these pools likely consumes much of the available O_2 during the night and prevents circulation that would mix the stratified layers. As a result, *Lymnaea* is exposed to conditions of severe hypoxia that are established throughout the water column during the night. Clearly, cutaneous respiration under these conditions is inadequate to meet metabolic O_2 demands. Fortunately, aerial respiration allows *Lymnaea* to breathe O_2 rich air, thus satisfying metabolic requirements for O_2 despite environmental hypoxia. Specifically, when exposed to acute hypoxia (30 minutes) ranging from 150 torr O_2 down to 5 torr O_2 (~8.5 to 0.3 mg/L), *Lymnaea* depends increasingly on aerial respiration over cutaneous respiration and is able to maintain normal levels of O_2 consumption even during very severe aquatic hypoxia (Jones, 1961; Figure 8a). Therefore, by employing aerial respiratory behaviour *Lymnaea* is able to sustain relatively normal levels of activity during acute hypoxia. This adaptation is highly advantageous, as it allows the animal to inhabit hypoxic environments, yet continue normal behaviours such as feeding. In addition, *Lymnaea* does not need to remain at the water surface near to O_2 rich air, and can spend most of its time hiding from predators among the vegetation.

The acute hypoxic response of *Lymnaea* is mediated largely by the O_2 chemoreceptors located within the osphradial ganglion, which drive activity in the rCPG to produce increased breathing frequency. These chemoreceptors appear to be quite sensitive to small changes in environmental O_2 and respond rapidly to hypoxia, implying that there is a fine degree of control over respiration even in these simple invertebrates. Specifically, *Lymnaea* exposed to O_2 levels above atmospheric (hyperoxia) in a closed system showed no change in the duration of respiratory behaviour compared with atmospheric levels. Conversely, animals exposed to hypoxia showed a logarithmic increase in the duration of aerial respiration, with a dramatic increase occurring at approximately 50 torr O_2 (~2.8 mg/L, elevation = 1,083m; Figure 8b) (Syed Lab, unpublished data). The observed changes in respiratory behaviour occurred within a few minutes of the animals being exposed to the hypoxic environment.

The finding that aerial respiration in adult *Lymnaea* increases at approximately 60 torr O_2 is interesting because this phenomenon has been documented for many other widely divergent species. Specifically, studies conducted using stage XXI tadpoles and goldfish report that the frequency of aerial respiration in these species also increases markedly at an environmental O_2 level of about 60 torr (Burggren, 1982; Crowder et al., 1998). Furthermore, the frequency of carotid sinus nerve activity recorded from cats increases dramatically when blood oxygenation drops to about 50 torr O_2 (Lihiri et al., 1993). Taken together, these results suggest that the activity of O_2 chemoreceptors and the resulting ventilation is maximized below a defined 'level' of hypoxia in many different animals.

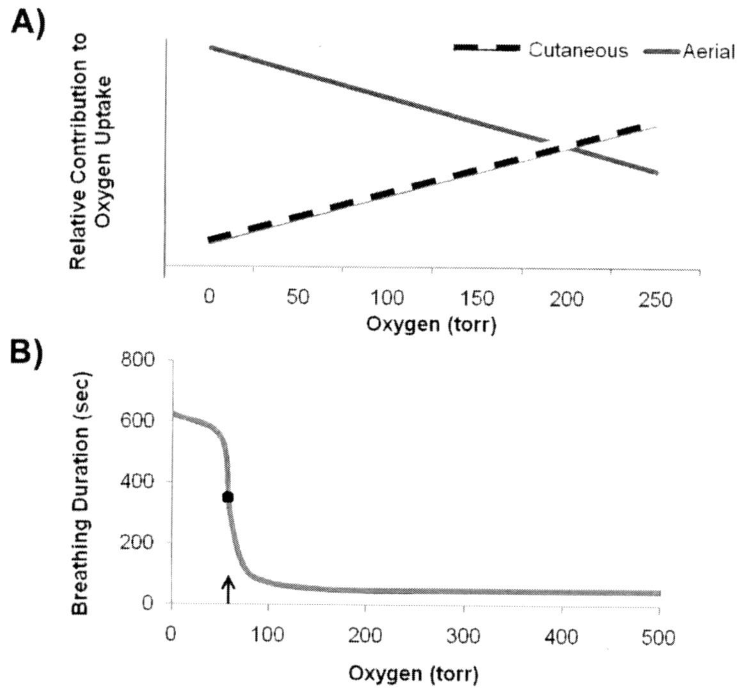

Figure 8. *Lymnaea* respiration is adapted to hypoxic habitats. A) Cutaneous respiration dominates under atmospheric (=150 torr O_2 at sea level) and hyperoxic (> 150 torr O_2) conditions. However, as the aquatic environment becomes more hypoxic, aerial respiration becomes increasingly important in maintaining O_2 uptake (Figure modified from Jones, 1961). B) The aerial respiratory behavior of adult *Lymnaea* is suppressed under conditions of normoxia and hyperoxia ($PO_2 \geq 159$ torr). Conversely, exposure of snails to increasingly hypoxic conditions ($PO_2 < 159$ torr) elicits a substantial increase in observed total breathing time. Note that respiration time increases dramatically around 60 torr O_2, a phenomenon that has been documented for many species (Syed Lab, unpublished data).

Respiratory Plasticity

Until recently, respiratory plasticity in response to hypoxia had only been studied in mammalian models. However, new evidence is emerging that respiratory plasticity can also be observed in invertebrates and may represent a fundamental and stereotypical response of the nervous system to a specific pattern of sensory input. Specifically, while exploring the characteristics of the synapse between *Lymnaea* osphradial chemoreceptors and RPeD1, Bell et al. (2007) observed short-term enhancement of synaptic strength in response to hypoxia. Specifically, perfusion of this synapse with acute hypoxia (10 minutes, 67 torr O_2, ~3.68 mg/L, elevation = 1,083m) produced increased excitatory post-synaptic potential (EPSP) amplitude recorded from RPeD1 (Bell et al., 2007; Figure 9a). However, this observed increase in synaptic strength was only transient, and the EPSP amplitude returned to control levels once normoxia was restored. Clearly, these results mimic the short-term potentiation of the carotid body-NTS synapse observed in mammals during acute hypoxia. This suggests then that hypoxia-induced short-term potentiation within sensory pathways represents a conserved mechanism by which the nervous system adapts respiratory behaviour to acute hypoxia (Figure 9b, Acute).

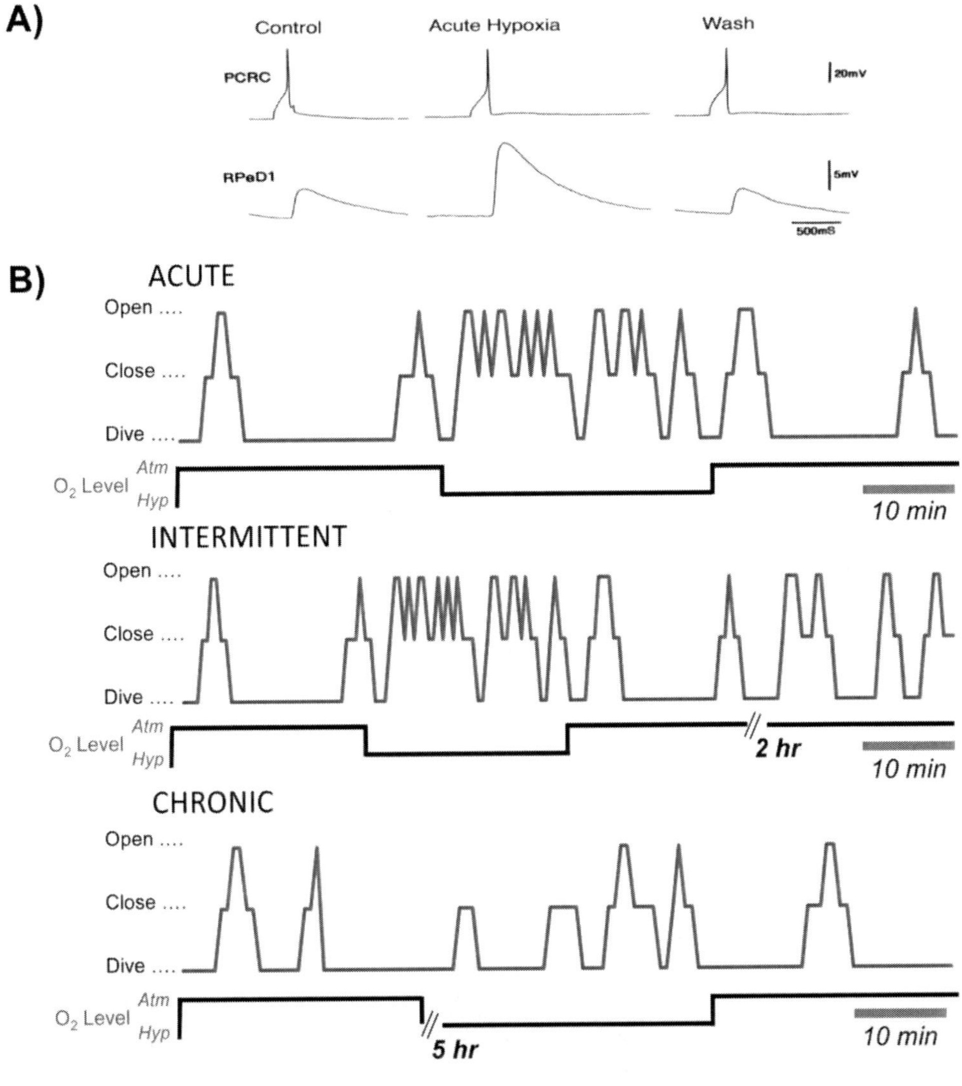

Figure 9. Model data depicting hypoxia–induced respiratory plasticity in *Lymnaea.* A) During normoxia, stimulation of an osphradial neuron produces an excitatory post-synaptic potential (EPSP) in its synaptic partner, RPeD1. However, perfusion of this synapse with hypoxia (65 torr O_2) transiently increases the EPSP amplitude, indicating that the synapse exhibits short-term synaptic potentiation. (Figure modified from Bell et al., 2007). B) Representative diagrams demonstrating that respiratory plasticity is apparent in adult *Lymnaea* exposed to different time-courses of hypoxia. Specifically, snails exposed to acute hypoxia spend more time at the surface performing aerial respiration, but return their respiration to baseline levels when atmospheric O_2 is restored. In contrast, following several hours of intermittent hypoxia animals continue to perform aerial respiration more frequently under atmospheric conditions than during baseline. Chronic hypoxia results in long-term depression of aerial breathing.

The finding that *Lymnaea* respiratory synapses could demonstrate short-term potentiation in response to acute hypoxia presented the possibility that chronic and intermittent hypoxia could also produce respiratory plasticity in this species. Interestingly, adult *Lymnaea* exposed to 5 hours of chronic hypoxia (6.7 torr O_2/~0.4 mg/L in a closed system, elevation = 1,083m) demonstrated long-term depression of aerial respiratory behaviour (Figure 9b, Chronic). Specifically, after 5 hours of chronic hypoxia the frequency of pneumostome openings was 80% lower than at the start of hypoxia. Furthermore, when the animals were returned to normoxic conditions, the frequency of aerial respiration remained depressed below control levels. In contrast, animals exposed to an intermittent hypoxia challenge (30 min of normoxia followed by 30 min of hypoxia at 6.7 torr O_2 for 4 hours) exhibited long-term increases in aerial respiration that persisted when the snails were returned to normoxic conditions (Figure 9b, Intermittent). These changes in aerial respiratory behaviour are similar to those reported for mammals and allude to the development of long-term plasticity between *Lymnaea* synapses following chronic and intermittent hypoxia. Interestingly, recent evidence suggests that chronic hypoxia does cause LTD between osphradial chemoreceptors and RPeD1, while intermittent hypoxia results in LTF at this synapse (Syed et al., in preparation).

Taken together, the finding that *Lymnaea* synapses conferring sensory information to the rCPG exhibit hypoxia-induced respiratory plasticity similar to that of mammals strongly suggests fundamental cellular and molecular mechanisms underlying this phenomenon.

Embryonic and Juvenile Responses to Hypoxia

Respiratory behaviour is influenced primarily by environment and metabolic demand but can also be affected by developmental state. Respiratory behaviour is not necessarily the same for a species at different stages of development. For example, juvenile *Lymnaea* do not perform aerial respiration as often as adults, presumably because their very large surface area to volume ratio makes cutaneous respiration highly effective at this developmental stage. Interestingly, there are differences between juveniles and adults in the connectivity of the rCPG and in the intrinsic properties of the component neurons which favours rhythmic activity in the adult rCPG only (McComb et al., 2003). One might predict then that the acute hypoxic response will also differ between adult and juvenile animals, with adults spending more time performing aerial respiration, and juveniles simply opting to remain at the water surface near to the O_2-rich air.

We can also consider the responses of embryonic *Lymnaea* to environmental hypoxia. O_2 acquisition in embryonic *Lymnaea* is necessarily restricted to cutaneous respiration as the animals develop within the egg mass. If environmental hypoxia occurs, the embryos will respond by initiating rapid, cilia-driven rotational behaviour within the egg resulting in mixing of the surrounding fluid and reduction of the diffusion barrier for O_2. This phenomenon, termed the "embryonic stir bar hypothesis," is enhanced by hypoxia and involves a pair of O_2 sensitive neurons, which are established early in development and regulate cilia-driven rotational behaviour (Kuang et al., 2002; Byrne et al., 2009). This embryonic respiratory behaviour is an essential adaptation because *Lymnaea* inhabits environments that are likely to become hypoxic.

As a whole, the respiratory system of *Lymnaea* is well adapted to allow these animals to survive in hypoxic environments and even under conditions of extreme hypoxia. This ability is due to the bimodal respiration these animals utilize as well as the precise sensory feedback provided by O_2 chemoreceptors. Because *Lymnaea* aerial respiratory behaviour can be modified by long-term hypoxia it would be interesting to determine how this plasticity might lead to respiratory adaptation over many generations. In addition, several lines of evidence suggest that there are fundamental mechanisms that operate within the nervous system of metazoans during hypoxic exposure. Therefore, these mechanisms could be elucidated using the *Lymnaea* model and the information applied to complex mammalian neural circuits.

Cumulative Stress: Effects of Hypoxia, Temperature and Heavy Metals

The idea of "stress" is currently a hot topic in medical research as we are just now beginning to understand the fundamental and far reaching affects it has on human health. However, stress is not limited to humans and affects all animals especially in this current time of climate change and habitat degradation. Stress is any factor that threatens an animal's physiological homeostasis and includes physical (e.g., injury), chemical (e.g., hypoxia, toxic metals) or perceived (e.g., presence of predators) stressors. Because some degree of stress is unavoidable, animals have evolved coping mechanisms to restore homeostasis following a stressful experience, including changes in hormone release, immunity, neural activity, metabolism, behaviour and cardiovascular/respiratory activity (Barton, 2002). Interestingly, acute stress is now thought to be beneficial for both working memory and for the immune system (Dhabhar and McEwen, 1997; Yuen et al., 2009). In contrast, there is no doubt that chronic or cumulative stress is detrimental and limits the ability of coping mechanisms to maintain physiological homeostasis and general health.

In aquatic ecosystems, invertebrates may be subjected to a multitude of stressors at any given time. In addition to fluctuations in O_2 levels, aquatic invertebrates are frequently subjected to dramatic changes in temperature. Specifically, water temperature at the surface can range from 35°C on hot summer days to 0°C during the winter months, but different species will be adapted to a specific optimal temperature range. A second biologically important stressor is heavy metal contamination from industrial activity. Heavy metals are extremely toxic for many species, and can accumulate in food webs resulting in poisoning of higher trophic levels. For example, acute lead exposure (72 hours) inhibits locomotion (and therefore aerial respiration), feeding and shell emergence in *Lymnaea stagnalis* (Pyatt et al., 2002). These behavioral changes are not surprising given the ability of lead, cadmium and mercury to alter neuronal growth and activity in *Lymnaea* (Szücs et al., 1994; Leong et al., 2001).

Unfortunately, hypoxia, high temperatures and heavy metals are occurring together more frequently in aquatic ecosystems. High temperatures confound hypoxic stress by further reducing water O_2 levels while raising the metabolic rates of invertebrates, and therefore their O_2 demands (Gillooly et al., 2001). However, the presence of heavy metals depresses respiration and locomotion, thus preventing the normal physiological mechanisms designed to maintain O_2 homeostasis. One can easily imagine *Lymnaea* (and many other species) becoming locally extinct under such extreme conditions.

Conclusion

Respiration is an essential, homeostatic behaviour that appeared early in metazoan evolution to facilitate aerobic metabolism. Even today, all metazoans share common features of the respiratory system and the underlying neural circuits producing respiratory behaviour. In this chapter, we have highlighted some of these fundamental characteristics and suggested they might best be elucidated using invertebrate models. For example, the pond snail, *Lymnaea* stagnalis, has been widely utilized as a model system in neuroscience for studies of synapse formation, regeneration, learning and respiration. *Lymnaea* respiratory behaviour is exquisitely adapted at all stages of development to the hypoxic environments which these animals inhabit. However, this chapter only scratches the surface in terms of adaptations of invertebrates to hypoxia as there are a plethora of cardiovascular and metabolic adaptations that are possible, but not yet fully understood. Some invertebrates have an incredible ability to adapt their respiratory behaviour to extreme environments, from the annelids inhabiting deep sea thermal vents (Hourdez and Lallier, 2007) to the snail, Physella johnsoni, found in the Banff National Park hot springs (Remigio et al., 2001). However, these species represent a minority, and a great number of invertebrates are facing extinction because they cannot adapt their respiratory behaviour to the rapidly changing environmental conditions currently occurring. Specifically, hypoxia is widely considered to be one of the most important threats to ecosystem health. While this is certainly true on a large scale, isolated and non-severe hypoxia may act to maintain diversity by forcing species to adopt different respiratory strategies and therefore ecological niches (Chapman et al., 2004). Clearly, conservation research must aim to understand what environmental stressors enhance biodiversity and adaptation versus more extreme stressors that have the opposite effect. On a final note, there is a great need for respiratory research in the field of medicine because of the serious diseases that affect millions of people worldwide. In particular, the neural circuits and cellular/molecular mechanisms governing respiratory behaviour have yet to be elucidated. To this end, multidisciplinary approaches may offer new perspectives necessary to answer these fundamental questions that surround respiration.

Evolution is an ongoing process for all animals (including humans) and is driven by the ability of animals to adapt to changing environments. Understanding how certain species have adapted to extreme environmental conditions provides unique insight into the fundamental properties of the nervous system, which controls many essential homeostatic behaviours, yet still remains poorly understood.

References

Alheid, G.F., and D.R. McCrimmon. 2008. The chemical neuroanatomy of breathing. *Respiratory Physiology & Neurobiology*. 164:3–11.

Barton, B.A. 2002. Stress in fishes: A diversity of responses with particular reference to changes in circulating corticosteroids. *Integrative and Comparative Biology*. 42:517–525.

Beall, C.M. 2006. Andean, Tibetan, and Ethiopian patterns of adaptation to high-altitude hypoxia. *Integrative and Comparative Biology*. 46:18–24.

Bell, H.J., T. Inoue, K. Shum, C. Luk, and N.I. Syed. 2007. Peripheral oxygen-sensing cells directly modulate the output of an identified respiratory central pattern generating neuron. *European Journal of Neuroscience*. 25:3537–3550.

Blain, G.M., C.A. Smith, K.S. Henderson, and J.A. Dempsey. 2009. Contribution of the carotid body chemoreceptors to eupneic ventilation in the intact, unanesthetized dog. *Journal of Applied Physiology*. 106:1564–1573.

Bradley, T.J. 2007. Control of the respiratory pattern in insects. *Advances in Experimental Medicine and Biology*. 618:211–220.

Brouwer, M., N.J. Brown-Peterson, P. Larkin, V. Patel, N. Denslow, S. Manning, and T.H. Brouwer. 2007. Molecular and whole animal responses of grass shrimp, *Palaemonetes pugio*, exposed to chronic hypoxia. *Journal of Experimental Marine Biology and Ecology*. 341:16–31.

Brown, T.G. 1911. The intrinsic factors in the act of progression in the mammal. *Proceedings of the Royal Society of London. Series B, Containing Papers of a Biological Character*. 84:308 –319.

Budha, P.B., Dutta, J. and B.A. Daniel. 2010. *Lymnaea stagnalis*. In: IUCN 2012. IUCN Red List of Threatened Species. Version 2012.2. <www.iucnredlist.org>. Downloaded on 27 October 2012.

Burggren, W.W., and K. Johansen. 1986. Circulation and respiration in lungfishes (dipnoi). *Journal of Morphology*. 190:217–236.

Burggren, W.W. 1982. "Air Gulping" improves blood oxygen transport during aquatic hypoxia in the goldfish *Carassius auratus*. *Physiological Zoology*. 55: 327-334.

Byrne, R.A., S.D. Rundle, J.J. Smirthwaite, and J.I. Spicer. 2009. Embryonic rotational behaviour in the pond snail *Lymnaea stagnalis*: influences of environmental oxygen and development stage. *Zoology*. 112:471–477.

Chapman, L.J., K.R. Schneider, C. Apodaca, and C.A. Chapman. 2004. Respiratory ecology of macroinvertebrates in a swamp–river system of East Africa. *Biotropica*. 36:572–585.

Cook, A.A., P. John D. Lambshead, L.E. Hawkins, N. Mitchell, and L.A. Levin. 2000. Nematode abundance at the oxygen minimum zone in the Arabian Sea. *Deep Sea Research Part II: Topical Studies in Oceanography*. 47:75–85.

Correa, A.C., J.S. Escobar, P. Durand, F. Renaud, P. David, P. Jarne, J.-P. Pointier, and S. Hurtrez-Boussès. 2010. Bridging gaps in the molecular phylogeny of the Lymnaeidae (Gastropoda: Pulmonata), vectors of Fascioliasis. *BMC Evolutionary Biology*. 10:381.

Costello, E.K., K. Stagaman, L. Dethlefsen, B.J.M. Bohannan, and D.A. Relman. 2012. The application of ecological theory toward an understanding of the human microbiome. *Science*. 336:1255–1262.

Crabb, E.D. 1929. Growth of a pond snail, *Lymnæa stagnalis appressa*, as indicated by increase in shell-size. *Biological Bulletin*. 56:41–63.

Crowder, W.C., Nie, M. and G.R. Ultsch. 1998. Oxygen uptake in bullfrog tadpoles (*Rana catesbeiana*). *Journal of Experimental Zoology*. 280:121-134.

De Castro F. 1928. Sur la structure et l'innervation du sinus carotidien de l'homme et des mammifères. Nouveaux faits sur l'innervation et la fonction du glomus caroticum. *Trav. Lab. Rech. Biol*. 25, 331–380.

Dempsey, J.A., S.C. Veasey, B.J. Morgan, and C.P. O'Donnell. 2010. Pathophysiology of sleep apnoea. *Physiological Reviews*. 90:47 –112.

de Paula, P.M. de, G. Tolstykh, and S. Mifflin. 2007. Chronic intermittent hypoxia alters NMDA and AMPA-evoked currents in NTS neurons receiving carotid body chemoreceptor inputs. *American Journal of Physiology: Regulatory, Integrative and Comparative Physiology*. 292:R2259–R2265.

Dhabhar, F.S., and B.S. McEwen. 1997. Acute stress enhances while chronic stress suppresses cell-mediated immunity in vivo: a potential role for leukocyte trafficking. *Brain, Behavior and Immunity*. 11:286–306.

Diaz, R.J., and R. Rosenberg. 2008. Spreading dead zones and consequences for marine ecosystems. *Science*. 321:926–929.

Dusenbery, D.B. 1983. Chemotactic behavior of Nematodes. *Journal of Nematology*. 15:168–173.

Eldridge, F.L. and D.E. Millhorn. 1986. Oscillation, gating, and memory in the respiratory control system. In: Cherniack, N.S., Widdicombe, J.G. (Eds.), Handbook of Physiology, section 3: The Respiratory System: Control of Breathing, part 1, vol. II. American Physiological Society, Washington, DC, pp. 93–114.

Engwall, M.J., L. Daristotle, W.Z. Niu, J.A. Dempsey, and G.E. Bisgard. 1991. Ventilatory afterdischarge in the awake goat. *Journal of Applied Physiology*. 71:1511–1517.

Feder, M.E., and W.W. Burggren. 1985. Cutaneous gas exchange in vertebrates: design, patterns, control and implications. *Biological Reviews*. 60:1–45.

Feldman, J.L., and C.A.D. Negro. 2006. Looking for inspiration: new perspectives on respiratory rhythm. *Nature Reviews Neuroscience*. 7:232–241.

Fredericks, C.A. 1976. Oxygen as a limiting factor in phototaxis and in intraclonal spacing of the sea anemone *Anthopleura elegantissima. Marine Biology*. 38:25–28.

Gillooly, J.F., J.H. Brown, G.B. West, V.M. Savage, and E.L. Charnov. 2001. Effects of size and temperature on metabolic rate. *Science*. 293:2248–2251.

Glanzman, D.L. 2010. Common mechanisms of synaptic plasticity in vertebrates and invertebrates. *Current Biology*. 20:R31–R36.

Graham, J.B. 1988. Ecological and evolutionary aspects of integumentary respiration: body size, diffusion, and the invertebrata. *American Zoologist*. 28:1031–1045.

Gröndahl, F. 1988. A comparative ecological study on the scyphozoans *Aurelia aurita, Cyanea capillata* and *C. lamarckii* in the Gullmar Fjord, western Sweden, 1982 to 1986. *Marine Biology*. 97:541–550.

Harrison, F. 2007. Microbial ecology of the cystic fibrosis lung. *Microbiology*. 153:917–923.

Harrison, J.F., A. Kaiser, and J.M. VandenBrooks. 2010. Atmospheric oxygen level and the evolution of insect body size. *Proceedings of the Royal Academy B*. 277: 1937–1946.

Heymans, C., Bouckaert, J.J. and P. Regniers. (1933) Le Sinus Carotidien. Ed. G. Doin, Paris.

Heymans, C., and J.J. Bouckaert. 1930. Sinus caroticus and respiratory reflexes. *Journal of Physiology*. 69:254–266.

Hodgkin, A.L., and A.F. Huxley. 1952. A quantitative description of membrane current and its application to conduction and excitation in nerve. *Journal of Physiology*. 117:500–544.

Hourdez, S., and F. Lallier. 2007. Adaptations to hypoxia in hydrothermal-vent and cold-seep invertebrates. *Reviews in Environmental Science and Biotechnology*. 6:143–159.

Huey, R.B., and P.D. Ward. 2005. Hypoxia, Global Warming, and Terrestrial Late Permian Extinctions. *Science*. 308:398–401.

Inoue, T., Z. Haque, K. Lukowiak, and N.I. Syed. 2001. Hypoxia-induced respiratory patterned activity in *Lymnaea* originates at the periphery. *Journal of Neurophysiology*. 86:156–163.

Jackson, D.C. 2002. Hibernating without oxygen: physiological adaptations of the painted turtle. *Journal of Physiology*. 543:731–737.

Janes, T.A. and N.I. Syed. 2012. Neuronal mechanisms of oxygen chemoreception: an invertebrate perspective. *Advances in Experimental Medicine and Biology*. 758: 7-17.

Jones, J.D. 1961. Aspects of respiration in *Planorbis corneus* (L.) and *Lymnaea stagnalis* (L.) (gastropoda: pulmonata). *Comparative Biochemistry and Physiology*. 4:1–29.

Kamardin, N.N. 1976. Structure and cellular organization of the osphradium of *Limnaea stagnalis* (L.) *Arkhiv Anatomii Gistologii I Embriologii*. 71:87–90.

Kanz, J.E., and W.D. Quast. 1992. Respiratory pumping behavior in the marine snail *Aplysia californica* as a function of ambient hypoxia. *Physiological Zoology*. 65:35–54.

Kline, D.D., A. Ramirez-Navarro, and D.L. Kunze. 2007. Adaptive depression in synaptic transmission in the nucleus of the solitary tract after in vivo chronic intermittent hypoxia: evidence for homeostatic plasticity. *Journal of Neuroscience*. 27:4663–4673.

Kuang, S., S.A. Doran, R.J.A. Wilson, G.G. Goss, and J.I. Goldberg. 2002. Serotonergic sensory-motor neurons mediate a behavioral response to hypoxia in pond snail embryos. *Journal of Neurobiology*. 52:73–83.

Leong, C.C., N.I. Syed, and F.L. Lorscheider. 2001. Retrograde degeneration of neurite membrane structural integrity of nerve growth cones following in vitro exposure to mercury. *Neuroreport*. 12:733–737.

Lihiri, S., Rumsey, W.L., Wilsin, D.F. and R. Iturriaga. 1993. Contribution of in vivo microvascular PO2 in the cat carotid body chemotransduction. *Journal of Applied Physiology*. 75:1035-1043.

López-Barneo, J., P. Ortega-Sáenz, R. Pardal, A. Pascual, J.I. Piruat, R. Durán, and R.Gómez-Díaz. 2009. Oxygen sensing in the carotid body. *Annals of the New York Academy of Sciences*. 1177:119–131.

Lotze, H.K., H.S. Lenihan, B.J. Bourque, R.H. Bradbury, R.G. Cooke, M.C. Kay, S.M. Kidwell, M.X. Kirby, C.H. Peterson, and J.B.C. Jackson. 2006. Depletion, degradation, and recovery potential of estuaries and coastal seas. *Science*. 312:1806–1809.

Luo, M., Liming, S. and J. Hu. 2009. Neural detection of gases- carbon dioxide, oxygen -in vertebrates and invertebrates. Current Opinion in Neurobiology. 19:354-361.

Marder, E., and D. Bucher. 2007. Understanding circuit dynamics using the stomatogastric nervous system of lobsters and crabs. *Annual Review of Physiology*. 69:291–316.

Marder, E. and D. Bucher 2001. Central pattern generators and the control of rhythmic movements. *Current Biology*. 11: R986-R996.

Massabuau, J.C., and B. Burtin. 1984. Regulation of oxygen consumption in the crayfishAstacus leptodactylus at different levels of oxygenation: role of peripheral O_2 chemoreception. *Journal of Comparative Physiology B*. 155:43–49.

McComb, C., R. Meems, N. Syed, and K. Lukowiak. 2003. Electrophysiological differences in the CPG aerial respiratory behavior between juvenile and adult *Lymnaea. Journal of Neurophysiology*. 90:983 –992.

McMahon, B.R. 2001. Respiratory and circulatory compensation to hypoxia in crustaceans. *Respiratory Physiology*. 128:349–364.

Menendez, A.A., T.J. Nuckton, J.E. Torres, and D. Gozal. 1999. Short-term potentiation of ventilation after different levels of hypoxia. *Journal of Applied Physiology*. 86:1478–1482.

Mifflin, S.W. 1997. Short-term potentiation of carotid sinus nerve inputs to neurons in the nucleus of the solitary tract. *Respiratory Physiology & Neurobiology*. 110:229–236.

Milsom, W.K., and M.L. Burleson. 2007. Peripheral arterial chemoreceptors and the evolution of the carotid body. *Respiratory Physiology & Neurobiology*. 157:4–11.

Morton, D.B., and A. Vermehren. 2007. Soluble guanylyl cyclases in invertebrates: targets for NO and O_2. *Advances in Experimental Biology*. 1:65–82.

Neubauer, J.A., and J. Sunderram. 2004. Oxygen-sensing neurons in the central nervous system. *Journal of Applied Physiology*. 96:367–374.

Noland, L.E., and M.R. Carriker. 1946. Observations on the biology of the snail *Lymnaea stagnalis appressa* during twenty generations in laboratory culture. *American Midland Naturalist*. 36:467–493.

Onimaru, H., and I. Homma. 2006. Point:Counterpoint: The parafacial respiratory group (pFRG)/pre-Bötzinger complex (preBötC) is the primary site of respiratory rhythm generation in the mammal. *Journal of Applied Physiology*. 100:2094–2098.

Prabhakar, N.R., T.E. Dick, J. Nanduri, and G.K. Kumar. 2007. Systemic, cellular and molecular analysis of chemoreflex-mediated sympathoexcitation by chronic intermittent hypoxia. *Experimental Physiology*. 92:39–44.

Prabhakar, N.R., and G.L. Semenza. 2012. Adaptive and maladaptive cardiorespiratory responses to continuous and intermittent hypoxia mediated by Hypoxia-Inducible Factors 1 and 2. *Physiological Reviews*. 92:967–1003.

Purcell, J.E., D.L. Breitburg, M.B. Decker, W.M. Graham, M.J. Youngbluth, and K.A. Raskoff. 2001. Pelagic cnidarians and ctenophores in low dissolved oxygen environments: A review. *Coastal Estuarine Studies*. 58:77–100.

Pyatt, A.J., F.B. Pyatt, and V.W. Pentreath. 2002. Lead toxicity, locomotion and feeding in the freshwater snail, *Lymnaea stagnalis* (L.). *Invertebrate Neuroscience*. 4:135–140.

Rabalais, N.N., R.E. Turner, and W.J. Wiseman. 2002. Gulf of Mexico Hypoxia, A.k.a. "the Dead Zone." *Annual Review of Ecology and Systematics*. 33:235–263.

Raven and Johnson. Biology. 6th ed. Boston (MA): McGraw-Hill; 2001.

Remigio E.A., Lepitzki D.A., Lee J.S., and Hebert P.D. 2001. Molecular systematic relationships and evidence for a recent origin of the thermal spring endemic snails *Physella johnsoni* and *Physella wrighti* (Pulmonata: Physidae). *Canadian Journal of Zoology*. 79:1941–1950.

Ribeiro, M., M. Schofield, I. Kemenes, P.R. Benjamin, M. O'Shea, and S.A. Korneev. 2010. Atypical guanylyl cyclase from the pond snail *Lymnaea stagnalis*: cloning, sequence analysis and characterization of expression. *Neuroscience*. 165:794–800.

Roman, M., J. Pierson, D. Kimmel, W. Boicourt, and X. Zhang. Impacts of hypoxia on zooplankton spatial distributions in the northern Gulf of Mexico. *Estuaries and Coasts*. 1–9.

Scott, K. 2011. Out of thin air: sensory detection of oxygen and carbon dioxide. *Neuron*. 69:194-202.

Selverston, A.I. 2010. Invertebrate central pattern generator circuits. *Philosophical Transactions of the Royal Society B*. 365:2329–2345.

Smith, J.C., H.H. Ellenberger, K. Ballanyi, D.W. Richter, and J.L. Feldman. 1991. Pre-Bötzinger complex: a brainstem region that may generate respiratory rhythm in mammals. *Science*. 254:726–729.

Sokolov, V. A. and N. N. Kamardin. 1977. The relation of impulse frequency in the osphradial nerve to the concentration of oxygen and inulin in liquid passing over the osphradium of the pond snail. *Vestnik Leningr. Univ Biol*. 1:87-90.

Strong, E.E., O. Gargominy, W.F. Ponder, and P. Bouchet. 2008. Global diversity of gastropods (Gastropoda; Mollusca) in freshwater. In Freshwater Animal Diversity Assessment. E.V. Balian, C. Lévêque, H. Segers, K. Martens, and H.J. Dumont, editors. Springer Netherlands. 149–166.

Syed, N.I., and W. Winlow. 1991b. Respiratory behavior in the pond snail *Lymnaea-Stagnalis* .2. neural elements of the central pattern generator (CPG). *Journal of Comparative Physiology*. 169:557–568.

Syed, N.I., D. Harrison, and W. Winlow. 1991a. Respiratory behavior in the pond snail *Lymnaea stagnalis*. *Journal of Comparative Physiology A*. 169.

Syed, N., A. Bulloch, and K. Lukowiak. 1990. *In vitro* reconstruction of the respiratory central pattern generator of the mollusk *Lymnaea*. *Science*. 250:282 –285.

Szücs, A., J. Salánki, and K. S.-Rózsa. 1994. Effects of chronic exposure to cadmium- or lead-enriched environments on ionic currents of identified neurons in *Lymnaea stagnalis* (L.). *Cellular and Molecular Neurobiology*. 14:769–780.

Tabata, M., H. Kurosawa, Y. Kikuchi, W. Hida, H. Ogawa, S. Okabe, Y. Tun, T. Hattori, and K. Shirato. 2001. Role of GABA within the nucleus tractus solitarii in the hypoxic ventilatory decline of awake rats. *American Journal of Physiology: Regulatory, Integrative and Comparative Physiology*. 281:R1411–1419.

Taylor, A.C., and A.R. Brand. 1975. A comparative study of the respiratory responses of the Bivalves *Arctica islandica* (L.) and *Mytilus edulis* (L.) to declining oxygen tension. *Proceedings of the Royal Society of London B*. 190:443 –456.

Vaquer-Sunyer, R., and C.M. Duarte. 2008. Thresholds of hypoxia for marine biodiversity. *Proceedings of the National Academy of Sciences*. 105:15452–15457.

Vermehren-Schmaedick, A., J.A. Ainsley, W.A. Johnson, S.-A. Davies, and D.B. Morton. 2010. Behavioral responses to hypoxia in *Drosophila* larvae are mediated by atypical soluble guanylyl cyclases. *Genetics*. 186:183–196.

West JB. Respiratory physiology—the essentials. 6th ed. Baltimore (MD), London (U.K.), and Los Angeles (CA): Williams & Wilkins; 1999.

Wetzel, R.G. 1983. Limnology, 2nd Edition. Saunders College Publishing, Philadelphia, PA.

Wignall, P.B., and R.J. Twitchett. 1996. Oceanic anoxia and the end Permian mass extinction. *Science*. 272:1155–1158.

Wohlgemuth, S.E., A.C. Taylor, and M.K. Grieshaber. 2000. Ventilatory and metabolic responses to hypoxia and sulphide in the lugworm *Arenicola marina* (L.). *Journal of Experimental Biology*. 203:3177–3188.

Worm, B., E.B. Barbier, N. Beaumont, J.E. Duffy, C. Folke, B.S. Halpern, J.B.C. Jackson, H.K. Lotze, F. Micheli, S.R. Palumbi, E. Sala, K.A. Selkoe, J.J. Stachowicz, and R. Watson. 2006. Impacts of biodiversity loss on ocean ecosystem services. *Science*. 314:787–790.

Wu, R.S.S. 2002. Hypoxia: from molecular responses to ecosystem responses. *Marine Pollution Bulletin*. 45:35–45.

Yuen, E.Y., W. Liu, I.N. Karatsoreos, J. Feng, B.S. McEwen, and Z. Yan. 2009. Acute stress enhances glutamatergic transmission in prefrontal cortex and facilitates working memory. *Proceedings of the National Academy of Sciences*. 106:14075–14079.

Zeis, B., T. Lamkemeyer, R.J. Paul, F. Nunes, S. Schwerin, M. Koch, W. Schütz, J. Madlung, C. Fladerer, and R. Pirow. 2009. Acclimatory responses of the *Daphnia pulex* proteome to environmental changes. I. Chronic exposure to hypoxia affects the oxygen transport system and carbohydrate metabolism. *BMC Physiology*. 9:7.

Zhang, W., F.R. Carreño, J.T. Cunningham, and S.W. Mifflin. 2008. Chronic sustained and intermittent hypoxia reduce function of ATP-sensitive potassium channels in nucleus of the solitary tract. *American Journal of Physiology: Regulatory, Integrative and Comparative Physiology*. 295:R1555–R1562.

Zhang, Z-Q. and W.A Shear. 2007. Linnaeus tercentenary and invertebrate taxonomy: an introduction. *Zootaxa*. 1668: 7-10.

Zimmer, M., J.M. Gray, N. Pokala, A.J. Chang, D.S. Karow, M.A. Marletta, M.L. Hudson, D.B. Morton, N. Chronis, and C.I. Bargmann. 2009. Neurons detect increases and decreases in oxygen levels using distinct guanylate cyclases. *Neuron*. 61:865–879.

In: Neuroecology and Neuroethology in Molluscs
Editors: Anna Di Cosmo and William Winlow

ISBN: 978-1-62948-970-4

Chapter 10

Epilogue: Molluscs in Neuroecology

***Anna Di Cosmo*[1,*] *and William Winlow*[2,†]**

[1]Professor of Zoology, Department of Biology, Univerità degli Studi di Napoli, Federico II, Napoli; Managing Editor of "Frontiers in Bioscience" Review Editor of "Frontiers in Experimental Endocrinology"

[2]Visiting Professor, Departimento di Biologia, Università degli Studi di Napoli, Federico II, 80126 Napoli, Italia; Professor Emeritus of Neuroscience, School of Pharmacy and Biomedical Sciences, University of Central Lancashire, Preston, Lancashire, UK; Honorary Research Fellow, Department of Veterinary Preclinical Sciences, University of Liverpool, UK

In their excellent prologue to this volume Derby and Zimmer made the point that neuroecology is the meeting point of ethology, neuroscience, evolution, ecology, physics and chemistry as visualised by them in Figure 1 of chapter 1. They also pointed out that although molluscan species are second only to arthropods in the number of extant species among the metazoans, many of the model species so far studied are not major contributors to structuring ecological communities. They also felt that these species functioned poorly in elucidating fundamental principles at higher levels of biological organisation beyond the level of whole behaving organisms.

Kirsten Benkendorff (chapter 2) has carried out an extensive metanaylysis of 828 publications on the current state of knowledge on molluscan marine natural products and found that most of the species studied were grazing opisthobranchs from subtidal rocky reefs from tropical and warm temperate locations and she outlines various chemical defence strategies and some interesting neurotoxic compounds. She makes the point that only 277 species were studied, representing only 0.53% of all named marine molluscan species and were unevenly distributed with few studies on polar and cold temperate species. However significant neuroecological progress has been made with a number of molluscan groups in

* Address for Correspondence: Department of Biology, Complesso Universitario Monte Sant'Angelo, Via Cinthia, Build 780126 Napoli. E-mail: mailto:dicosmo@unina.it.

† Address for Correspondence: 32 Hill Crescent, Newton, Preston, PR4 3TR, Lancashire, UK. E-mail: bill@winlow.myzen.co.uk.

terms of neurotoxic peptides, complex defensive ink alarm pheromones, bioaccumulation of shellfish toxins and the evolution of chemical defence mechanisms in species with reduced physical protection. Thus key concepts in neuroecology are emerging from this heterogeneous group of mollusks.

Leading on from the chemical diversity of molluscan communities, Cummins and Wyeth (chapter 3) consider olfaction in three distantly related aquatic gastropods and three more closely related terrestrial gastropods. As a group gastropod molluscs inhabit a wide range of environments and their perceptions of the outside world are largely chemically driven, leading to odorant driven behavioural responses – their sense of smell is paramount and appears to require substantial peripheral processing with some behavioural responses requiring no CNS participation. Thus many molluscs inhabit a world of olfactory cues, important to both molluscan predators and prey and this area is explored by Nusnbaum (Chapter 4) where he explores the predator-prey dynamics in molluscs which are so important to species survival. He makes the valid point that "By studying the signals through which molluscs identify and avoid threats and locate and consume food, researchers can connect the dots between stimuli, neural pathways, behaviour, and ecology", i.e., a significant contribution to neuroecology. Cephalopod molluscs have excellent vision, but also rely on distance chemoreception particularly those that live in the deep sea or are nocturnal. Given that more needs to be known about the specific cues that alert predatory cephalopods and then initiate their search behaviour, this would seem to be a fruitful area for future neuroecological research as Benkendorff (Chapter 2) also indicates.

Continuing with the theme of predator and prey molluscs Wilson and Weissburg (Chapter 5) examine the ecological consequences of chemically mediated prey perception in marine molluscan communities. In this they point out three significant gaps in our current understanding: a lack of knowledge of identified chemical deterrents which prevents comprehension of concentrations emitted from multiple predators; how changes in perceptual abilities in different flow environments controls how well prey can detect predators and *vice versa*; and finally the effects of invasive predatory species on prey communities. A better understanding the interactions of chemical cues between predator and prey molluscs will be important in future studies in the area of chemical ecology.

The plethora of signalling molecules found in molluscan communities is further considered in chapter 6 where Di Cosmo and Polese set cephalopods in a neuroecological framework, looking at them not only as animal models for neuroscience and neuroethology but also demonstrating their significance in affecting marine populations and communities. In aquatic systems, chemical cues determine feeding, defence, habitat and mating choices affecting not only individual behaviour but also community organisation and ecosystem function. Cephalopods, besides being equipped with excellent vision and direct and rapid neural control of the chromatophores that allow them unparalleled dynamic camouflage, also use tactile and olfactory cues to detect chemical signals. Oddly to date little attention has been given to the role played by the olfactory organ in the detection of chemical stimuli and chemical communication that strongly affect life processes in the sea, but in this volume this is addressed in both chapters 3 and 6. The studies presented in chapter 6 offer new perspectives on the Wells and Wells model for *O. vulgaris* reproduction introducing a concert of molecules, neuropeptides, steroid receptors, and neurotransmitters involved in the control and modulation of the optic gland activity through the olfactory lobe and the olfactory organ. In this new model, the olfactory lobe and olfactory organ play a crucial role in the chemical perception and integration of odour signals that allow the animal to balance their energy allocation of growth versus reproduction. This new perspective adds weight to the view that cephalopods, occupying a wide range of ecological niches as they do, are critical for understanding marine systems, but their omnipresence and impact are inadequately recognized.

Ottaviani (Chapter 7) explores the immune and neuroendocrine systems of molluscs and concludes that "the signaling molecules (i.e., neuropeptides, cytokines) as well as the large amount of molecular data now available on the major protein superfamilies with different activities, reveal that these primary signaling molecules are exceptionally well-conserved". Indeed the pool of molecules used is similar to those in advanced vertebrates. This conservation of molecular systems is important in neuroecology and underlies the conservation of neural mechanisms underlying behaviours as organ systems alter under evolutionary pressure as implied by Winlow and Polese (Chapter 8).

In Chapter 8 Winlow and Polese describe a neuroplastic network that underlies behavioural change in the great pond snail *Lymnaea stagnalis*. They discuss the evolution and reorganisation of the brain in relation to torsion and detorsion and show that many behavioural networks remain virtually intact during these processes, although reflecting the evolutionary changes that adapt *Lymnaea* to its environment. The network underlying behaviour is plastic and underlies many behaviours and its synaptic connectivity is described. Knowledge of these cellular interactions has allowed them to demonstrate seasonal changes within the respiratory central pattern generator of *Lymnaea,* with a loss of connectivity as winter approaches and a restoration of synaptic connectivity and the behavioural programme during the spring. Substantial further detail of *Lymnaea* respiratory behaviour is given by Janes and Syed (Chapter 9) and they conclude that "a great number of invertebrates are facing extinction because they cannot adapt their respiratory behaviour to the rapidly changing environmental conditions currently occurring" particularly where hypoxia threatens the health of ecosysystems due to human activity. Because *Lymnaea* utilises both skin respiration and pulmonary ventilation at the water surface, it is able to survive in stagnant environments, but makes more visits to the water surface in hypoxic conditions, making it more open to predation by surface predators. We suggest that *Lymnaea* will be a useful and sensitive model for studying the neuroecological effects of climate change with particular respect to invasive freshwater species such as predators, freshwater plans or microbial species.

In closing we would comment that up to the present Derby and Zimmer (Chapter 1) are correct in stating that molluscs have, with some notable exceptions, have functioned poorly in elucidating fundamental principles of neuroecology, but the reviews presented here suggest that this situation is changing rapidly. The areas of chemical ecology (chapter 2) and predator prey relationships have been well covered (chapters 3, 4, 5) as have neuroethology (chapters 3, 6, 8, 9) and evolution (chapters 7, 8, 9). The challenge now must be to utilise molluscan species research to illuminate the effects of environmental change on the neuroecology of these species, using them wherever possible to evaluate the degree of change taking place.

Index

A

access, 60, 96, 97, 98, 102
accessibility, 22
accounting, 26
acetylcholine, 58, 66, 194
acetylcholinesterase, 66
Achatina, 45, 55, 56, 57, 60, 61, 63, 69, 83, 151
Achatina fulica, 56, 63, 69, 83, 151
acid, 22, 29, 66, 86, 171
acidity, 29
A-cluster neurons, 158, 160
A-cluster rhythm, 159, 160
acquired immunity, 137
ACTH, 136, 137, 139, 141, 142
ACTH molecule, 137
action potential, 4, 151, 154, 160, 165, 166, 170, 180, 186, 188, 190
Active defenses, 78, 80
acute stress, 203
adaptability, 167, 170
adaptation(s), 9, 40, 49, 73, 74, 75, 76, 79, 82, 145, 151, 177, 180, 181, 182, 189, 191, 199, 202, 203, 204, 207
adductor, 75
adenine, 47
adrenal gland, 138
adulthood, 128
adults, 48, 100, 202
advancements, 3
Aeolidoidea nudibranchs, 29
aerial, 167, 169, 173, 177
aerial respiration, 167, 169, 177
  frequency of, 169
Africa, 34, 56, 205
agar, 143
age, 109, 115
aggregation, 45, 76, 81, 100, 102, 112, 113
aggressive behavior, 75
air, 53, 56, 60, 158, 161, 163, 181, 183, 184, 185
  air-breathing, 53, 158, 185
  air-breathing aquatic pulmonate, 53
airways, 187
alanine, 48
alarm pheromones, vii, 13, 18, 30
alertness, 82
algae, 18, 35, 53, 101, 196, 198
algal grazers, 24, 26
alien species, 67
alkaloids, 4, 18, 26, 29, 40
alters, 8, 85, 89, 94, 108, 157, 168, 206
alveoli, 184
amine(s), 39, 139, 141, 165, 172
amino, 22, 29, 48, 54, 55, 74
amino acid(s), 22, 29, 48, 54, 55, 74
amphibians, 91, 97, 142, 184, 185
Amphibious origins of basommatophoran pulmonates, 151
amplitude, 78, 79, 190, 200, 201
amputation, 65
Anapsidean sea hares, 13
Anaspidea, 29, 30, 38, 39
anatomy, 180, 182
ancestors, 28, 146
angiogenesis, 180
animal behavior, 82
animal cognition, 3
annual review, 13, 14, 15, 17, 18, 20, 21, 23, 26, 30
ANOVA, 79
anoxia, 180, 182, 196, 197, 198, 209
Antarctic marine invertebrates, 21, 37
Antarctic molluscs, 21
Antarctic opisthobranchs, 21
Antarctica, 21, 34
antibacterial activity, 19, 32, 132
anticancer activity, 26

anti-fouling activity, 19
antimicrobial activity, 18, 19, 39, 120
Antipredator behavior of mud crabs, 102
antipredator behaviors, 100, 104, 106
chemically-mediated, 90, 101, 105, 108
intense, 21, 47, 104, 106
odor-mediated, 90, 101, 108, 115
antipredator defenses, 106
effective, 2, 6, 46, 74, 75, 76, 79, 80, 89, 97, 99, 105, 158, 184
antipredator responses, 91, 93, 101, 106
illicit, 91, 94
antipredator strategies, 95, 102
Aplysia, 5, 6, 17, 18, 28, 36, 37, 38, 39, 40, 45, 46, 47, 48, 49, 50, 51, 55, 59, 60, 61, 62, 63, 64, 66, 67, 68, 69, 70, 83, 84, 85, 86, 91, 112, 173
Aplysia californica, 5, 37, 39, 40, 48, 50, 62, 63, 67, 68, 70, 85, 86, 112
Aplysia kurodai, 28
Aplysia punctata, 46, 49, 70
attraction pheromones, 51
candidate chemoreceptors in, 50
candidate chemoreceptors in draft genome, 48
candidate chemoreceptors in ink, 18, 29, 31, 76, 77, 78, 80, 85, 86, 91, 119, 120
candidate chemoreceptors in mate attraction, 18, 51, 64
candidate chemoreceptors in olfactory organs, vii, 46, 49, 51, 56, 66, 118, 120, 128
candidate chemoreceptors in peripheral sensory regions, 47
aplysioviolin, 78
apoptosis, 34
apparency, 97, 99, 100
APV, 78, 79
aquaculture, 37, 60
aquarium, 17
aquatic, 9, 32, 37, 45, 46, 51, 53, 59, 60, 74, 75, 76, 82, 83, 93, 109, 110, 120, 146, 158, 177, 179, 181, 182, 184, 185
aquatic habitats, 109, 177, 181
aquatic systems, 74, 120, 212
Archidoris pseudoargus, 52, 70
arousal, 59, 155
artery, 187
arthropods, 6, 60, 155, 211
Asia, 53, 168, 196
assessment, 16, 17, 32, 60, 90, 93, 112
atmosphere, 193, 197
atmospheric O_2, 182, 184
ATP, 210
attachment, 3, 37, 84, 99, 112
attractant, 101, 129
autonomy, 13, 29
aversion, 18, 56, 65, 66
aversive conditioning, 155
avoidance, 55, 67, 74, 75, 76, 78, 83, 84, 86, 110, 111, 114, 127, 187
avoidance behavior, 67, 111
awareness, 62
axons, 5, 6, 53, 56, 158, 164

B

bacteria, 18, 19, 22, 26, 35, 39, 93, 181
bacterial infection, 19
barnacles, 3, 98, 104
basal lamina, 49
base, 46, 47, 50, 184
basommatophoran pulmonates, 150, 151, 158
behavioral change, 96, 203
behaviorally-mediated trophic cascades, 101, 102, 104
behaviorists, 3
behaviors, 5, 6, 18, 67, 73, 74, 75, 80, 81, 82, 83, 85, 90, 96, 97, 99, 100, 104, 106, 113
behavioural, vii, 64, 85, 117, 119, 126, 130, 145, 146, 155, 156, 157, 161, 162, 163, 165, 167, 170, 171, 175, 176, 178, 179, 180, 181
adaptability, 167, 170
programme, 146, 167
repertoire, 120
responses, vii, 2, 5, 6, 45, 52, 54, 55, 58, 60, 61, 64, 65, 66, 70, 73, 76, 77, 78, 79, 82, 85, 89-94, 96, 98, 99, 100, 104, 106, 107, 108, 110, 111, 114, 120, 123, 130, 137, 139, 140, 155, 157, 160, 167, 168, 169, 177, 180
benefits, 5, 74, 178
benthic invertebrates, 32
bias, 28
Bilateral, 173
bile, 79
bioaccumulation, vii, 212
biochemistry, 39
biodiversity, 6, 7, 33, 110, 111, 178, 179, 198, 204, 209
biofilms, 19, 110
biogenic amines, 39, 139, 141, 172
biological activity, 24, 25, 27, 30
biological control, 110
biological processes, 99
biomass, 4, 93, 94, 111
biomaterials, 119
biosensors, 120
biosynthesis, 18, 34, 36, 37
biosynthetic origin, 13, 26

biotic, 3, 6, 19, 177, 181
biotic factor, 177, 181
bipedal, 120
birds, 80
bivalve prey, 82, 100, 101, 114
  large, 5, 6, 22, 31, 45, 47, 48, 51, 81, 94, 98, 100, 103, 111, 118, 119, 139, 158, 160, 164, 179, 184, 185
  preferred, 101
bivalve species form aggregations, 97
bivalves, 5, 6, 19, 21, 22, 23, 26, 28, 30, 75, 76, 98, 100, 105, 106, 108, 109
  chemosymbiotic, 30
  filter feeding, 26, 28, 30
blends, 90, 106
blood, 121, 142, 174, 183, 184, 188, 199, 205
blood flow, 183, 184
blood vessels, 121, 183
blue crabs, 93, 96, 97, 100, 101, 102, 103, 109, 111
  predatory, 4, 7, 13, 18, 21, 23, 24, 26, 28, 30, 31, 51, 52, 68, 73, 74, 80, 81, 82, 83, 84, 85, 93, 96, 97, 98, 99, 100, 101, 104, 106, 107, 108, 109
blueprint, 182
body patterning, 75, 81
body size, 111, 178, 184, 185, 193, 206
body wall, 52, 159, 163, 164, 184
body wall neurons, 159, 164
brain, 3, 6, 63, 66, 119, 120, 128, 129, 132, 140, 141, 145, 146, 148, 155, 165, 166, 171, 173, 175, 213
brain size, 6
brainstem, 209
branchial chamber, 184
branching, 56
Brazil, 131
breakdown, 60
breathing, 53, 158, 162, 163, 169, 185, 188, 191, 194, 199, 200, 201, 204
breeding, 167
bromination, 36
by-products, 90

C

Ca^{2+}, 130, 137, 194, 195
cadmium, 203, 209
caenogastropod, 21
Caenorhabditis elegans, 3, 180
  nematode worm, 3, 180
Cajal staining, 122
calcium, 47, 142, 165, 166, 170
Callinectes sapidus, 93, 109, 110
Cambrian, 118, 150
cancer, 32, 34
cancer cells, 32, 34
Cancer sp, 91
Candidate chemoreceptors in *Aplysia*, 59
candidates, 7, 37, 55
Cantareus, 45, 56, 57, 58, 60, 61
  olfactory search behavior, 56
  tentacle sensory neurons, 58
Cantareus, 56
capsule, 120
carapace, 103, 184
carbohydrate, 210
carbohydrate metabolism, 210
carbon, 59, 65, 178, 207, 208
carbon dioxide, 59, 178, 207, 208
carbon monoxide, 65
Carcinas menas, 93
Caribbean, 84
carnivores, 7, 24, 30, 81
carotenoids, 26
carotid sinus, 189, 199, 208
cascades, 101, 102, 104, 108, 111, 113, 114
catfish, 40
cation, 194
cell culture, 193
cell death, 136
cell groups, 151
cell line(s), 86, 138
cell membranes, 178
cell migration, 136, 138, 141, 142
cell motility, 136, 141
cell surface, 141
Cellana grata, 22, 37
cells, 1, 3, 4, 5, 32, 34, 46, 48, 52-64, 66, 67, 68, 70, 71, 119, 121, 122, 123, 125, 130, 136, 137, 138, 139, 141, 142, 143, 147, 149, 151, 153-161, 163, 164, 165, 167, 169, 170, 171, 173, 175, 176, 178
  cerebral, 48, 50, 51, 56, 57, 58, 65, 147, 148, 150, 154, 159, 164, 165, 169, 173, 176
  ciliated, 46, 49, 122, 175
  excitable, 3, 5
  giant dopamine, 157, 170, 176
  growth hormone secreting, 166
  mobile-immune, 139
  mucus, 13, 17, 18, 32, 37, 46, 49, 76, 81, 86, 159
  muscle, 1, 4, 26, 75, 100, 124, 126, 162, 185
  neuroendocrine, vii, 1, 123, 125, 135, 136, 138, 139, 140, 141, 142, 143
  neurosecretory, 67, 155, 159, 171, 172, 174, 175
  osphradial, 55, 66, 149, 163
  osphradial ganglion, 163
  paired cerebral giant, 147
  sensitive peripheral chemoreceptor, 149, 163

cellular energy, 178
cellular O_2 levels, 178
cellular/molecular mechanisms, 180
central nervous system (CNS), 47, 63, 68, 69, 70, 118, 146, 147, 150, 163, 187, 189, 192, 193, 194, 195, 208
Cephalaspidea, 23, 30
Cephalaspideans, 13, 18, 30
cephalic sensory organs, 46, 47, 48, 50, 52, 53, 54, 55, 59, 61, 63, 65, 70
cephalic tentacles, 48, 56, 58
cephalopods, vii, 5, 13, 18, 23, 76, 80, 81, 82, 86, 87, 117, 118, 119, 120, 121, 123, 128, 129, 132
cerebellum, 131
cerebral ganglia, 48, 56, 58, 65, 147, 148, 150, 164, 165, 169
cerebro-pedal connectives, 148
cerrata, 29
  autonomised, 29
challenges, 33, 63, 177, 179, 181, 182, 191, 196, 199
changing environment, viii, 179, 194, 204, 213
channel blocker, 4
chemical communication, 4, 17, 18, 128
Chemical Cue Transport, 94
chemical cues, vii, 18, 75, 84, 89, 90, 93, 94, 96, 100, 101, 107, 108, 109, 110, 111, 115, 118, 120, 126, 127, 128
chemical defense, 4, 7, 9, 13, 17, 21, 28, 29, 30, 31, 34, 38, 40, 41, 75, 76, 85, 86, 129
chemical ecology, vii, 4, 5, 8, 9, 17, 18, 28, 32, 33, 36, 37, 38, 41, 83, 86, 108, 112
chemical interaction(s), 18, 19, 22, 23, 29, 31
chemical properties, 39, 124
Chemically-mediated antipredator responses of molluscs display, 99
Chemoreception, v, 62, 63, 117
chemoreceptors, 31, 46, 48, 50, 51, 52, 54, 55, 59, 60, 62, 66, 82, 163, 183, 187, 188, 189, 191, 194, 195, 199, 200, 202, 203, 205, 208
  close-range, 46
  contact, 48, 51, 52, 53, 54, 81, 96, 99, 106, 157, 163
  identifying, 17, 51, 55, 59, 182
  osphradial, 55, 66, 149, 163
  primary, vii, 4, 22, 23, 28, 30, 34, 46, 52, 53, 60, 61, 66, 69, 96, 99, 100, 114, 119, 139
  unidentified, 52, 125
chemosensory, 4, 48, 51, 52, 53, 54, 55, 56, 59, 60, 63, 64, 65, 68, 80, 81, 82, 86, 90, 112, 115, 120, 163
chemosensory cells, 54, 55, 59, 60
  primary, vii, 4, 22, 23, 28, 30, 34, 46, 52, 53, 60, 61, 66, 69, 96, 99, 100, 114, 119, 139
chemosensory systems, 81, 82, 90
chemotaxis, 81, 127, 136, 138, 141
chemotherapy, 34
Chicago, 84
choline, 24, 26, 31
Chromodoridadae nudibranchs, 30
chronic hypoxia, 189, 191, 202, 205
cilia, 46, 49, 60, 158, 159, 175, 202
  long, 14, 21, 24, 46, 57, 59, 63, 64, 67, 76, 80, 81, 98, 108, 114, 118, 128, 137, 146, 157, 167, 169, 173, 174, 177, 178, 179, 180, 181, 182
  motile, 60
  pedal, 50, 68, 147, 148, 150, 157, 158, 159, 160, 164, 171, 173, 174, 175
ciliary activity, 60
ciliary flow, 60
ciliary glider, 159
ciliomotoneurons, 159
circuits, 5, 59, 60, 62, 146, 171, 180
circulation, 181, 199
Cladobranchia, 29, 39
Clams, 102
clarity, 74
Class Scaphopoda, 30
classes, 6, 22, 148
classical conditioning, 174
classification, 193
cleaning, 123
climate(s), viii, 8, 13, 20, 35, 36, 37, 146, 168, 169, 170, 196, 203, 213
climate change, viii, 35, 36, 37, 146, 168, 169, 170, 203, 213
Clione limacina, 47, 68
cloning, 208
closure, 163, 164
clusters, 54, 103, 145, 146, 148, 151, 153, 154, 155, 157, 158, 159, 160, 164, 169
  neuronal, 47, 66, 67, 70, 119, 131, 145, 146, 154, 157, 164, 167, 169, 171, 172, 173, 174, 175, 176
  paired, 21, 56, 147, 148, 150, 151, 157, 158, 159, 164, 170, 173, 174
CMCPG, 156, 157, 159
Cnidaria, 178, 179, 185
cnidarians, 29, 81, 182
CNS, 56, 57, 58, 61, 68, 123, 127, 150, 165, 167, 169, 212
CO_2, 55, 178, 181, 184
coding, 78
co-evolution, 75, 146
cognition, 2
cognitive impairment, 191
Collisella dorsuosa, 22

colonization, 7, 19, 104
color, 6, 78, 80, 192, 193
colorectal cancer, 40
colours, 30, 118
combined effect, 93
command neuron peptides (CNPs), 58
commercial, 19
commissure, 147, 148
communication, 4, 5, 6, 17, 18, 26, 34, 60, 74, 82, 120, 128, 140, 212
communication systems, 60
community(s), vii, 1, 2, 3, 4, 5, 6, 7, 8, 9, 14, 15, 17, 19, 21, 22, 24, 26, 30, 36, 38, 39, 40, 51, 85, 89, 90, 100, 104, 106, 108, 109, 111, 112, 115, 117, 118, 119, 120, 127, 128, 179, 181, 198, 211, 212
  chemical interactions shaping rocky intertidal, 22
  structuring, vii, 7, 15, 21, 24
comparative analysis, 130
compensation, 86, 207
competition, 21, 84, 89, 101, 110, 111, 112, 113, 139
competitive abilities, 99
competitive advantage, 106
competitive interactions, 17, 18, 101, 108, 127
competitors, 7, 80, 82, 96, 100, 113
complexity, 8, 9, 29, 53, 62, 111
components, 3, 31, 48, 57, 58, 61, 62, 64, 66, 78, 80, 81, 85, 108, 130, 163, 172
composition, 18, 24, 34, 49, 109, 129, 174, 198
compounds, 9, 13, 14, 15, 16, 18, 19, 21, 25, 26, 27, 28, 29, 30, 34, 38, 74, 76, 78, 82, 85, 93, 120, 211
  deterrent, 17, 28, 29, 39, 76, 78, 80, 82, 85
  dietary derived, 26, 29, 31
  neurotoxic, vii, 13, 24, 26, 28, 30
  toxic, 13, 17, 26, 30, 76, 81, 108, 180
comprehension, 212
computer, 137, 143
condensation(s), 150, 151, 171
conditioning, 7, 64, 83, 155, 167, 169, 172, 173
conduction, 5, 6, 206
configuration, 146, 167
Congress, 70
connections, 5, 56, 73, 119, 154, 156, 157, 160, 162, 163, 164, 167, 168, 170, 172, 185
  interneuronal, 56
connectivity, 146, 167, 168, 202, 213
conopeptides, 24, 26
consciousness, 117, 118
consensus, 181
conservation, vii, 3, 74, 178, 179, 180, 204, 213
consolidation, 155, 157, 173
conspecific odor, 92, 93
conspecifics, 6, 18, 71, 81, 86, 87, 91, 93, 98, 100, 106, 132
  crushed, 91
  injured, 91, 92, 93, 106, 107, 114
  threatened, 18, 91, 177
constituents, 140
consumers, 4, 17, 73, 75, 82, 96
consumption, 101, 103, 104, 105, 113, 199
contamination, 60, 179, 203
context, 51, 61, 68, 73, 75, 89, 94, 101, 105, 118, 139, 155, 162
  environmental, vii, viii, 1, 5, 19, 48, 73, 75, 84, 89, 94, 112, 113, 120, 121, 126, 148, 161, 163, 164, 168, 169, 174, 177, 178, 179, 181, 182, 185
contingency, 141
convergence, 35, 131
cooperation, 135
coordination, 2, 123, 126, 155, 163, 173
copulation, 118
correlation, 22, 166
correlation coefficient, 166
corticosteroids, 204
Corticotrophin-Realising Factor (CRF), 126
corticotropin, 131, 141, 142
cortisol, 142
cost, 18, 28, 97, 99, 104
courtship, 4, 45, 62
covering, 117, 118
CPG functions, 185
CPGs, 145, 155, 157
  activated, 164, 165, 166, 185
  feeding, 4, 5, 9, 17, 21, 26, 28, 32, 53, 54, 64, 67, 68, 75, 80, 81, 85, 93, 97, 101, 106, 108, 109, 112, 114, 118, 120, 127, 146, 155, 157, 161, 163, 164, 167, 169, 170, 171, 174, 176
  locomotor, 83, 145, 156, 157, 158, 159, 161, 162, 163, 164, 175, 185
  presumed ciliomotor, 157
  respiratory, viii, 48, 51, 53, 62, 145, 148, 149, 152, 155, 158, 159, 161, 162, 163, 164, 167, 169, 170, 171, 172, 173, 174, 175, 177, 178, 179, 180, 181, 182, 184, 185
  shell movement, 157, 159, 161
crabs, 76, 81, 82, 83, 91, 93, 96, 97, 98, 100, 101, 102, 103, 109, 111, 207
  green, 93, 101, 107, 111, 159, 165, 171, 172
  predatory, 4, 7, 13, 18, 21, 23, 24, 26, 28, 30, 31, 51, 52, 68, 73, 74, 80, 81, 82, 83, 84, 85, 93, 96, 97, 98, 99, 100, 101, 104, 106, 107, 108, 109
cracks, 97
Crassostrea gigas, 6, 39
Crassostrea spp, 6
Crassostrea virginica, 9, 100, 110

crayfish, 76, 83, 91, 92, 93, 97, 104, 110
crayfish predator cues, 76
creep, 196
CRF, 124, 126, 127, 141
CRH, 136, 137, 138, 139, 140, 142
human, viii, 14, 18, 32, 34, 71, 138, 139, 142, 177, 179, 180, 181, 182
crop, 62
CT, 173, 174
Cue Characteristics, 90
cues, vii, 4, 8, 17, 18, 29, 36, 37, 46, 51, 59, 73, 74, 75, 76, 81, 82, 84, 85, 86, 89, 90, 91, 92, 93, 94, 96, 97, 98, 99, 100, 101, 102, 103, 107, 108, 109, 110, 111, 112, 114, 115, 118, 120, 126, 127, 128, 157, 168, 212
chemosensory, 4, 48, 51, 52, 53, 54, 55, 56, 59, 60, 63, 64, 65, 68, 80, 81, 82, 86, 90, 112, 115, 120, 163
communication, 5, 34, 60, 74, 82, 120, 140
environmental, vii, viii, 1, 5, 19, 48, 73, 75, 84, 89, 94, 112, 113, 120, 121, 126, 148, 161, 163, 164, 168, 169, 174, 177, 178, 179, 181, 182, 185
olfactory, vii, 8, 9, 10, 45, 46, 47, 48, 49, 51, 52, 55, 56, 58, 59, 60, 61, 62, 63, 64, 65, 66, 67, 68, 69, 70, 82, 85, 112, 118, 120, 121, 122, 123, 124, 125, 126, 127, 128, 129, 130, 131, 132
reliable, 81, 90, 91, 92, 93, 166
settlement, 17, 18, 100, 109, 110, 111, 113
culture, 31, 180, 208
currents, 46
Cyanea capillata, 185
cyanide, 195
cycles, 161, 165, 168, 186, 192, 193
Cyclical behaviours, 155
cycling, 108
cystic fibrosis, 181, 206
cytokines, 135, 137, 138, 139, 142, 213
cytoskeleton, 137
cytotoxicity, 136, 138, 143

D

danger, 10, 101, 112
data collection, 119
database, 67
decay, 95
decoding, 34, 119
decomposition, 181, 196, 197, 198
defence, vii, 18, 38, 39, 85, 86, 117, 118, 120, 211, 212
defense mechanisms, 28, 29, 34
antipredator, 26, 29, 84, 90, 91, 93, 95, 97, 99, 100, 101, 102, 104, 105, 106, 108, 112, 113
defense strategies, 29
defenses, 6, 9, 17, 21, 32, 38, 74, 75, 76, 80, 82, 83, 84, 85, 86, 89, 99, 105, 106, 107, 111, 112, 114
anti-predator, 6, 74, 84, 89, 110
anti-predatory, 6
behavioral, 1, 3, 6, 7, 45, 51, 55, 58, 60, 61, 62, 69, 70, 73, 74, 75, 78, 80, 82, 85, 93, 96, 97, 99, 104, 105, 111, 114, 169
passive, 75, 76
defensive behaviors, 90
defensive ink, vii, 13, 30
complex, vii, 2, 3, 6, 17, 19, 30, 55, 58, 61, 73, 78, 80, 81, 82, 118, 119, 120, 131, 137, 138, 139, 145, 146, 148, 155, 157, 161, 163, 170, 171, 180, 185
degradation, 182, 203, 207
demography, 89, 100
dendrites, 53, 56
depolarisation, 165, 166
depolarization, 188
depression, 190, 201, 202, 207
depth, 13, 14, 96, 196
derivatives, 22, 24, 29, 32
desmosome, 49
detectable, 18
detection, 6, 8, 18, 39, 48, 51, 55, 56, 59, 68, 74, 82, 85, 86, 90, 94, 101, 105, 169, 174, 207, 208, 212
chemically-mediated predator, 105
odor-mediated predator, 101, 115
deterrence, 21, 28, 38, 76
deterrents, 28, 76, 80, 108
deviation, 137, 165
diatoms, 33
diet, 8, 17, 24, 34, 38, 39, 53, 76, 78, 81, 84, 91, 92, 93, 98, 114, 115
diffusion, 94, 178, 183, 184, 185, 191, 196, 197, 202, 206
diffusion rate, 183, 184, 185
dimorphism, 8
direct action, 163
discharges, 55
disease(s), 19, 34, 36, 37, 177, 178, 179, 180, 181, 182, 204
human, viii, 14, 18, 32, 34, 71, 138, 139, 142, 177, 179, 180, 181, 182
dishabituation, 68
dissociation, 187
dissolved O_2, 181, 185
dissolved oxygen, 208
distinctness, 24

distribution, 13, 17, 19, 20, 23, 24, 26, 27, 30, 38, 40, 48, 60, 63, 86, 94, 97, 100, 104, 105, 118, 159, 168, 171
Disturbance-released alarm cues, 91
divergence, 3
diversity, 3, 6, 8, 13, 14, 17, 21, 22, 24, 29, 30, 32, 35, 40, 53, 55, 64, 75, 105, 106, 178, 179, 204, 209, 212
DNA, 39, 40, 178
DNA damage, 40
DOI, 111, 115, 173
dominance, 6, 23
dopamine, 58, 69, 120, 139, 157, 165, 167, 168, 170, 176, 193
 content, 54, 58, 59, 165, 167, 171, 181
 receptors, 38, 48, 49, 51, 60, 62, 63, 80, 112, 124, 139, 141, 143, 161, 167, 168, 170, 173
dopaminergic, 65, 168
dorsal pedal commissure, 147, 148
Doryteuthis, 5, 6
double bonds, 148
down-regulation, 168, 170
draft, 48
drainage, 199
Drosophila, 3, 63, 69, 209
Drosophila melanogaster, 3, 69
Dual inhibition of VD4, 162
dyes, 53
dynamics, 4, 7, 35, 65, 73, 119
 predator-prey, 7, 29, 74, 75, 95, 105, 110, 111, 113
 prey population, 7, 8, 85, 99, 100, 112

E

eastern oyster, 102
echinoderms, 74
ecological communities, vii, 3, 6, 9
 contributors structuring, 6
 structuring, vii, 7, 15, 21, 24
Ecological Consequences of Prey Responses, 99
ecological interactions, 1, 6, 83, 89, 90, 105
ecological niches, 14, 117, 118
ecological roles, 7, 26
ecology, vii, 1, 3, 4, 5, 7, 8, 9, 17, 18, 26, 28, 31, 32, 33, 36, 37, 38, 41, 62, 73, 74, 83, 84, 86, 108, 110, 111, 112, 113, 119, 179, 181, 205, 206, 211, 212, 213
ecosystem(s), 3, 5, 6, 8, 14, 17, 18, 19, 21, 36, 73, 108, 114, 118, 119, 120, 127, 169, 178, 179, 204, 209, 212
 health, viii, 14, 19, 179, 181
 services, 108, 110, 179
 freshwater, viii, 8, 46, 53, 61, 64, 71, 74, 83, 84, 86, 97, 106, 110, 114, 115, 141, 142, 146, 148, 151, 161, 168, 169, 170, 171, 172, 173, 175, 176, 182, 184
 natural, 1, 4, 6, 13, 14, 15, 17, 18, 19, 20, 21, 22, 23, 24, 26, 28, 30, 31, 32, 33, 35, 36, 51, 53, 82, 85, 94, 96, 100, 103, 112, 119, 120, 130, 138, 141, 161, 178, 179, 182
editors, 209
effluent, 84, 111
egg, 18, 19, 22, 24, 26, 32, 37, 39, 40, 51, 118, 120, 123, 126, 169, 202
ejaculation, 124
electron, 46, 47, 50, 122
electron microscopy, 122
elevated PCO_2, 149
e-mail, 145
embryogenesis, 62
encoding, 48
endocrine, 123, 130, 140, 141, 169
endocrine glands, 123, 169
endocrine system, 123
endocrinology, 172
endorphins, 141
energy, vii, 4, 28, 74, 83, 89, 90, 100, 105, 111, 114, 123, 126, 127, 136, 178, 181, 189, 213
energy transfer, 114
enlargement, 125
environment(s), vii, 9, 14, 19, 22, 23, 38, 45, 46, 59, 74, 75, 80, 83, 84, 89-91, 93, 94, 95, 97, 102, 105, 106, 108, 115, 119, 120, 127, 131, 145, 146, 148, 151, 155, 157, 161, 169, 171, 177-182, 184, 185, 187, 191, 199, 200, 202, 204, 208, 209, 212, 213
 aquatic, 9, 32, 37, 45, 46, 51, 53, 59, 60, 74, 75, 76, 82, 83, 93, 109, 110, 120, 146, 158, 177, 179, 181, 182, 184, 185
 ever-changing, 179
 flow, 51, 60, 68, 70, 71, 83, 87, 89, 94, 95, 97, 105, 108, 109, 110, 113, 183, 184
 freshwater, viii, 8, 46, 53, 61, 64, 71, 74, 83, 84, 86, 97, 106, 110, 114, 115, 141, 142, 146, 148, 151, 161, 168, 169, 170, 171, 172, 173, 175, 176, 182, 184
 high risk, 106
 hypoxic, 177, 179, 181, 182
environmental challenges, 179, 182
environmental change, viii, 169, 179, 181, 210, 213
environmental characteristics, 84
environmental conditions, viii, 75, 161, 178, 181, 182, 191, 204
 changing, viii, 3, 113, 118, 127, 159
 extreme, 30
environmental factors, 112, 185, 194

environmental hypoxia, 177, 179, 181, 185
environmental O_2, 181
environmental stimuli, 1
environmental stress, 19, 148, 163, 204
enzyme(s), 29, 40, 178, 187
Eogastropoda, 22
epidermis, 46, 171
epinephrine, 139
epithelia, 48, 49, 53, 56, 60, 64
epithelium, 40, 46, 47, 48, 56, 57, 58, 60, 64, 121, 122, 123, 175, 184
sensory, 1, 2, 5, 7, 18, 29, 45, 46, 47, 48, 49, 50, 51, 52, 53, 54, 55, 56, 57, 58, 59, 60, 61, 62, 63, 64, 65, 68, 69, 70, 71, 74, 89, 105, 114, 115, 120, 122, 123, 125, 148, 155, 157, 172, 178
erythrocytes, 138
escape, 5, 6, 18, 29, 52, 65, 73, 74, 75, 80, 84, 85, 89, 90, 96, 97, 105, 112
rapid, 5, 47, 65, 96, 105, 161
responses, vii, 2, 5, 6, 45, 52, 54, 55, 58, 60, 61, 64, 65, 66, 70, 73, 76, 77, 78, 79, 82, 85, 89-94, 96, 98, 99, 100, 104, 106, 107, 108, 110, 111, 114, 120, 123, 130, 137, 139, 140, 155, 157, 160, 167, 168, 169, 177, 180
esophagus, 185
ester, 24, 31
estuarine systems, 108
ethology, vii, 3, 119, 211
EU, 117, 118, 129, 131
EU Directive, 117, 118, 131
Euglandina, 81, 84
Europe, 53, 56, 129, 168, 196
European Parliament, 129
evaporation, 196
evidence, 6, 18, 21, 29, 40, 48, 51, 52, 54, 55, 58, 61, 85, 89, 122, 130, 135, 136, 141, 142, 151, 166, 170, 186, 187, 200, 202, 203, 207, 208
Evidence of Seasonal Changes, 165
evolution, vii, 1, 2, 3, 8, 14, 21, 28, 31, 34, 50, 64, 67, 69, 74, 75, 82, 84, 86, 106, 111, 128, 130, 135, 136, 137, 139, 141, 142, 145, 146, 148, 150, 151, 170, 171, 174, 177, 180, 204, 206, 208, 211, 212, 213
metazoan, 178, 179, 180
olfactory system, 45, 49, 51, 55, 56, 58, 59, 60, 61, 65, 66, 69, 123
evolution of chemical defense, 28, 31
evolutionary origin, 135, 142, 151
evolutionary pressures, 148
strong, 4, 7, 21, 24, 28, 54, 73, 74, 77, 89, 91, 95, 99, 104, 105, 108, 120, 127, 157, 164, 168
evolutionary relationships, 61, 62
evolutionary time, vii, 6, 146
excitability, 5, 169, 174, 190
excitation, 4, 5, 58, 152, 163, 187, 193, 194, 195, 206
exclusion, 81, 101
excretion, 178
excurrent, 97, 110
exoskeleton, 6
experimental design, 93
expiration, 161, 162, 164, 183
exposure, 80, 81, 86, 98, 99, 106, 152, 187, 189, 190, 200, 203, 209, 210
external environment, 148, 157
extinction, viii, 106, 178, 179, 182, 204, 209, 213
global, 8, 13, 19, 67, 178, 179, 182
extracts, 17, 18, 19, 32, 78, 84, 169

F

families, 39
family Aplysiidae, 46
family Conidae, 24
fatty acids, 32, 33, 37
fauna, 14
FCPG, 146, 156, 157
fear, 94, 112, 114
feeding, 4, 5, 9, 17, 21, 26, 28, 32, 53, 54, 64, 67, 68, 75, 80, 81, 85, 93, 97, 101, 106, 108, 109, 112, 114, 118, 120, 127, 146, 155, 157, 161, 163, 164, 167, 169, 170, 171, 174, 176
behavior, 1, 2, 4, 5, 7, 8, 18, 23, 26, 45, 51, 53, 55, 56, 59, 62, 65, 67, 69, 70, 73, 74, 75, 76, 80, 82, 83, 84, 87, 93, 97, 99, 100, 101, 102, 103, 104, 105, 106, 108, 109, 110, 111, 112, 113, 114, 129, 146, 171, 172, 179
behaviour, vii, viii, 8, 63, 65, 69, 70, 84, 86, 111, 117, 118, 119, 120, 123, 124, 126, 127, 128, 132, 137, 146, 148, 151, 155, 157, 161, 163, 164, 165, 168, 169, 170, 171, 174, 175, 176, 177, 178, 180, 181, 182
cycle, 18, 74, 127, 129, 157, 163, 167, 179, 183
deterrence, 21, 28, 38, 76
deterrent activity, 17
motor neuron B4, 155
motor neurons, 152, 154, 156, 159, 161, 162, 163, 164, 170, 175
motor program, 54, 119, 131
network, 58, 66, 145, 146, 148, 155, 157, 164, 168, 169, 170, 171, 174
system, 3, 7, 18, 40, 50, 52, 59-70, 73, 85, 93, 94, 108, 111, 118, 119, 120, 123, 124, 128, 129, 131, 139, 140, 145, 146, 148, 150, 151, 152,

155, 156, 157, 159, 161, 162, 164, 169, 171, 176, 177, 179, 180, 182, 184
fertilization, 118, 129
fertilizers, 179
fibers, 58, 63
Fick, 184, 185
fictive feeding motor pattern, 155
filter feeders, 26, 27, 28
filtration, 100
fires, 186
fish, 8, 17, 38, 74, 76, 77, 78, 80, 82, 85, 91, 92, 97, 98, 104, 107, 111, 112, 113, 114, 131, 181, 184, 185, 198
fish predators, 38, 80, 104, 112
fisheries, 119, 179
fitness, 18, 28, 74, 76, 123, 169
flexibility, 169, 170, 178, 189
flight, 74
flooding, 196
fluctuating O_2 conditions, 181
fluctuations, 113, 127, 129, 181, 187, 203
fluid, 67, 89, 105, 108, 202
fluid mixing, 105
fluorescence, 131
FMRFamide, 48, 58, 67, 122, 123, 125, 126, 127, 129, 131
FMRFamide and GnRH neurons, 126
FMRF-related peptides, 47
food, 4, 6, 9, 17, 30, 46, 56, 59, 62, 63, 64, 65, 67, 69, 70, 73, 75, 80, 81, 93, 96, 97, 98, 99, 104, 105, 106, 109, 113, 114, 118, 119, 121, 123, 126, 127, 166, 167, 169, 179, 203, 212
food chain, 114, 179
food intake, 99, 118, 121, 127, 169
food source, 17, 30, 75, 81, 119
food web, 4, 113, 203
force, 99, 107, 159
Ford, 19, 34, 35
formation, 9, 19, 150, 169, 173, 204
formula, 137
fouling, 17, 18, 19, 39
foundation species, 3, 6, 7
fragments, 137, 138, 141
France, 33
freshwater, viii, 8, 46, 53, 61, 64, 71, 74, 83, 84, 86, 97, 106, 110, 114, 115, 141, 142, 146, 148, 151, 161, 168, 169, 170, 171, 172, 173, 175, 176, 182, 184, 187, 193, 196, 197, 198, 208, 209, 213
freshwater snail(s)
  natural environment, 94, 161
freshwater species, viii, 146, 169, 213
frontal lobe, 120
function, 3, 4, 6, 14, 18, 19, 48, 54, 55, 56, 59, 60, 61, 62, 63, 76, 80, 81, 82, 91, 93, 94, 96, 101, 103, 105, 109, 114, 117, 118, 120, 122, 127, 131, 137, 139, 140, 141, 178, 184, 185
  immune, vii, 18, 36, 120, 135, 136, 138, 139, 140, 141, 142, 143
  immune-neuroendocrine, 135, 140
  neuroendocrine, vii, 1, 123, 125, 135, 136, 138, 139, 140, 141, 142, 143
  neurotoxins, 4, 28
functional analysis, 69
functional roles, 26, 146
fungi, 18
fusion, 2, 6, 150, 151

G

GABA, 58, 67, 173, 189, 209
Galanin, 124, 126, 127, 131
gametogenesis, 124
ganglia, 52, 56, 57, 58, 63, 65, 120, 126, 145, 146, 147, 148, 150, 151, 153, 155, 158, 160, 172, 174
  buccal, 68, 147, 148, 150, 173, 174, 184
  circumsoesophageal, 146
  crab stomatogastric, 180
  left parietal, 147, 150, 153, 174
  osphradial, 55, 66, 149, 163
  paired pleural, 147, 148
  rhinophore, 46, 47, 48, 49, 50, 52, 70
  subintestinal, 150
  supraintestinal, 150
ganglion, 46, 47, 50, 51, 56, 57, 58, 120, 129, 147, 148, 150, 151, 152, 163, 174, 194, 195, 199
ganglionic fusion, 151
  parietal/visceral, 151
gas diffusion, 184
gas exchange, 178, 183, 184, 185
  efficient, 74, 120, 178, 183, 184
gas exchange surfaces, 183, 184, 185
  ideal, 184
gastropod, vii, 7, 24, 34, 35, 38, 39, 45, 47, 56, 59, 60, 61, 62, 63, 64, 80, 83, 84, 85, 86, 91, 101, 106, 109, 110, 111, 112, 113, 122, 141, 142, 145, 148, 149, 171, 173
  chemoreceptor gene sequences, 59
  olfaction, 45, 46, 48, 50, 52, 53, 55, 56, 58, 59, 60, 61, 62, 63, 68, 69, 127, 130
gastropods, 5, 13, 18, 19, 22, 23, 24, 26, 28, 32, 34, 38, 39, 45, 46, 48, 52, 55, 57, 58, 59, 60, 61, 65, 76, 81, 82, 84, 85, 86, 96, 98, 110, 115, 150, 171, 184

aquatic, 9, 32, 37, 45, 46, 51, 53, 59, 60, 74, 75, 76, 82, 83, 93, 109, 110, 120, 146, 158, 177, 179, 181, 182, 184, 185
grazing, 13, 24, 26, 28, 90, 108
predatory marine, 24
terrestrial, 17, 36, 45, 49, 55, 56, 58, 59, 60, 61, 62, 64, 65, 66, 68, 69, 81, 83, 151, 181, 184, 185
general knowledge, 94
genes, 3, 33, 48, 59, 60, 62, 64, 127, 136, 180
genetics, 3, 177, 180
genome(s), 6, 9, 48, 49
genomics, 132
genus, 38
geological history, 118
Georgia, ix, x, 1, 73, 89, 106, 115
geotaxis, 161
Geukensia demissa, 100
giant neurons, 145, 148, 151, 157, 175
Gibbula umbilicalis, 101
gill, 6, 63, 149
gills, 48, 148, 150, 151, 178, 183, 184, 185
internal, 28, 121, 123, 129, 147, 148, 153, 155, 168, 169, 184
gland, 26, 35, 37, 39, 40, 69, 78, 120, 124, 125, 126, 127, 128, 129, 130, 212
glia, 48, 49
global warming, 182
glomeruli, 47, 52, 56, 57, 60, 69
putative, 53, 54, 56, 57, 58, 68, 172
glomerulus, 47, 49, 51, 56, 57, 60, 67
glucose, 66, 178
glutamate, 58, 63, 65, 66
GnRH immunoreactivity, 123, 129
GnRH mRNA level, 124
GnRH neurons, 126
GnRH-ir fibres, 123
gonadotropin-releasing hormone (GnRH), 129
gonads, 37, 125, 126, 169
gradients, 21, 51, 105, 108, 112
grants, 140
granules, 120
grass(s), 3, 205
gravity, 172
gray matter, 191
grazers, 24, 26, 47, 78, 101, 106
grazing, 13, 24, 26, 28, 90, 108, 211
growth, vii, 18, 21, 34, 74, 85, 91, 96, 99, 100, 111, 112, 118, 121, 123, 126, 128, 136, 166, 169, 172, 196, 198, 199, 203, 207, 213
reduced, vii, 13, 29, 30, 31, 76, 79, 90, 97, 98, 99, 100, 165, 166
reduced prey, 100
growth factor, 136
growth hormone, 166
growth rate, 21, 96
guidance, 115
guiding principles, 59
Gulf of Mexico, 20, 198, 208

H

habitat(s), 4, 7, 8, 13, 17, 22, 23, 30, 45, 51, 53, 74, 76, 89, 94, 96, 97, 99, 100, 101, 104, 109, 111, 112, 113, 114, 118, 120, 127, 151, 168, 177, 179, 181, 182, 184, 196, 198, 200, 203, 212
freshwater, viii, 8, 46, 53, 61, 64, 71, 74, 83, 84, 86, 97, 106, 110, 114, 115, 141, 142, 146, 148, 151, 161, 168, 169, 170, 171, 172, 173, 175, 176, 182, 184
harassment, 80
health, viii, 19, 179, 181, 203, 204, 213
heavy metals, 203
Helisoma trivolvis, 169, 176
Helix, 55, 56, 61, 62, 63, 64, 65, 66, 67, 68, 70
Helix aspersa, 56, 62, 63, 64, 66, 68
heme, 188
hemisphere, 196
Hermissenda crassicornis, 6, 62, 81, 83
herpes, 19, 34, 39
herpes simplex, 34
herpes virus, 19, 39
Heterobranchia, 27, 28, 64
Heterobranchs, 31
heterogeneity, 67, 104
heterospecifics, 91, 92
HIF-1, 180
high blood pressure, 191
HIJK cells, 151, 152
histochemistry, 53, 69
histology, 31
history, 14, 24, 61, 75, 90, 94, 96, 106, 108, 119, 130, 169, 180
homeostasis, 178, 182, 185, 187, 198, 203
homeostatic behaviour, 177
hormone(s), 67, 121, 124, 125, 127, 128, 129, 136, 139, 143, 166, 169, 172, 176, 203
host, 4, 18, 32, 109
hot springs, 204
hotspots, 32
human, viii, 14, 18, 32, 34, 71, 138, 139, 142, 177, 179, 180, 181, 182, 198, 203, 205, 213
human activity, viii, 198, 213
human body, 181, 182
human condition, 177
human health, 14, 203

humans, 177, 180, 182, 184
hunting, 80
hybrid, 50
hydroid, 81
Hygrophila, 61
hyperpolarisation, 165, 166
hypertrophy, 125
hyperventilate, 190
hypothesis, 21, 58, 106, 107, 112, 167, 202
hypoxia, viii, 149, 152, 163, 177, 179, 180, 181, 182, 185, 187, 188, 189, 190, 191, 194, 195, 196, 197, 198, 199, 200, 201, 202, 203, 204, 205, 206, 207, 208, 209, 210, 213
  acute, 40
  aquatic, 9, 32, 37, 45, 46, 51, 53, 59, 60, 74, 75, 76, 82, 83, 93, 109, 110, 120, 146, 158, 177, 179, 181, 182, 184, 185
  chronic, 181
hypoxic, 177, 179, 181, 182
  conditions, 75, 95, 99, 148, 161, 166, 177, 178, 179, 180, 181, 182
  exposure, 80, 81, 86, 98, 99, 106, 152
  stress, 9, 19, 105, 136, 138, 139, 140, 141, 142, 148, 163, 169, 174, 181

I

ideal, 184
identification, 3, 17, 24, 37, 50, 54, 59, 60, 64, 70, 86, 112, 175, 193
identity, 51, 76, 90, 94, 98, 100, 104, 114, 187
IFS (invasive freshwater species)
  insidious, 169
IL-8, 136, 137
image, 137
image analysis, 137
immune function, 136, 140
immune response(s), 36, 139, 142
immune system, 18, 120, 139, 140, 143, 203
immunity, 34, 142, 203, 206
immunobiology, 141
immunocytes, 137, 138, 139, 141, 142
  invertebrate, 8, 30, 32, 38, 51, 111, 119, 132, 138, 141, 142, 178, 179, 180, 182, 185
Immunocytochemical comparisons of cellular distribution patterns, 48
immunohistochemical evidence, 58
immunohistochemical localization, 58
immunohistochemistry, 53, 123
immunolocalization, 47, 129
immunoreactivity, 47, 58, 63, 69, 122, 123, 129, 130
in situ hybridization, 136
in vitro, 68, 192, 194, 207
in vitro exposure, 207
in vivo, 68, 180, 192, 193, 206, 207
incidence, 19, 94, 106, 177
India, 39
indirect effect, 29, 89, 105, 108, 113
individuals, 2, 4, 75, 81, 90, 91, 97, 104, 118
Indo-Pacific molluscs, 21
Indo-Pacific species, 21
inducer, 125
Inducible defenses, 97
induction, 83, 111
industrialization, 179
inefficiency, 184
Infaunal bivalves, 96
infection, 22, 39, 40
inflation, 193
information exchange, 95
information processing, 2, 65
ingestion, 6, 76, 77
inhibition, 5, 8, 152, 157, 159, 162, 170, 186
initiation, 13, 64, 172
injure, 93
injury, 76, 89, 109, 180, 203
ink, 18, 29, 31, 76, 77, 78, 80, 85, 86, 91, 119, 120
  cephalopod, vii, 80, 81, 121, 128, 131
  purple, 26, 31, 32, 34, 37, 40, 78
  releasing, 76, 129, 135, 136, 140, 176
  squid, 5, 9, 18, 80, 83, 118, 119, 120, 122, 130, 131, 180
ink gland, 78, 120
ink sac, 80, 119, 120
ink secretion, 78, 85
Inking, 80
innate immunity, 38
insects, 47, 49, 184, 185, 187, 205
insulin, 128
integration, vii, 3, 45, 60, 82, 128, 163, 213
integrity, 207
interface, 7
interference, 3, 8
Interleukin (IL)
  IL-1, 137, 139, 141
  IL-6, 136
interleukin-8, 142
interneuron(s), 57, 58, 65, 147, 157, 159, 162, 163, 164, 172, 174, 175, 176, 193
interneuron VD4, 164
Interneurons L/RPeD11, 164
interspecific communication, 6, 26
intimidation, 113
invasion resistance, 106
invasions, 113

invertebrate(s), 8, 17, 19, 21, 30, 32, 35, 37, 38, 51, 66, 111, 112, 113, 117, 118, 119, 128, 132, 136, 138, 139, 140, 141, 142, 169, 177, 178, 179, 180, 181, 182, 184, 185, 187, 188, 191, 198, 199, 200, 203, 204, 206, 207, 208, 213
Invertebrate neural circuits function, 180
investment, 100, 102
ion channels, 5, 7, 38, 188
ionotropic receptors (IRs), 49
ions, 188
ipsilateral, 159, 164
irrigation, 196
isolation, 17, 34, 185, 194
issues, 7, 132
Italy, ix, 135, 140

J

juveniles, 169, 202

K

key species, 5
keystone species, 3, 4, 9, 41
keystone species concept, 4, 9, 41
kill, 24, 76, 196
kleptocnides, 13, 29

L

L/RPeA cluster neurons, 159
L/RPeD1, 157, 164
L/RPeD10, 157
L/RPeD11, 164
L/RPeD2, 157, 160
L/RPeD3, 157
L/RPIB cells, 164
labeling, 47, 54, 58, 84, 109, 111
laboratory studies, 51
lakes, 45, 94, 196, 197, 198
laminar, 71
landscape, 94, 115
larva, 111
larvae, 138, 148, 184, 209
latitudinal gradients, 21
laws, 136
layering, 196
LCPG, 156, 157, 158, 159
lead, viii, 19, 21, 60, 90, 167, 190, 197, 203, 209
learning, 1, 2, 5, 6, 9, 56, 58, 59, 65, 66, 68, 69, 83, 113, 120, 130, 146, 155, 169, 170, 171, 172, 173, 174, 204
legal protection, 118
leptin, 128
lesions, 51, 85, 125
LGC activity, 165, 166
LGC activity pattern, 165
life cycle, 18, 198
light, 3, 81, 121, 123, 126, 127, 165, 172, 174, 177, 196, 198
Limax, 45, 55, 56, 57, 58, 60, 61, 63, 65, 66, 67, 69, 70
Limax marginatus, 56, 65, 66, 67, 69
Limax maximus, 56, 65
lineages, 3, 28, 185
lipids, 37
lips, 53, 54, 55, 81
Littoraria irrorata, 98
Littoraria sp, 96, 101
Littorina littorea, 100
Littorina obtusata, 98
Littorina sp, 100, 107
livestock, 196, 198
living environment, 181
lobsters, 31, 40, 76, 86, 100
local field potentials, 57, 58
localization, 45, 58, 62, 68
locomotion, 74, 81, 85, 120, 137, 146, 149, 155, 157, 159, 161, 163, 164, 170, 172, 175
locomotor, 83, 145, 156, 157, 158, 159, 161, 162, 163, 164, 175, 185
locomotor behaviour, 159, 163
locomotor CPG, 157, 185
  cat, 185
locomotor motoneurones, 164
Loligo, 5, 117, 119, 130
  Loligo pealeii, 119
  Loligo vulgaris, 117, 119
long term memory (LTM), 169
long-term memory, 63, 67, 157, 173, 174
Louisiana, 198
love, 62
low O_2 conditions, 181
low risk, 106
LP1, 147, 151, 152
LPaV1, 151
LPeD11, 147, 164
LPeD5, 157
LTD, 190, 202
LTM formation, 169
lung cavity, 152, 161, 183
lung ventilation, 149, 155, 157, 161, 163, 170, 183
Lung Ventilation in *Lymnaea*, 161
lungs, 178, 181, 183, 184, 185
Luo, 9, 188, 207

lying, 158
Lymnaea, v, vi, vii, 6, 45, 47, 53, 54, 55, 60-71, 76, 84, 85, 91, 110, 145, 146, 147, 148, 149, 150, 151, 152, 153, 155, 156, 157, 158, 161, 163, 164, 165, 166, 167, 168, 169, 170, 171, 172, 173, 174, 175, 176, 177, 179
aerial respiration, 167, 169, 177
aerial respiratory behaviour, 169, 173
behaviour, vii, viii, 8, 63, 65, 69, 70, 84, 86, 111, 117, 118, 119, 120, 123, 124, 126, 127, 128, 132, 137, 146, 148, 151, 155, 157, 161, 163, 164, 165, 168, 169, 170, 171, 174, 175, 176, 177, 178, 180, 181, 182
behavioural repertoire, 146, 163, 170
brain of, 128, 129, 148, 155, 171, 175
cephalic sensory organs, 46, 47, 48, 50, 52, 53, 54, 55, 59, 61, 63, 65, 70
cerebral ganglia, 48, 56, 58, 65, 147, 148, 150, 164, 165, 169
chemosensory responses, 52, 55
CNS, 47, 56, 57, 58, 61, 68, 123, 127, 150, 165, 167, 169
distance chemoreception, vii, 46, 52, 53, 55
feeding CPG, 146
feeding system, 67, 155, 157, 163, 164, 176
lips, 53, 54, 55, 81
model, 6, 19, 24, 36, 45, 52, 56, 57, 59, 61, 74, 118, 121, 125, 126, 127, 128, 129, 130, 131, 139, 141, 145, 146, 169, 170, 179, 180, 185
nervous system, 3, 5, 47, 59, 60, 61, 63, 65, 66, 67, 68, 69, 70, 82, 117, 119, 128, 129, 145, 146, 147, 148, 150, 151, 155, 157, 163, 169, 170, 171, 173, 174, 177, 178, 180, 182, 185
osphradial neurons, 66
respiratory behaviour, viii, 148, 155, 159, 161, 163, 164, 169, 173, 175, 177, 178, 179, 180, 181, 182
synapses, 5, 52, 54, 56, 60, 61, 167, 185
Lymnaea stagnalis, v, vi, vii, 6, 47, 53, 54, 64, 65, 66, 68, 69, 70, 71, 76, 84, 85, 91, 110, 145, 146, 147, 150, 151, 168, 169, 170, 171, 172, 173, 174, 175, 176, 177, 179
gastropod mollusc, vii, 34, 63, 64, 80, 141, 142, 145, 171, 173
pond snail, 6, 64, 65, 69, 70, 71, 76, 170, 172, 173, 175, 177, 179
pulmonate pond snail, 76
lymphocytes, 138
lysis, 138

M

MAAs, 22, 24
magnitude, 33, 78, 91, 100, 111, 184, 185
majority, 13, 14, 21, 23, 48, 94, 117, 119, 166, 185
mammals, 48, 80, 135, 138, 180, 181, 182, 184, 186, 187, 189, 191, 193, 201, 202, 205, 208, 209
diving, 22, 181
mammalian cells, 138
man, 142
management, 106, 108, 179
mantle, 29, 30, 48, 53, 121, 122, 149, 150, 152, 156, 161, 162, 169, 170, 184, 193
mantle cavity, 53, 122, 149, 150, 156, 161, 162, 169, 170, 184
mapping, 175
marine, vii, 7, 8, 9, 13, 14, 15, 17, 18, 19, 20, 21, 23, 24, 25, 26, 27, 28, 29, 30, 31, 32, 33, 34, 35, 36, 37, 38, 39, 40, 41, 46, 51, 65, 69, 74, 80, 83, 84, 85, 86, 109, 110, 111, 112, 113, 114, 118, 119, 120, 127, 128, 129, 179
ecosystems, 3, 5, 8, 14, 18, 21, 36, 73, 118, 119, 127, 169, 178, 179
invertebrates, 17, 19, 30, 32, 35, 38, 66, 112, 113, 117, 118, 119, 128, 136, 138, 139, 140, 141, 142, 169, 177, 178, 179, 180, 181, 182, 184, 185
molluscan communities, 89, 90, 108
molluscan species, viii, 5, 13, 15, 21, 30, 32, 37, 39, 89, 94, 108, 109, 136
pathogens, 18, 19, 22
marine environment, 9, 19, 110
marine species, 14, 21
marriage, 181
marsh, 84, 101, 112, 114, 115
masking, 17
mass, 6, 19, 51, 80, 96, 182, 202, 209
maternal care, 127, 128
mating, 4, 5, 45, 51, 81, 100, 112, 118, 120, 128
matter, 38, 140
measurement(s), 94, 137, 166
mechanistic explanations, 17
median, 54, 147, 148, 153
mediation, 113, 128
medical, 177, 180, 182, 203
medical science, 180, 182
medicine, 5, 191, 204
Mediterranean, 20, 34, 196
medulla, 186, 189
melanin, 18, 86, 120
membranes, 5
memory, 1, 2, 5, 6, 9, 63, 66, 67, 120, 130, 155, 157, 169, 171, 173, 174, 206
Mercenaria, 6, 92, 93, 109, 110, 113, 114
Mercenaria mercenaria, 6, 92, 93, 109, 110, 113, 114

hard clam, 93, 95, 97, 102, 109, 110
mercury, 203, 207
messages, 90
messengers, 123, 124, 127, 137
metabolic, 74, 90, 113, 128, 178, 182, 184
demands, 178, 182
rates, 4, 21, 22, 61, 97, 99, 100, 101, 102, 107, 114, 121, 179, 183
state, 13, 137, 143, 146, 178
metabolic responses, 209
metabolism, 21, 36, 83, 126, 128, 173, 177, 178, 181, 182, 189, 198, 203, 204
aerobic, 178, 182
metabolite(s) 13, 14, 15, 16, 17, 21, 24, 26, 28, 29, 30, 31, 32, 34, 35, 36, 37, 39, 74, 76, 84, 91, 93, 95, 97, 103, 169
release, 18, 74, 78, 80, 90, 95, 97, 101, 102, 120, 125, 138, 139
defensive, vii, 13, 17, 18, 21, 29, 30, 31, 32, 37, 38, 39, 74, 76, 84, 86, 90, 93, 106, 112, 114, 145, 155, 174
dietary derived, 26, 29, 31
metabotropic G-protein, 48
metacerebrum, 57, 58
metals, 203
metamorphosis, 81
metazoans, 6, 178, 182, 183, 185
MET-enkephalin, 136
meter, 102
Mexico, 198
Mg^{2+}, 194, 195
mice, 3
microorganisms, 18, 120, 181
microscope, 46, 174
microscopy, 123, 137
migration, 113, 136, 137, 138, 141, 142, 181, 198
mimicry, 131
Mississippi River, 198
mixing, 94, 95, 99, 105, 183, 198, 202
model system(s), 45, 52, 56, 57, 59, 61, 74, 130, 145, 146, 192, 204
models, viii, 3, 5, 6, 17, 31, 36, 46, 105, 106, 117, 119, 180, 200, 204, 212
invertebrate, 8, 30, 32, 38, 51, 111, 119, 132, 138, 141, 142, 178, 179, 180, 182, 185
modifications, 115, 137, 167
modulation of, 67, 127, 141, 175, 178
Lymnaea, v, vi, vii, 6, 45, 47, 53, 54, 55, 60-71, 76, 84, 85, 91, 110, 145, 146, 147, 148, 149, 150, 151, 152, 153, 155, 156, 157, 158, 161, 163, 164, 165, 166, 167, 168, 169, 170, 171, 172, 173, 174, 175, 176, 177, 179
molecular biology, 112
molecular mechanisms, 2, 70
conserved, vii, 135, 137, 139, 180, 182
molecular weight, 183
molecules, vii, 3, 4, 5, 6, 7, 8, 9, 17, 29, 31, 49, 55, 60, 93, 110, 118, 120, 128, 129, 135, 136, 137, 138, 139, 140, 141, 142, 212, 213
mollusc prey, 89, 106
potential, 5, 6, 14, 19, 24, 28, 31, 37, 40, 52, 58, 62, 75, 76, 80, 89, 90, 91, 96, 97, 99, 100, 101, 104, 105, 108, 112, 117, 119, 151, 160, 165, 166, 175
molluscan, vii, viii, 5, 6, 13, 14, 15, 16, 17, 18, 19, 21, 22, 24, 30, 32, 37, 39, 76, 81, 89, 90, 91, 94, 100, 104, 105, 106, 108, 113, 128, 136, 137, 138, 139, 141, 142, 170, 176
communities, vii, 1, 4, 6, 7, 8, 9, 14, 15, 19, 21, 22, 24, 26, 30, 36, 38, 39, 85, 106, 108, 111, 115, 119, 128
groups, vii, 26, 30, 75, 92, 154, 170, 182, 184, 185
immunocytes, 137, 138, 139, 141, 142
metabolites, 13, 14, 17, 18, 21, 22, 26, 28, 29, 30, 34, 35, 36, 37, 39, 74, 76, 91, 93, 103, 169
prey, vii, 7, 8, 13, 18, 24, 29, 45, 51, 52, 53, 55, 71, 73, 74, 75, 76, 80, 81, 82, 83, 84, 85, 87, 89, 90, 91, 92, 93, 94, 95, 96, 97, 98, 99, 100, 101, 102, 103, 104, 105, 106, 107, 108, 109, 110, 111, 112, 113, 114, 115, 118, 120, 128
molluscs model organisms, 109
mollusk-hunters, 81
mollusks, 32, 34, 35, 66, 212
cephalopod, vii, 80, 81, 121, 128, 131
early non-torted, 150
larval, 6, 100, 110
marine, vii, 7, 8, 9, 13, 14, 15, 17, 18, 19, 20, 21, 23, 24, 25, 26, 27, 28, 29, 30, 31, 32, 33, 34, 35, 36, 37, 38, 39, 40, 41, 46, 51, 65, 69, 74, 80, 83, 84, 85, 86, 109, 110, 111, 112, 113, 114, 118, 119, 120, 127, 128, 129, 179
shelled, 13, 28, 98, 110
Moon, 85
morphine, 143
morphological defenses, 90, 97, 98, 99, 104, 105, 107
inducible, 19, 85, 86, 97, 98, 111, 112, 114
morphology, 96, 97, 98, 107, 109, 110, 112, 120, 131, 138, 151, 169, 171, 173
mortality, 19, 33, 39, 74, 95, 181
motor activity, 130, 180, 185, 192, 193
motor commands, 164
motor control, 58, 61, 155
motor neurons, 152, 154, 156, 159, 161, 162, 163, 164, 170, 175, 192, 193, 207

motor system, 85, 170
movements, 51, 52, 56, 60, 69, 80, 94, 96, 120, 159, 161, 163, 164, 172
mRNA, 124, 127, 136, 141
mucosa, 123
mucus, 13, 17, 18, 32, 37, 46, 49, 76, 81, 86, 159
mucus trails, 18, 81
  conspecific, 18, 46, 51, 83, 84, 91, 92, 93, 111
mud crabs, 102, 103, 111
multicellular organisms, 121
multifunctional, 4, 13, 146, 155, 163, 164, 170, 174
  interneurons, 57, 65, 157, 163, 164, 174, 176
  network, 58, 66, 145, 146, 148, 155, 157, 164, 168, 169, 170, 171, 174
multivariate analysis, 102
Murcidae species, 19
Muricidae, 18, 24, 26, 32, 40
Muricidae family, 18, 24
Muricids, 24
Mus musculus, 3
muscles, 75, 119, 159, 161, 163, 164, 172, 183, 185, 188, 193
  adductor, 75
  columellar, 147, 156, 159, 162, 164
mussels, 3, 33, 37, 75, 84, 97, 99, 104, 109, 110
mutant, 9
Mytilus edulis, 9, 75, 85, 86, 110, 112, 114, 138, 143
  odor, 46, 51, 52, 55, 57, 58, 59, 60, 66, 67, 69, 75, 80, 81, 90, 91, 92, 94, 95, 97, 98, 101, 102, 103, 106, 107, 108, 113, 115
Mytilus galloprovincialis, 136, 137
Mytilus spp, 6

N

Na^+, 130
Nassarius sp, 91
National Academy of Sciences, 35, 69, 209, 210
National Research Council, 172
native species, 179
natural assemblages, 18
natural habitats, 4, 96
natural killer cell, 141
natural products, 6, 13, 14, 15, 17, 18, 19, 20, 21, 22, 23, 24, 26, 28, 30, 31, 32, 33, 35, 36, 82
  chemistry, vii, 6, 13, 14, 21, 22, 24, 28, 31, 35, 93
  products research, 13, 14, 22, 24, 30
natural resources, 178, 179
natural selection, 1, 6, 14, 182
Navanax, 81
necrosis, 136
Nematoda, 179, 185
nematode(s), 3, 178, 180, 182
nematode worms, 3
Neogastropods, 18, 19, 24
  predatory marine, 24
nerve, 1, 4, 5, 9, 47, 48, 49, 50, 52, 55, 57, 65, 120, 125, 128, 130, 132, 147, 153, 159, 165, 189, 195, 199, 206, 207, 208, 209
nerve fibers, 48
nervous system, 3, 5, 47, 59, 60, 61, 62, 63, 65, 66, 67, 68, 69, 70, 82, 117, 119, 128, 129, 145, 146, 147, 148, 150, 151, 155, 157, 163, 169, 170, 171, 173, 174, 177, 178, 180, 182, 185, 187, 189, 191, 200, 203, 204, 207
  peripheral, 5, 6, 47, 48, 52, 53, 54, 58, 60, 61, 62, 63, 64, 65, 66, 67, 68, 69, 126, 131, 146, 163, 167, 169, 171, 172, 179
nervous system of metazoans, 182
Netherlands, 199, 209
network, 58, 66, 145, 146, 148, 155, 157, 164, 168, 169, 170, 171, 174
  behavioural, vii, 64, 85, 117, 119, 126, 130, 145, 146, 155, 156, 157, 161, 162, 163, 165, 167, 170, 171, 175, 176, 178, 179, 180, 181
  motor, 54, 58, 61, 68, 85, 119, 120, 126, 130, 131, 146, 151, 152, 154, 155, 156, 159, 161, 162, 163, 164, 170, 172, 174, 175, 180, 185
  peripheral odor processing, 58
  procerebral, 58
  sensory, 1, 2, 5, 7, 18, 29, 45, 46, 47, 48, 49, 50, 51, 52, 53, 54, 55, 56, 57, 58, 59, 60, 61, 62, 63, 64, 65, 68, 69, 70, 71, 74, 89, 105, 114, 115, 120, 122, 123, 125, 148, 155, 157, 172, 178
network interactions, 146
neural circuits, 3, 59, 60, 146, 178, 180, 185
  specialized, 30, 178, 180, 185
neural control, 53, 73, 171
neural function, 189
neural mechanisms, vii, 5, 9, 52
neural network(s), 132, 145, 146, 157, 164, 168, 169, 170, 177, 185, 186, 189, 191
  complex, vii, 2, 3, 6, 17, 19, 30, 55, 58, 61, 73, 78, 80, 81, 82, 118, 119, 120, 131, 137, 138, 139, 145, 146, 148, 155, 157, 161, 163, 170, 171, 180, 185
  multiganglionic, 145, 148, 155, 157, 168, 169, 170, 175, 176
  pathways, 3, 48, 52, 56, 57, 58, 63, 64, 67, 73, 131, 137, 142, 182
  plasticity, 62, 75, 84, 97, 106, 107, 108, 110, 111, 113, 114, 167, 169, 170, 171, 174, 175, 183
  recognition, 4, 24, 45, 59, 113
neural system, 146
neuroanatomy, 47, 52, 53, 54, 55, 61, 67, 70

functional, 3, 4, 8, 17, 26, 29, 47, 48, 60, 69, 119, 136, 142, 146, 151, 163, 169, 172
sensory, 1, 2, 5, 7, 18, 29, 45, 46, 47, 48, 49, 50, 51, 52, 53, 54, 55, 56, 57, 58, 59, 60, 61, 62, 63, 64, 65, 68, 69, 70, 71, 74, 89, 105, 114, 115, 120, 122, 123, 125, 148, 155, 157, 172, 178
neurobiology, 3, 45, 68, 130
neuroecological approach, 2, 3, 4, 5, 82, 120
neuroecological effects of climate change, viii, 146, 169, 170
neuroecology, vii, viii, 1, 2, 5, 6, 7, 8, 9, 15, 17, 30, 34, 82
neuroendocrine cells, 138
neuroendocrine responses, 136, 139
neuroendocrine system(s), vii, 135, 136, 140, 141, 142, 213
neuroethology, vii, 1, 2, 3, 59, 73
neurohormonal, 169
neuromodulators, 58, 124
neurones, 127, 147, 151, 153, 155, 157, 159, 162, 163, 164, 165, 170, 171, 172, 173, 174, 175, 176
neurons, 3, 5, 6, 47, 48, 49, 52-58, 61, 64-69, 119, 123, 125, 126, 130, 131, 145-148, 151, 152, 154-170, 173, 175, 179, 180, 185, 186, 187, 188, 189, 190, 191, 192, 193, 194, 195, 202, 206, 208, 209
active, 16, 17, 18, 19, 24, 26, 28, 39, 74, 75, 80, 81, 82, 86, 137, 146, 161, 165
key, 5, 22, 45, 108, 118, 119, 120, 125, 126, 139, 147, 170, 183
multifunctional, 4, 13, 146, 155, 163, 164, 170, 174
osphradial, 55, 66, 149, 163
osphradial ganglion, 163
pacemaker, 185
paired, 21, 56, 147, 148, 150, 151, 157, 158, 159, 164, 170, 173, 174
peripheral, 5, 6, 47, 48, 52, 53, 54, 58, 60, 61, 62, 63, 64, 65, 66, 67, 68, 69, 126, 131, 146, 163, 167, 169, 171, 172, 179
procerebral, 58
respiratory CPG, 146, 152, 167, 171
rhinophore ganglion, 46, 47
neuropeptides, 38, 70, 118, 121, 125, 126, 129, 139, 212, 213
neurophysiology, 18, 55, 61
neuroscience, vii, 3, 5, 57, 59, 74, 119, 204, 211, 212
neurosecretory, 67, 155, 159, 171, 172, 174, 175
neurotransmission, 68, 189
neurotransmitter(s), 48, 53, 54, 55, 58, 66, 118, 124, 125, 126, 139, 212
New England, 9
New Zealand, 21, 35, 67, 168, 196
NH_2, 47
nicotinamide, 47
nitric oxide, 47, 54, 58, 63, 65, 68, 69, 142, 187
nitric oxide synthase, 63, 69
nitrogen, 198
Nobel Prize, 5
norepinephrine, 139
North Africa, 168, 196
North America, 21, 31, 53, 168, 196
NPY, 123, 124, 126, 127, 128
Nucella, 7, 8, 91, 93, 94, 97, 98, 100, 101, 104, 112, 115
foraging, 8, 64, 74, 76, 86, 99, 101, 104, 111
Nucella lapillus, 8, 104, 112, 115
Nucella spp, 7
nucleic acid, 29
nucleotides, 74
nucleus, 49, 189, 207, 208, 209, 210
nucleus tractus solitarius, 189
Nudibranch molluscs
Hermissenda crassicornis, 6, 62, 81, 83
Melibe leonine, 47
Tritonia diomedea, 6, 51, 62, 63, 64, 65, 67, 68, 69, 70, 71, 80, 87
Nudibranchia, 30, 35, 63, 64, 70
nudibranchs, 13, 27, 29, 31, 38, 51, 61, 80, 81, 84
aeolid, 13
carnivourous grazing, 27
Nudipleura, 52
nutrient(s), 108, 165, 168, 198

O

O_2, 55, 178, 179, 181, 182, 183, 184, 185
low, 21, 74, 79, 80, 82, 95, 102, 104, 105, 106, 107, 167, 177, 179, 181
zero, 182
oceanic areas, 198
oceans, 20, 21, 117, 118, 198
octopus(s), 26, 40, 75, 80, 82, 83, 86, 119, 122, 123, 125, 127, 128, 130, 131, 132
blue ringed, 26
octopus predators, 80
Octopus vulgaris, v, 33, 117, 119, 121, 122, 123, 128, 129, 131, 132
Octopus brain, 119, 120
Octo-Sap, 127
odor cues, 46, 97, 101, 102, 103
predatory green crab, 101, 104
Odor detection, 48
odorants, 45, 48, 49, 60, 149
odors, 46, 51, 52, 54, 56, 58, 60, 62, 63, 64, 80, 84, 91, 93, 97, 98, 99, 100, 101, 104, 106, 107, 113

injured conspecific, 91, 92, 93, 106, 107, 114
noxious, 56, 58, 163
predatory, 4, 7, 13, 18, 21, 23, 24, 26, 28, 30, 31, 51, 52, 68, 73, 74, 80, 81, 82, 83, 84, 85, 93, 96, 97, 98, 99, 100, 101, 104, 106, 107, 108, 109
predatory asteroid, 106
predatory Cancer productus, 98
predatory green crab, 101, 104
olfaction, v, 7, 45, 46, 48, 50, 51, 52, 53, 55, 56, 58, 59, 60, 61, 62, 63, 68, 69, 127, 130, 212
proximate mechanisms of, 52, 59, 60, 62
olfactory, vii, 8, 9, 10, 45, 46, 47, 48, 49, 51, 52, 55, 56, 58, 59, 60, 61, 62, 63, 64, 65, 66, 67, 68, 69, 70, 82, 85, 112, 118, 120, 121, 122, 123, 124, 125, 126, 127, 128, 129, 130, 131, 132
capabilities, 55, 96, 117, 118, 127, 174
inputs, 2, 52, 53, 55, 62, 67, 120, 151, 155, 156, 158, 159, 160
mechanisms, vii, 1, 2, 4, 5, 7, 19, 28, 29, 34, 52, 59, 60, 61, 62, 70, 80, 82, 86, 99, 111, 123, 130, 139, 171, 178, 180, 182
neural pathways, 56, 63, 73, 82, 123
organs, 9, 48, 50, 52, 53, 56, 60, 61, 64, 66, 68, 100, 119, 120, 122, 128, 138, 139, 185
pits, 82
processing, 2, 3, 6, 47, 49, 51, 52, 56, 58, 59, 61, 62, 65, 68, 142, 143, 148
responses, vii, 2, 5, 6, 45, 52, 54, 55, 58, 60, 61, 64, 65, 66, 70, 73, 76, 77, 78, 79, 82, 85, 89-94, 96, 98, 99, 100, 104, 106, 107, 108, 110, 111, 114, 120, 123, 130, 137, 139, 140, 155, 157, 160, 167, 168, 169, 177, 180
sensory epithelium, 48, 57, 58, 64
sensory organs, 46, 48, 51, 52, 59, 60, 61, 63, 65, 70, 123
olfactory lobe (OL), 121, 128
olfactory nerve, 122, 125
Olfactory Receptor Neurons (ORNs), 123
olfactory systems, 45, 49, 56, 59, 60, 61, 65
gastropod, vii, 7, 24, 34, 35, 38, 39, 45, 47, 56, 59, 60, 61, 62, 63, 64, 80, 83, 84, 85, 86, 91, 101, 106, 109, 110, 111, 112, 113, 122, 141, 142, 145, 148, 149, 171, 173
opaline, 18, 29, 77, 78, 91
opaline secretions, 18
opisthobranchs, 13, 21, 26, 33
grazing, 13, 24, 26, 28, 90, 108
opportunities, 6, 55, 59, 62, 83
optic nerve, 125, 127, 147
optical microscopy, 122
Optimal defense theory, 28
Order Nudibranchia, 29
organ(s), vii, 9, 46, 47, 48, 49, 50, 51, 52, 53, 54, 55, 56, 58, 59, 60, 61, 63, 64, 65, 66, 68, 70, 118, 119, 120, 121, 122, 123, 125, 126, 127, 128, 129, 130, 131, 132, 138, 139, 185, 187, 194, 195, 212, 213
organism, 2, 17, 50, 74, 75, 80
organize, 59
oscillatory activity, 58, 67
osphradium, 48, 52, 53, 55, 61, 66, 70, 122, 148, 163, 175
overlap, 91, 136
oviduct, 124
ox, 148, 184
oxygen, 62, 97, 148, 159, 161, 164, 167, 170, 171, 172, 177, 178, 184, 195, 196, 205, 206, 207, 208, 209, 210
oxygen consumption, 207
oyster(s), 6, 8, 9, 19, 34, 35, 39, 101, 102, 103, 106, 108, 110, 111, 114
Pacific, 6, 21, 39, 40
populations, vii, 1, 8, 14, 19, 36, 55, 100, 106, 107, 108, 110, 112, 128, 169, 181, 182
reefs, 13, 14, 32, 101, 102, 111, 114
released, 51, 74, 76, 83, 91, 93, 94, 102, 109, 125, 128, 132, 165, 169
oyster reef system, 102

P

Pacific, 6, 21, 39, 40
pain, 117, 118
parallel, 17, 56, 102, 151
parapodia, 30
parasite(s), 22, 35, 37
parental care, 121
parietal ganglia, 146, 147, 148, 150, 151, 170
paired, 21, 56, 147, 148, 150, 151, 157, 158, 159, 164, 170, 173, 174
partial pressure differences, 185
participants, 105
pathogens, 18, 19, 22
pathology, 191
pathways, 3, 48, 52, 56, 57, 58, 63, 64, 67, 73, 82, 123, 131, 137, 142, 182, 187, 189, 201, 212
molecular, vii, 1, 2, 3, 4, 5, 6, 17, 32, 34, 46, 48, 59, 70, 74, 83, 94, 112, 120, 126, 128, 137, 139, 140, 157, 177, 182, 183
polysynaptic, 58
pattern generators, 53, 145, 155, 167, 170, 180, 183, 185
patterned activity, 185
PCR, 39
PCRCs, 149, 163

PDGF, 136, 137, 139, 141, 142
PDGF-AB, 137, 139
pedal, 50, 68, 147, 148, 150, 157, 158, 159, 160, 164, 171, 173, 174, 175
pedal ganglia, 147, 148, 150, 158, 173, 174
paired, 21, 56, 147, 148, 150, 151, 157, 158, 159, 164, 170, 173, 174
penis, 147, 162
peptide(s), vii, 18, 19, 24, 27, 29, 30, 33, 37, 38, 47, 58, 65, 66, 93, 110, 120, 122, 123, 126, 127, 128, 129, 132, 135, 136, 138, 141, 142, 169, 212
neurotoxic, vii, 13, 24, 26, 28, 30
perception, vii, 6, 58, 66, 94, 99, 102, 105, 108, 127, 172
perceptual abilities, 105, 108
perceptual range, 95, 105, 108
perfusion, 194, 200, 201
periodicity, 155
peripheral nervous system, 61, 65, 187, 195
peripheral neural circuits, 179
peripheral sensory cells, 52, 53, 62
peripheral synapses, 52, 54
periphery, 63, 68, 137, 146, 164
permission, 92, 98, 107
peroxide, 31
Persian Gulf, 20
personality, 117, 118, 131
pests, 62
pH, 178
phagocytic activity, 138
phagocytosis, 136, 138, 141, 142, 143
pharmacology, 34, 35, 37, 176
Phe-Met-Arg-Phe-NH2, 47
phenotypes, 75, 113
Phenotypic plasticity, 75, 114
pheromones, 4, 6, 18, 34, 40, 46, 51, 59, 60, 64, 67, 69, 120
Phestilla, 60, 63
Philadelphia, 209
phosphate, 47
phosphorus, 198
phosphorylation, 130
photographs, 95
photosynthesis, 196, 197, 198
phototaxis, 206
phylum, 14, 74, 117
Physa acuta, 91
Physa sp, 96
Physella gyrina, 97
physical environment, 177
physical properties, 46
physical protection, vii, 13, 27, 28, 31
reduced, vii, 13, 29, 30, 31, 76, 79, 90, 97, 98, 99, 100, 165, 166
physics, vii, 211
Physid snails, 76
Physiological, 53, 54, 63, 69, 70, 205, 206, 207, 208
physiological mechanisms, 203
physiology, 5, 26, 59, 68, 73, 75, 82, 180, 182, 209
phytoplankton, 111
pigmentation, 148
Placida dendritica, 29
Placopecten magellanicus, 96
Planorbarius corneus, 67, 136, 141, 142
plant growth, 198
plants, 53, 83, 84, 108, 111, 181, 196
plastic, vii, 75, 107, 165, 167, 168
plastic phenotypes, 75
plasticity, 62, 75, 84, 97, 106, 107, 108, 110, 111, 113, 114, 167, 169, 170, 171, 174, 175, 183, 189, 190, 191, 200, 201, 202, 203, 207
induced, 8, 19, 40, 51, 59, 84, 86, 108, 109, 110, 112, 119, 137, 139, 141, 142, 155, 163, 169, 174
long-term, 63, 67, 114, 157, 173, 174, 177, 178, 179, 181, 182
phenotypic, 84, 97, 114, 169
seasonal, vii, 66, 146, 148, 165, 167, 168, 170, 181
synaptic, 52, 59, 60, 145, 151, 155, 157, 159, 160, 163, 167, 168, 171, 172, 174, 177, 185
Platyhelminthes, 178, 185
playing, 118
PIB, 147, 156, 157
Pleurobranchaea californica, 6, 47, 68
PM, 129, 131, 170, 174
pneumostome, 53, 148, 149, 151, 152, 156, 159, 161, 162, 163, 164, 170
pneumostome closure, 163, 164
point mutation, 8
polar, 21, 30, 211
polar regions, 21
pollinators, 9
polychaetes, 32, 100
polychete worms, 184
polypeptides, 120
Polyplacophora, 22, 30
class, 22, 24, 30, 65, 117, 118
polyunsaturated fat, 18
polyunsaturated fatty acids, 18
POMC, 136, 138, 139, 140, 141
POMC genes, 136
POMC mRNA, 136
POMC products, 139
POMC-derived peptides, 138

ponds, 94, 170, 196
pools, 114, 199
population, 3, 7, 17, 54, 55, 100, 111, 120, 127, 186, 191, 193, 198
population growth, 198
population size, 121
population structure, 127
positive feedback, 179
positive relationship, 91
potassium, 195, 210
prayer, 173
predation, 6, 7, 8, 10, 21, 30, 32, 39, 51, 74, 76, 80, 81, 82, 83, 85, 86, 91, 93, 96, 97, 98, 99, 100, 101, 102, 103, 105, 106, 107, 108, 109, 110, 111, 112, 113, 114, 115, 117, 118, 121, 169, 198, 213
predation cue, 93
predation events, 74, 106
predation intensity, 91, 105, 108
predation pressure, 21, 106, 107, 114
predation risk, 10, 74, 93, 100, 101, 106, 111, 112, 113, 114
predator abundance, 99
predator adaptations, 74, 82
predator avoidance, 55, 86, 110, 114
predator cues, 76, 96, 100
predator diet, 91, 93, 98
predator identity, 90, 98
predator odor cues, 96, 97, 99, 100, 101
predator odor plumes, 51
predator odors, 51, 80, 90, 94, 96, 97, 98, 99, 100, 109
predator offense, 82
predator searches, 74
predator species, 93, 96, 106
  invasive, viii, 9, 56, 67, 106, 110, 146
  multiple, 4, 17, 21, 46, 53, 58, 59, 61, 67, 76, 93, 108, 109, 115, 128, 131, 146, 154, 170, 185
predator threat, 89
Predator-dependent alterations, 100
predator-induced habit shifts, 101
predator–prey, 17, 18
predator-prey communication, 74
predator-prey duet, 105
predator-prey interactions, 7, 29, 75, 111, 113
predators, vii, viii, 4, 7, 8, 9, 17, 18, 21, 24, 26, 28, 29, 34, 36, 38, 46, 47, 51, 63, 71, 73-76, 78, 80, 82, 84-115, 118, 123, 146, 148, 169, 174, 199, 203, 212, 213
  active, 16, 17, 18, 19, 24, 26, 28, 39, 74, 75, 80, 81, 82, 86, 137, 146, 161, 165
  blue crab, 93, 96, 97, 100, 101, 102, 103, 109, 110, 111
  detection of, 8, 51, 68, 86, 90, 105
  multiple, 4, 17, 21, 46, 53, 58, 59, 61, 67, 76, 93, 108, 109, 115, 128, 131, 146, 154, 170, 185
  native, 56, 106, 111, 113, 179
  non-native, 106, 111
  potential, 5, 6, 14, 19, 24, 28, 31, 37, 40, 52, 58, 62, 75, 76, 80, 89, 90, 91, 96, 97, 99, 100, 101, 104, 105, 108, 112, 117, 119, 151, 160, 165, 166, 175
  snail, 8, 36, 56, 59, 62, 63, 64, 65, 66, 67, 68, 69, 70, 71, 76, 81, 83, 84, 85, 86, 91, 92, 93, 97, 99, 101, 108, 112, 114, 141, 142, 163, 168, 171, 172, 173, 174, 175, 176
  surface, 19, 22, 30, 34, 37, 39, 46, 48, 56, 60, 92, 117, 118, 123, 141, 147, 148, 149, 151, 159, 161, 162, 163, 167, 169, 183, 184, 185
  sympatric, 169, 174
predators feeding, 100
predatory behaviors, 81
predatory Cephalopods, 26, 30
Predatory decapod crustaceans, 75
predatory Neogastropoda, 30
  neogastropods, 26
predatory species, 7, 24
  invasive, viii, 9, 56, 67, 106, 110, 146
prefrontal cortex, 210
preparation, 31, 52, 62, 157, 194, 195, 202
pressure gradient, 112, 184
prey, vii, 7, 8, 13, 18, 24, 29, 45, 51, 52, 53, 55, 71, 73, 74, 75, 76, 80, 81, 82, 83, 84, 85, 87, 89, 90, 91, 92, 93, 94, 95, 96, 97, 98, 99, 100, 101, 102, 103, 104, 105, 106, 107, 108, 109, 110, 111, 112, 113, 114, 115, 118, 120, 128
  apparency, 97, 99, 100
  conspecific, 18, 46, 51, 83, 84, 91, 92, 93, 111
  consumed, 89, 91, 102, 103, 104
  control, 3, 53, 54, 58, 60, 69, 73, 77, 79, 106, 107, 108, 110, 118, 119, 120, 121, 124, 125, 126, 127, 128, 129, 131, 132, 138, 152, 155, 164, 172, 173, 174, 178
  defenses, 6, 9, 17, 21, 32, 38, 74, 75, 76, 80, 82, 83, 84, 85, 86, 89, 99, 105, 106, 107, 111, 112, 114
  determined, 47, 51, 91, 93, 94, 127, 137, 184
  focal, 96, 99
  foundation species, 2, 3, 6, 7, 75, 81, 90, 91, 97, 104, 118
  heterospecific, 91
  non-native, 106, 111
  perception, vii, 6, 58, 66, 94, 99, 102, 105, 108, 127, 172
  populations, vii, 1, 8, 14, 19, 36, 55, 100, 106, 107, 108, 110, 112, 128, 169, 181, 182

potential, 5, 6, 14, 19, 24, 28, 31, 37, 40, 52, 58, 62, 75, 76, 80, 89, 90, 91, 96, 97, 99, 100, 101, 104, 105, 108, 112, 117, 119, 151, 160, 165, 166, 175
search, 4, 14, 55, 56, 73, 76, 80, 81, 82, 93, 136
sensitivity, 91, 94, 95, 106, 108, 114
species abundance, 6, 101
to Predators, 96, 100
principles, vii, 1, 2, 6, 46, 59, 136, 155, 211, 213
probability, 74
probe, 24
procerebrum, 56, 57, 58, 59, 66, 67, 68
producers, 4, 82, 101
project, 48, 53, 54, 56, 57, 58, 61, 159, 160
proliferation, 19, 136
propagation, 172
prostaglandins, 18, 29
protection, vii, 13, 22, 27, 28, 30, 31, 59, 77, 98, 100, 117, 118, 120, 212
protein kinase C, 137
protein superfamilies, 139
proteins, 19, 24, 26, 27, 33, 35, 37, 48, 51, 60
proteome, 210
Ptlosarcus gurneyi, 51
Pulmonata, 22, 68, 69, 70, 141, 142, 172
pulmonate carnivores, 81
pulmonates, vii, 21, 27, 61, 68, 149, 150
freshwater, viii, 8, 46, 53, 61, 64, 71, 74, 83, 84, 86, 97, 106, 110, 114, 115, 141, 142, 146, 148, 151, 161, 168, 169, 170, 171, 172, 173, 175, 176, 182, 184
herbivorous, 27, 35, 64
purpura, 35
Pycnopodia helianthoides, 51

Q

quantification, 39
Queensland, ix, 45

R

race, 74, 82, 105, 110
radiation, 180
radula, 157, 163
RCPG (*Lymnaea* respiratory central pattern generator)
activity, viii, 14, 17, 18, 19, 24, 25, 26, 27, 30, 34, 36, 58, 59, 61, 63, 66, 67, 69, 74, 76, 89, 90, 96, 100, 101, 104, 124, 125, 126, 127, 130, 138, 146, 149, 151, 154, 155, 157, 159, 163, 164, 165, 167, 168, 172, 173, 180, 185
discharge, 155, 163
neurons, 3, 5, 6, 47, 48, 49, 52, 53, 54, 55, 56, 57, 58, 61, 64, 65, 66, 67, 68, 69, 119, 123, 125, 130, 131, 146, 147, 151, 152, 154, 155, 156, 157, 158, 159, 160, 161, 162, 163, 164, 165, 166, 167, 169, 170, 173, 175, 179, 180, 185
silent, 165
RE, 109, 132
reactions, 19, 94, 99, 141, 155
reactive oxygen, 178
reactive oxygen species, 178
reactivity, 68
recall, 155, 157, 173
receptive field, 52
receptors, 38, 48, 49, 51, 60, 62, 63, 80, 112, 124, 139, 141, 143, 161, 167, 168, 170, 173, 189, 190, 212
recognition, 4, 24, 45, 59, 113
reconstruction, 59, 175, 209
recovery, 207
Red List, 205
redistribution, 137
redundancy, 136
reflexes, 3, 155, 206
reform, 192
refuge habitats, 104, 111
refuges, 76, 100, 102, 111
regeneration, 13, 29, 204
regression, 165, 166
regression line, 165, 166
rejection, 77
relative size, 151
relevance, 82
reliability, 54
relief, 31, 104
remediation, 108
remotely operated vehicles, 22
repair, 136
reproduction, vii, 26, 46, 59, 74, 99, 100, 111, 117, 118, 121, 123, 124, 125, 126, 127, 128, 129, 131, 169, 176, 199, 212
reproductive behavior, 129
reproductive organs, 100
requirements, 46, 130, 199
researchers, 5, 14, 21, 29, 46, 61, 73, 81, 177, 181, 212
resilience, 104
resistance, 4, 105, 106, 115, 184
resolution, 91
resource allocation, 28, 99
resources, 3, 6, 32, 73, 75, 89, 90, 96, 99, 100, 101, 104, 114, 178, 179, 198
basal, 49, 101, 102, 104, 105, 120

natural, 1, 4, 6, 13, 14, 15, 17, 18, 19, 20, 21, 22, 23, 24, 26, 28, 30, 31, 32, 33, 35, 36, 51, 53, 82, 85, 94, 96, 100, 103, 112, 119, 120, 130, 138, 141, 161, 178, 179, 182
respiration, 55, 146, 151, 155, 159, 161, 163, 164, 167, 169, 171, 172, 177, 178, 179, 180, 181, 182, 184, 185, 187, 188, 190, 191, 192, 193, 194, 196, 197, 198, 199, 200, 201, 202, 203, 204, 205, 206, 207, 213
invertebrate, 8, 30, 32, 38, 51, 111, 119, 132, 138, 141, 142, 178, 179, 180, 182, 185
skin, 21, 71, 76, 118, 119, 122, 151, 161, 178, 184
respiratory, viii, 48, 51, 53, 62, 145, 148, 149, 152, 155, 158, 159, 161, 162, 163, 164, 167, 169, 170, 171, 172, 173, 174, 175, 177, 178, 179, 180, 181, 182, 184, 185
CPG, 146, 157, 159, 180, 185
homeostasis, 178, 182, 185
invertebrate, 8, 30, 32, 38, 51, 111, 119, 132, 138, 141, 142, 178, 179, 180, 182, 185
mammalian, 138, 139, 141, 180
neurons, 3, 5, 6, 47, 48, 49, 52, 53, 54, 55, 56, 57, 58, 61, 64, 65, 66, 67, 68, 69, 119, 123, 125, 130, 131, 146, 147, 151, 152, 154, 155, 156, 157, 158, 159, 160, 161, 162, 163, 164, 165, 166, 167, 169, 170, 173, 175, 179, 180, 185
plasticity, 62, 75, 84, 97, 106, 107, 108, 110, 111, 113, 114, 167, 169, 170, 171, 174, 175, 183
rhythm, 155, 162, 167, 178
strategies, 13, 39, 70, 89, 105, 108, 109, 111, 115, 117, 118, 119, 136
structures, 30, 47, 52, 53, 54, 56, 57, 60, 68, 69, 74, 82, 125, 158, 177, 178, 184, 185
surfaces, 120, 148, 184, 185
respiratory behaviour, viii, 148, 155, 159, 161, 163, 164, 169, 173, 175, 177, 178, 179, 180, 181, 182
response, 7, 18, 19, 22, 29, 32, 33, 39, 40, 51, 58, 66, 75, 76, 78, 79, 80, 82, 83, 84, 90, 91, 93, 94, 96, 97, 98, 100, 104, 108, 110, 111, 113, 114, 138, 141, 154, 155, 168, 174, 178, 180, 185, 191, 194, 195, 198, 199, 200, 202, 207
responsiveness, 62, 106
restoration, 110, 111, 146, 167, 213
reticulum, 49
RH, 47, 130, 138
rhinophores, 46, 47, 48, 52, 67, 68, 70
nudibranch *Phestilla sibogae*, 52
single *Aplysia punctata*, 48
rhodopsin, 48
rhythm, 155, 159, 160, 162, 167, 178, 186, 188, 189, 192, 193, 194, 206, 208, 209
rhythmic activity, 185
rhythmic behaviour, 167
RIE, 111
right parietal ganglion, 150, 151
risk(s), 10, 36, 74, 77, 85, 90, 93, 94, 99, 100, 101, 102, 104, 106, 107, 111, 112, 113, 114
risk perception, 94
RNA, 3, 8, 194
room temperature, 167
rotifers, 97
roughness, 113
Royal Society, 128, 130, 131, 205, 208, 209
RPeD1, 147, 157, 162, 163, 165, 166, 167, 168, 169, 170
activity of, 17, 27, 67, 90, 101, 104, 125, 127, 138, 157, 165, 185
follower cells of, 167
projections of, 56, 173
RPeD11, 147, 157, 162, 164, 169, 174, 175
runoff, 198

S

Saccoglossa, 29, 30
shell-less, 21, 28, 29
salinity, 181
salivary gland(s), 26
salts, 31, 79
scaling, 1, 3
scallops, 100, 109
scanning electron microscopy, 53
S-cells, 52
scent, 10, 18, 101, 106, 108, 112
Schwerin, 210
science, 5, 181
scope, 189
SCPG, 156, 157, 158, 159
sea hare, 5, 6, 18, 28, 30, 31, 36, 37, 38, 39, 40, 78, 83, 84, 85, 86, 112
sea level, 181, 182, 200
sea star, 21, 51, 97, 109, 110
seasonal changes(s), 146, 148, 167, 168, 213
in *Lymnaea* Behaviour, 167
secondary metabolism, 21, 36
secondary metabolites, 13, 14, 15, 16, 17, 21, 24, 26, 28, 29, 30, 31, 32, 35, 39, 84, 97
secrete, 166
secretion(s), 17, 18, 19, 26, 32, 37, 39, 76, 78, 84, 85, 86, 112, 142
cloudy, 78
purple, 26, 31, 32, 34, 37, 40, 78
sediment(s), 23, 24, 109, 181, 187, 196, 197, 198
segregation, 113
selection, vii, 1, 6, 9, 14, 45, 52, 59, 67, 74, 75, 76, 82, 87, 109, 111, 127, 176, 182

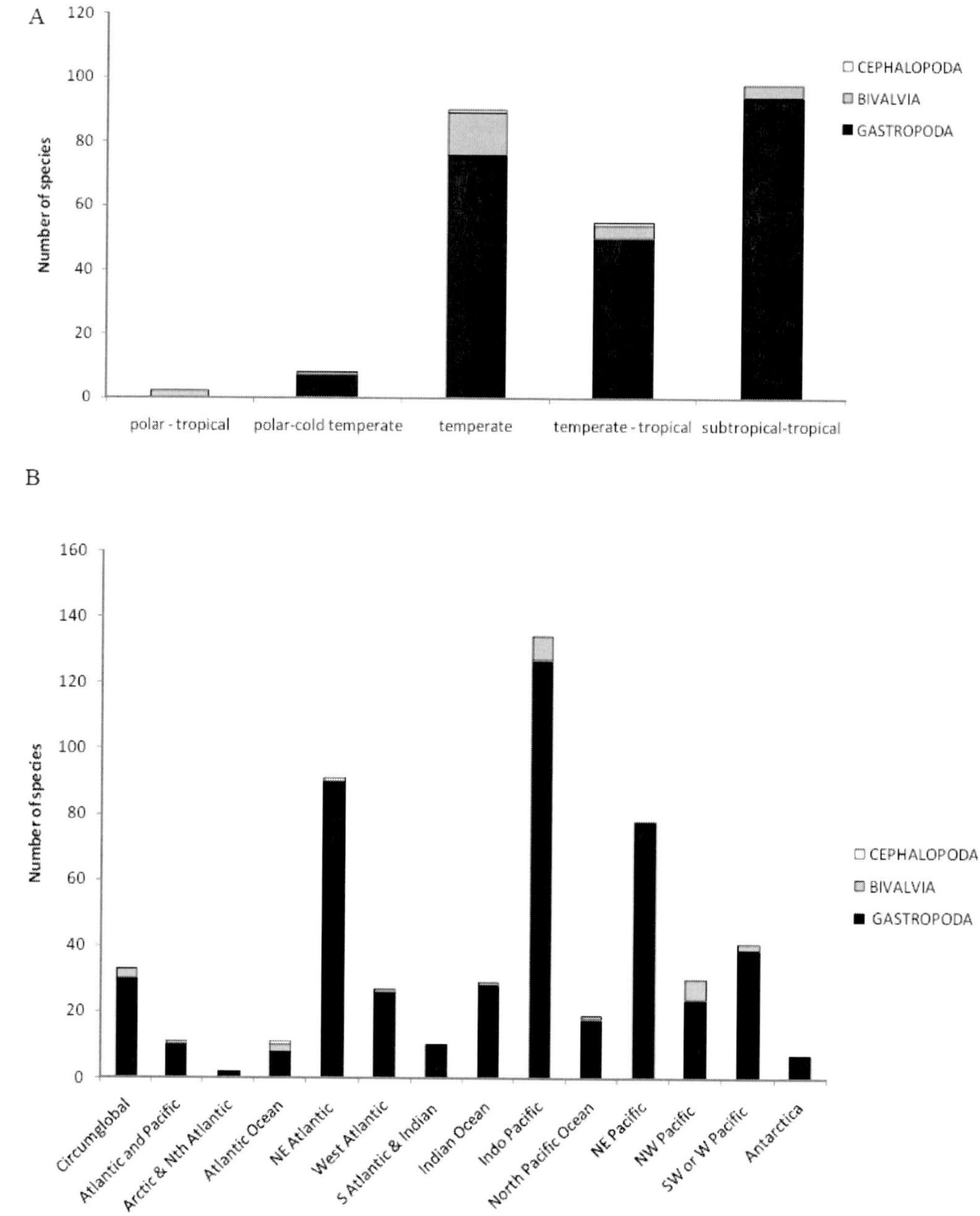

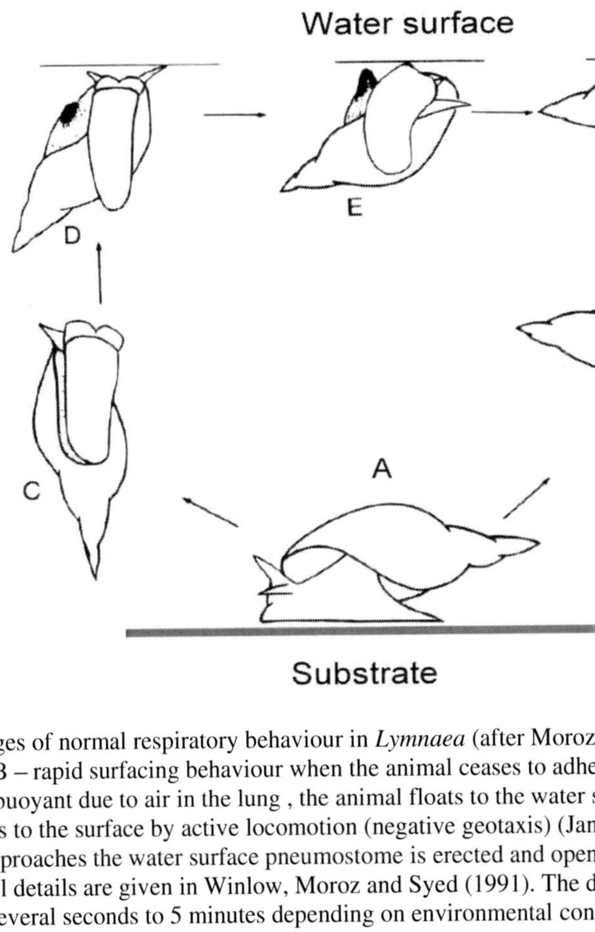

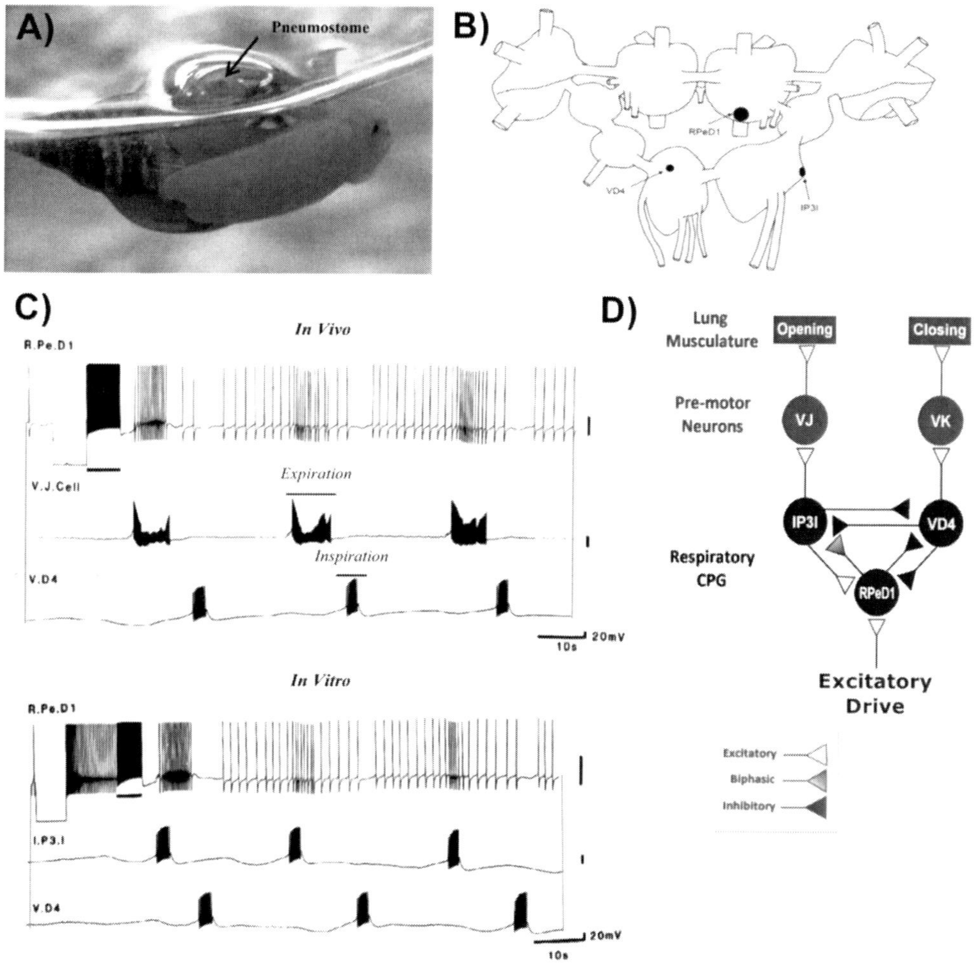

narrow, 52, 178
reciprocal, 82
selective pressure(s), 21, 73, 74, 75
self-adaptation, 79
semi-intact preparations, 159, 174
senses, 5, 7, 55, 75, 80, 85
sensing, 48, 62, 105, 170, 187, 194, 205, 207, 208
sensitivity, 91, 94, 95, 106, 108, 114, 187
sensors, 188
sensory, 1, 2, 5, 7, 18, 29, 45, 46, 47, 48, 49, 50, 51, 52, 53, 54, 55, 56, 57, 58, 59, 60, 61, 62, 63, 64, 65, 68, 69, 70, 71, 74, 89, 105, 114, 115, 120, 122, 123, 125, 148, 155, 157, 172, 178
anteroceptive, 148
bipolar, 54, 55, 56, 57, 58
epithelia, 48, 49, 53, 56, 60, 64
inputs, 2, 52, 53, 55, 62, 67, 120, 151, 155, 156, 158, 159, 160
neurons, 3, 5, 6, 47, 48, 49, 52, 53, 54, 55, 56, 57, 58, 61, 64, 65, 66, 67, 68, 69, 119, 123, 125, 130, 131, 146, 147, 151, 152, 154, 155, 156, 157, 158, 159, 160, 161, 162, 163, 164, 165, 166, 167, 169, 170, 173, 175, 179, 180, 185
systems, vii, 1, 3, 17, 45, 55, 57, 59, 60, 61, 65, 66, 68, 70, 71, 74, 82, 91, 94, 100, 101, 108, 110, 120, 123, 126, 139, 143, 145, 146, 156, 157, 159, 162, 170, 172, 177, 180
sensory modalities, 46, 55, 59
sensory projection, 52
sensory systems, 1, 55, 70, 71, 74
Sepia officinalis, 62, 81, 85, 117, 119, 125, 128, 129, 130, 132
Sepioteuthis sepioidea, 80
sequencing, 3, 59, 70
serotonin, 47, 48, 58, 69, 131, 159, 171
services, 108, 110, 179, 209
sewage, 179, 198
sex, 51, 52, 69, 118, 120, 126, 127
sex differences, 52
sex steroid, 118, 127
sexual behaviour, 123, 169
shape, 76, 119, 121, 136, 137, 141, 142, 166, 181, 189
shell, 8, 9, 21, 22, 24, 28, 29, 38, 75, 82, 83, 96, 97, 98, 100, 106, 107, 110, 112, 148, 150, 157, 159, 161, 163, 169
internal, 28, 121, 123, 129, 147, 148, 153, 155, 168, 169, 184
length, 93, 98, 100, 107, 167, 168
movements, 51, 52, 56, 60, 69, 80, 94, 96, 120, 159, 161, 163, 164, 172
rhythmic, 159, 167, 185
shelled predatory caenogastropods, 13
shellfish, vii, 14, 26, 28, 30, 40, 212
shellfish toxins, vii, 14, 26, 28, 30
shelter, 76, 109
shock, 33
shoots, 96
shores, 6, 7, 30
showing, 2, 49, 104, 130, 136, 149, 153, 195
shrimp, 77, 198, 205
signal transduction, 5, 64, 142
signaling molecules, 60, 139
signaling pathway, 131, 137, 142
signalling, vii, 118, 131, 136, 137, 212
signals, vii, 4, 8, 36, 45, 51, 73, 74, 75, 76, 80, 81, 82, 84, 93, 94, 95, 102, 104, 108, 114, 118, 119, 120, 123, 126, 128, 137, 212
Silent inactivatable neurons, 165
Silent neurons, 165
Singapore, 143
siphon, 95, 123, 193
Siphonaria diemenensis, 19, 39
gelatinous egg masses, 18, 39
Siphonaria spp, 22
Siphonaridae, 22
skin, 21, 71, 76, 118, 119, 122, 151, 161, 178, 184, 187, 191, 213
slugs, 6, 28, 34, 51, 52, 70
sea, 3, 5, 6, 8, 18, 21, 22, 28, 30, 31, 34, 36, 37, 38, 39, 40, 45, 51, 52, 63, 64, 67, 68, 78, 80, 83, 84, 85, 86, 97, 109, 110, 111, 112, 117, 118, 128, 181, 182
snails, 26, 28, 53, 54, 56, 63, 70, 71, 74, 76, 81, 82, 86, 90, 91, 92, 97, 98, 104, 107, 110, 113, 114, 161, 165, 166, 167, 168, 169, 171, 172, 176
pond, 6, 64, 65, 69, 70, 71, 76, 111, 167, 170, 172, 173, 175, 177, 179
social behaviour, 118
sodium, 4, 7, 8, 9
solubility, 183, 197
solution, 187, 194
somata, 6, 47, 48, 52
South Africa, 21
South America, 21, 34
SP, 110, 113
Spain, 131
Spartina alterniflora shoots, 96, 101
Spartina sp, 108
spatial memory, 9
species, vii, viii, 1, 3-9, 13-30, 32, 34, 35, 37-39, 41, 46, 48, 51, 55, 58-62, 67, 73-78, 81, 86, 89-91, 93, 94, 96, 97, 99-102, 104-106, 108-110, 112, 114, 117-120, 126, 136, 138, 139, 146, 150, 158, 169, 177-179, 181, 182, 184, 185, 187, 198-200, 202-204, 211-213

aquatic, 9, 32, 37, 45, 46, 51, 53, 59, 60, 74, 75, 76, 82, 83, 93, 109, 110, 120, 146, 158, 177, 179, 181, 182, 184, 185
extirpation, 179, 182
invasive, viii, 9, 56, 67, 106, 110, 146
microbial, viii, 17, 18, 19, 22, 146, 169, 181
survival, 3, 46, 73, 74, 82, 89, 96, 97, 103, 110, 111, 121, 137, 142, 163, 177, 179, 180, 181, 182
species richness, 22, 28, 32, 179
spending, 202
sperm, 118, 127, 128
spermatogenesis, 124
spiders, 18
sponge, 21, 29, 31, 32, 33, 36, 40, 84
squid giant axons, 5
squids, 6, 82, 120
conspecific, 18, 46, 51, 83, 84, 91, 92, 93, 111
stability, 7, 8, 178, 179
standard deviation, 166
stars, 21, 97, 109
Starved predators, 94
state(s), 13, 137, 143, 146, 178, 184, 187, 202, 211
sterols, 29, 32
stimulation, 46, 48, 57, 127, 130, 152, 157, 159, 164, 169, 189, 194, 201
Stimulation of gametogenesis, 124
Stimulation of spermatogenesis, 124
stimulus, 2, 69, 79, 81, 137, 138, 155, 163
stock, 148
stomatogastric ganglion, 180
storage, 4, 28, 126, 127
Stramonita haemastoma, 7
stratification, 181, 196, 197
stress, 9, 19, 105, 136, 138, 139, 140, 141, 142, 148, 163, 169, 174, 181, 197, 199, 203, 206, 210
biotic, 3, 6, 19, 177, 181
environmental, vii, viii, 1, 5, 19, 48, 73, 75, 84, 89, 94, 112, 113, 120, 121, 126, 148, 161, 163, 164, 168, 169, 174, 177, 178, 179, 181, 182, 185
induced, 8, 19, 40, 51, 59, 84, 86, 108, 109, 110, 112, 119, 137, 139, 141, 142, 155, 163, 169, 174
performing, 138
response, 7, 18, 19, 22, 29, 32, 33, 39, 40, 51, 58, 66, 75, 76, 78, 79, 80, 82, 83, 84, 90, 91, 93, 94, 96, 97, 98, 100, 104, 108, 110, 111, 113, 114, 136, 138, 139, 141, 142, 154, 155, 168, 174, 178, 180, 185
stress response, 136, 138, 139, 141, 142
stressors, 19, 66, 114, 203, 204
environmental, vii, viii, 1, 5, 19, 48, 73, 75, 84, 89, 94, 112, 113, 120, 121, 126, 148, 161, 163, 164, 168, 169, 174, 177, 178, 179, 181, 182, 185
extreme, 30
striped blennies, 80
structure, 3, 8, 9, 14, 19, 21, 26, 36, 37, 58, 59, 60, 62, 63, 64, 96, 101, 104, 111, 112, 117, 119, 123, 127, 128, 179, 184, 194, 205
structuring, vii, 6, 7, 15, 21, 24, 211
STX, 4
style, 120
substrate(s), 6, 19, 23, 30, 52, 75, 82, 99, 108, 137, 155, 157, 159, 161, 163, 175, 176
succession, 104
sucrose, 65, 173
summer, 19, 39, 165, 167, 168
suppression, 108
surface area, 184, 202
surface layer, 123
survival, 3, 46, 73, 74, 82, 89, 96, 97, 103, 110, 111, 121, 137, 142, 163, 177, 179, 180, 181, 182, 212
susceptibility, 39, 106, 113, 155, 198
Sweden, 206
symmetry, 173
symptoms, 181
synapse(s), 5, 52, 53, 54, 56, 60, 61, 167, 185, 186, 194, 200, 201, 202, 204
synaptic connections, 163, 167, 168, 171, 174, 185
normal, 51, 161, 162, 163, 167, 178
synaptic connectivity, 146
synaptic inputs, 145, 151, 155, 157, 160, 163, 171
common, 5, 6, 22, 48, 50, 61, 85, 90, 91, 99, 100, 101, 106, 114, 120, 126, 135, 139, 140, 142, 143, 151, 155, 179
wide-acting, 145, 151, 155, 157, 171
synaptic plasticity, 191, 206
synaptic strength, 189, 190, 200
synaptic transmission, 194, 207
synthesis, 3, 4, 5, 9, 28, 29, 33, 34, 45, 59

T

tactics, 83
tactile stimuli, 69, 157, 173
Taiwan, 33
Tambja ceutae, 138, 141
tambjamine family, 138
tangles, 48
target, 6, 24, 57, 120, 128
target organs, 120
taste aversion, 173

taxa, 16, 57, 59, 60, 61, 74, 94, 97, 100, 135, 139, 198
  euthyneuran, 61, 67
  primary, vii, 4, 22, 23, 28, 30, 34, 46, 52, 53, 60, 61, 66, 69, 96, 99, 100, 114, 119, 139
  vertebrate, 60, 135, 138, 141, 142, 155, 180
taxonomy, 210
techniques, 3, 47, 52, 122, 123, 167
teeth, 91, 98
temperate, 13, 19, 21, 30, 32, 112
  cold, 21, 30, 185
  regions, 19, 20, 21, 48, 51, 107
temperature, 21, 34, 111, 121, 123, 165, 167, 168, 169, 181, 196, 203, 206
temporal variation, 196
tension, 209
tentacle ganglion, 56, 57, 58
tentacles, 6, 46, 47, 48, 50, 53, 54, 55, 56, 58, 62, 63, 66, 67, 68, 81, 185
  oral, 46, 47, 48, 50, 51, 52, 62
terminals, 165
terpenes, 27, 29
terrestrial animals, 45
terrestrial gastropod olfaction, 56
terrestrial slugs, 56
testing, 28, 30, 38, 61, 93
texture, 117, 118
TGF, 136, 137, 139, 141, 142
threats, 73, 179, 204, 212
thymus, 142
Tibetans, 182
time periods, 106, 189
tissue, 29, 76, 131, 178, 180, 181, 183, 184
TNF, 136, 137, 139
tooth, 98
torsion, 145, 146, 148, 150, 151, 169, 213
toxic metals, 180, 203
toxicity, 76, 208
toxicology, 26
toxin, 8
Toxin-mediated processes, 4
trade, 28, 36, 113, 114
trade-off, 28, 36, 113, 114
trafficking, 206
trails, 76, 81
traits, 6, 7, 73, 90
transcription, 180
transduction, 4, 48, 82, 137
transforming growth factor, 136
transmission, 53, 56, 94, 194, 210
transport, 3, 28, 71, 94, 96, 109, 126, 178, 205, 210
transport processes, 96
treatment, 90, 92, 98, 102, 107, 180, 182
trial, 69, 173
Tridacna squamosa, 94, 98, 112
trigeminal nerve, 78
triggers, 168
Tritonia, 6, 45, 51, 52, 60, 62, 63, 64, 65, 67, 68, 69, 70, 71, 80, 87
Tritonia diomedea, 6, 51, 62, 63, 64, 65, 67, 68, 69, 70, 71, 80, 87
Tritonia rhinophores, 51, 52
trophic niche, 13, 24, 25, 28
TTX, 4, 9, 10
tumor, 138
tumor cells, 138
tumour necrosis factor (TNF), 136
tunicates, 81
turbulence, 94, 198
turbulent flow, 51, 84
turbulent mixing, 94, 95
turbulent transport processes, 96
turgor, 123
turnover, 47
turtle, 207
Tyrosinase, 80
tyrosine, 47
tyrosine hydroxylase, 47

U

UK, x, 62, 145, 167, 171, 172, 211
ultrastructure, 71, 132
uniform, 138
unipolar nitrergic cells, 53, 54, 55
United, 35, 130
United Nations, 130
United States, 35
urban, 196
Urosalpinx cinerea, 7
USA, ix, x, 70, 83, 106
UV, 22
UV radiation, 22

V

valve, 184
variables, 105, 181
vascular system, 159
vasculature, 178
vasoactive intestinal peptide, 58
vector, 70
vegetation, 179, 196, 199
velocity, 95, 97

ventilation, 149, 151, 155, 157, 161, 163, 170, 180, 183, 185, 186, 187, 189, 191, 200, 205, 208, 213
vertebrates, vii, 47, 49, 60, 118, 123, 136, 138, 139, 140, 169, 172, 180, 187, 206, 207, 213
vesicle, 49
virus infection, 33
viruses, 18
visceral ganglia, 150, 151
medial, 50, 147, 148, 158
modern, 32, 60, 150, 151, 170
single median, 148
unpaired, 151, 170
vision, vii, 46, 56, 81, 212
visual stimuli, 85
visual stimulus, 80
voracious carnivores, 7
vulnerability, 89

W

Washington, 206
waste, 97, 178
water, 17, 23, 38, 45, 46, 48, 51, 60, 64, 69, 71, 74, 75, 76, 80, 82, 85, 87, 92, 93, 94, 95, 108, 113, 127, 148, 149, 151, 159, 161, 162, 165, 167, 169, 178, 179, 181, 183, 184, 185, 187, 192, 193, 196, 197, 198, 199, 202, 203, 213
bottom, 46, 128, 167
chemistry, vii, 6, 13, 14, 21, 22, 24, 28, 31, 35, 93
column, 76, 159, 164
conditioned, 81, 94, 167, 169, 170
conditions, 75, 95, 99, 148, 161, 166, 177, 178, 179, 180, 181, 182
contamination, 60, 179
currents, 46
fresh, 17
jets, 80
layers, 113, 126
level, vii, 1, 2, 15, 17, 28, 79, 89, 90, 99, 102, 104, 105, 107, 108, 118, 120, 123, 124, 126, 169, 181, 182
water chemistry, 93
water O_2 levels
oxygenation, 36, 165
quality, 54, 93, 104, 108, 128
reducing
oxygenated, 148, 184
surface, 19, 22, 30, 34, 37, 39, 46, 48, 56, 60, 92, 117, 118, 123, 141, 147, 148, 149, 151, 159, 161, 162, 163, 167, 169, 183, 184, 185
temperatures, 22, 113, 167
water quality, 108
watershed, 196
wealth, 180, 187, 191
Western Europe, 56
wetlands, 179
whole-body withdrawal motoneurones, 163
winter, 146, 165, 166, 167, 168, 170, 182
animals, viii, 2, 3, 6, 9, 32, 37, 46, 51, 52, 56, 59, 60, 62, 81, 83, 90, 94, 96, 112, 118, 120, 121, 140, 157, 158, 163, 167, 168, 169, 170, 177, 178, 179, 180, 181, 182, 185
conditioning, 7, 64, 83, 155, 167, 169, 172, 173, 174
hibernation, 182
snails, 26, 28, 53, 54, 56, 63, 70, 71, 74, 76, 81, 82, 86, 90, 91, 92, 97, 98, 104, 107, 110, 113, 114, 161, 165, 166, 167, 168, 169, 171, 172, 176
withdrawal, 6, 58, 59, 63, 106, 145, 146, 151, 154, 155, 156, 157, 159, 162, 163, 164, 170, 172, 174
behaviour, vii, viii, 8, 63, 65, 69, 70, 84, 86, 111, 117, 118, 119, 120, 123, 124, 126, 127, 128, 132, 137, 146, 148, 151, 155, 157, 161, 163, 164, 165, 168, 169, 170, 171, 174, 175, 176, 177, 178, 180, 181, 182
whole-body, 146, 163, 172
workers, 13, 20, 52
working memory, 203, 210
worldwide, 55, 196, 204
worms, 3, 184, 198

X

xenografts, 138

Y

Yellow Cells, 159
Yellow-green Cells, 159
yield, 2, 81

Z

zooplankton, 110, 198, 208